SO-AIH-278

David Lloyd and Ernest L. Rossi (Eds.)

Ultradian Rhythms in Life Processes

An Inquiry into Fundamental Principles of Chronobiology and Psychobiology

With 143 Figures

Springer-Verlag
London Berlin Heidelberg New York
Paris Tokyo Hong Kong
Barcelona Budapest

D. Lloyd, BSc, PhD, DSc
Professor of Microbiology, School of Pure and Applied Biology,
University of Wales College of Cardiff, PO Box 915, Cardiff CF1
3TL, Wales

E. L. Rossi PhD
Clinical Psychologist, C. G. Jung Institute of Los Angeles, 23708
Harbor Vista Drive, Malibu, CA 90265, USA

Cover Illustration: Ch. 12, Fig. 12. Day-chart of sleep–waking behaviour
and stereotyped activity of the mentally impaired boy "Ulf".

ISBN 3-540-19746-X Springer-Verlag Berlin Heidelberg New York
ISBN 0-387-19746-X Springer-Verlag New York Berlin Heidelberg

British Library Cataloguing in Publication Data
Ultradian Rhythms in Life Processes:
Inquiry into Fundamental Principles of
Chronobiology and Psychobiology
  I. Lloyd, David
  II. Rossi, Ernest Lawrence
  574. 1882
ISBN 3-540-19746-x

Library of Congress Cataloging-in-Publication Data
Ultradian rhythms in life processes: an inquiry into fundamental
  principles of chronobiology and psychobiology / David Lloyd and
  Ernest L. Rossi (eds.).
    p. cm.
  includes index.
  ISBN 3-540-19746-X (alk. paper). — ISBN 0-387-19746-X (alk.
paper)
    1. Ultradian rhythms. 2. Life cycles (Biology)
  [DNLM: 1. Activity Cycles—genetics. 2. Chronobiology—genetic.
3. Circadian Rhythm—genetics. 4. Sleep. 5. Wakefulness. QT 167
U465]
QP84.6.U44   1992
612".022—dc20
DNLM/DLC   92-2298
for Library of Congress   CIP

Apart from any fair dealing for the purposes of research or private study, or criticism or review, as
permitted under the Copyright, Designs and Patents Act 1988, this publication may only be
reproduced, stored or transmitted, in any form or by any means, with the prior permission in
writing of the publishers, or in the case of reprographic reproduction in accordance with the terms
of licences issued by the Copyright Licensing Agency. Enquiries concerning reproduction outside
those terms should be sent to the publishers.

© Springer-Verlag London Limited 1992
Printed in the United States of America

First published 1992

The use of registered names, trademarks etc. in this publication does not imply, even in the
absence of a specific statement, that such names are exempt from the relevant laws and regulations
and therefore free for general use.

Product liability: The publisher can give no guarantee for information about drug dosage and
application thereof contained in this book. In every individual case the respective user must check
its accuracy by consulting other pharmaceutical literature.

Typeset by Best-set Typesetter Ltd, Hong Kong
12/3830-543210 Printed on acid-free paper

Preface

Profound progress has been made in the fields of chronobiology and psychobiology within the past decade, in theory, experiment and clinical application. This volume integrates these new developments on all levels from the molecular, genetic and cellular to the psychosocial processes of everyday life. We present a balanced variety of research from workers around the globe, who discuss the fundamental significance of their approach for a new understanding of the central role of ultradian rhythms in the self-organizing and adaptive dynamics of all life processes.

The years since the publication of *Ultradian rhythms in physiology and behavior* by Schultz and Lavie in 1985 have seen a burgeoning realization of the ubiquity and importance of ultradian rhythms within and between every level of the psychobiological hierarchy. The experimental evidence lies scattered through a disparate literature, and this volume attempts, albeit in a highly selective manner, to bring together some of the different strands. The editors are very conscious of the omission of many important current aspects; e.g. we have not included any of the fascinating and indeed long- and well-established experiments with plants (Bünning 1971, 1977; Guillaume and Koukkari 1987; Millet et al. 1988; Johnsson et al. 1990) that are widely regarded as having initiated the whole field of chronobiology (De Mairan 1729). Neither have we reviewed recent developments on glycolytic oscillations, since a great deal of the seminal work was already completed by 1973 (Chance et al. 1973). Cell signalling systems have been covered by Goldbeter's (1989) multiauthor volume, as has chaos in biology (the subject of a prescient meeting here in Cardiff in 1986 (Degn et al. 1987)), the evolution of biological complexity (Mosekilde 1991) and self-modifying systems (Kampis 1991).

The common denominator of all the contributors to this volume is their agreement on the evidence for the all-pervasiveness of rhythmic organization, its evolutionary conservation from amoeba to human, and the continuum of its nature as a hallmark of life itself. In the Epilogue we attempt to utilize this new consensus by outlining an emergent unifying hypothesis of chronobiology and psychobiology.

References

Bünning E (1971) The adaptive value of leaf movements. In: Menaker M (ed) Biochronometry. National Academy of Sciences, Washington, DC, pp 203–211

Bünning E (1977) Fifty years of research in the wake of Wilhelm Pfeffer. Ann Rev Plant Physiol 28:1–22

Chance B, Pye EK, Ghosh AK, Hess B (eds) (1973) Biological and biochemical oscillators. Academic Press, London

De Mairan (1729) Observation botanique. Histoire de l'Académie Royale des Sciences, Paris, vol 35

Degn H, Holden AV, Olsen LF (eds) (1987) Chaos in biological systems. Plenum Press, New York

Goldbeter A (ed) (1989) Cell to cell signalling: from experiments to theoretical models. Academic Press, London

Guillaume FM, Koukkari WL (1987) Two types of high frequency oscillations in Glycine max (L.) Merr. In: Pauly JE, Scheving LE (eds) Advances in chronobiology, pt A. Alan R Liss, New York, pp 47–57

Johnsson A, Engelmann W, Antkowiak B (1990) Leaf movements in nyctinastic plants as hands of the biological clock. In: Satter RL, Gorton HL, Vogelmann TC (eds) The pulvinus. American Society for Plant Physiology, Rockville, MD, pp 79–100

Kampis G (1991) Self-modifying systems in biology and cognitive science. A new framework for dynamics, information and complexity. Pergamon Press, Oxford

Millet B, Botton A-M, Hayoum C, Koukkari WL (1988) An experimental analysis and comparison of three rhythms of movements in bean (*Phaseolus vulgaris* L.). Chronobiol Int 3:187–193

Mosekilde E (ed) (1991) Complex dynamics and biological evolution. Plenum Press, New York

Schultz H, Lavie P (eds) (1985) Ultradian rhythms in physiology and behavior. Springer, Berlin Heidelberg New York

David Lloyd and Ernest L. Rossi
Cardiff, Wales, November 1991

Acknowledgements

The editors acknowledge the loyal assistance of many members of the faculty staff and student body of the School of Pure and Applied Biology and the University of Wales, Cardiff. In particular we recognize the enthusiastic help of Juliette Thomas and Margaret Ryan, who have helped this volume to make its appointed chronobiological rounds.

Contents

Contributors

R. Abraham
Division of Natural Sciences, University of California, Santa Cruz, CA 95064, USA

G. Brandenberger
Laboratoire de Physiologie et de Psychologie Environmentales CNRS/INRS, 21 rue Bequerel, 67087 Strasbourg Cedex, France

V. Y. Brodsky
Institute of Developmental Biology, Academy of Sciences of the USSR, 26 Vavilov Street, Moscow 117808, Russia

H. B. Dowse
Department of Zoology, University of Maine, Orono, ME 04469, USA

A. Garfinkel
Brain Research Institute, 1804 Life Sciences Building, University of California, Los Angeles, CA 90024-1527, USA

D. A. Gilbert
Department of Medical Biochemistry, Medical School, PO Wits, Johannesburg 2050, South Africa

Mary L. Greenacre
Department of Genetics, University of Leicester, Adrian Building, University Road, Leicester LE1 7RH, UK

J. C. Hall
Department of Biology, Brandeis University, Waltham, MA 02254, USA

Toke Hoppenbrouwers
Rm 9L19, Women's Hospital, 1240 Mission Road, Los Angeles, CA
90033, USA

R. R. Klevecz
Division of Biology, Beckman Research Institute, City of Hope,
1450 East Duarte Road, Duarte, CA 91010, USA

C. P. Kyriacou
Department of Genetics, University of Leicester, Adrian Building,
University Road, Leicester LE1 7RH, UK

P. Lavie
Sleep Laboratory, Faculty of Medicine, Technion-Israel Institute of
Technology, Technion City, Haifa 32000, Israel

B. M. Lippincott
John F. Kennedy University, Graduate School of Professional
Psychology, 1, West Campbell #67, Campbell, CA 95008, USA

D. Lloyd
Microbiology Group, School of Pure and Applied Biology,
University of Wales, PO Box 915, Cardiff CF1 3TL, UK

Heather MacKinnon
Division of Food Science and Technology, CSIR, Pretoria, South
Africa

A. Meier-Koll
Fachgrüppe   Psychologie,   Sozialwissenschaftliche   Fakultät,
Universität Konstanz, Postfach 5560, D-7750 Konstanz I, Germany

A. A. Peixoto
Department of Genetics, University of Leicester, Adrian Building,
University Road, Leicester LE1 7RH, UK

J. M. Ringo
Department of Zoology, University of Maine, Orono, ME 04469,
USA

M. G. Ritchie
Department of Genetics, University of Leicester, Adrian Building,
University Road, Leicester LE1 7RH, UK

E. L. Rossi
Psychological Perspectives, 23708 Harbor Vista Drive, Malibu, CA
90265, USA

G. Shiels
Department of Genetics, University of Leicester, Adrian Building,
University Road, Leicester LE1 7RH, UK

Helen C. Sing
Department of Behavioral Biology, Walter Reed Army Institute of Research, Forest Glen Annex Building 189, Washington, DC 20307-5100, USA

M. Stupfel
Mécanismes Physiopathologiques des Nuisances de l'Environment, Institut National de la Santé et de la Recherche Medicale, 44 chemin de Ronde, 78110 le Vésinet, France

J. D. Veldhuis
Division of Endocrinology, Department of Internal Medicine, University of Virginia Health Sciences Center, and the NSF Science Center for Biological Timing, Charlottesville, VA 22908, USA

R. A. Wever
Max Planck Institut für Psychiatrie, D-8138 Erling-Andechs, Germany

Part I

The Molecular–Genetic–Cellular Level

There is nothing more powerful than an idea whose time has come.

Victor Hugo

Life is an ensemble of oscillators; when plotted on a time scale measured in seconds, its component reactions span at least 22 decades (Fig. I.1).

Relaxation times <100 ms are characteristic of intramolecular events, solid state membrane-associated phenomena, lateral mobilities of membrane phospholipids and proteins, transport processes, energy conservation, electron transport and cytoskeletal dynamics. These are all rapid reactions in the *fast metabolic* time domain.

Intermediary metabolism involves a complex network of diffusion rate-limited molecular interactions, and employs (largely) "soluble" enzymes; here (in the *metabolic* time domain) isotopic labelling experiments indicate relaxation times within the range 0.2–300 s.

The *epigenetic* time domain encompasses all those reactions that stem from information encoded and controlled by a pre-existing genome (the product of the previous replication) plus those remarkable properties inherent in, and emergent from, the forces and fluxes of the complex system itself (i.e. "self-organization"). It is not yet established exactly how much of the information content of an organism ($>10^9$ bits for a bacterium, $>10^{28}$ bits for a human) is "hardwired", i.e. actually resides physically in DNA; it may be rather a small proportion of the total. In this time domain we are concerned with transcription, translation (and the attendant control systems requiring DNA-binding proteins, initiation, elongation and termination factors, etc.), protein folding and processing, membrane assembly, organelle assembly, membrane and cell wall extension and the elaboration of extracellular structures.

Next we have the *cell cycle* domain. *Vibrio natriegens*, a marine bacterium, sets the record with a mean generation time of 600 s. Each and every molecule of this organism has to be duplicated within this time in order to produce two daughters identical with (or rather very similar to) the mother cell when the latter was newly divided. This is in fact an underestimate of biosynthetic requirement because it neglects the extensive turnover of macromolecules essential for the continued maintenance of life, even when conditions are ideal and growth is rapid. Degradation is as important as synthesis in determining the levels of the inherently unstable constituents of a living organism. Even DNA itself, so chemically inert when outside a cell that its sequence survives, certainly on an archaeological time scale, is prone in vivo to damage and requires continual repair. In metazoa, higher plants and other organisms living in the natural environment, circadian control dominates all life processes: the approximately 24-h periods of the *circadian* domain that coordinate the organism with its environment have been studied for a quarter of a millenium.

In Chap. 1, Lloyd describes the discovery of an ultradian clock which is responsible for intracellular time keeping in lower eukaryotes. He suggests that multicellular organization also uses this timekeeper for the synchronization of events in tissues and organs, and that suitable coupling may result in the generation of circadian rhythms. Brodsky (Chap. 2) emphasizes that the entrainable hourly rhythms of protein synthesis that he first observed 30 years ago occuring in cell-free systems as well as in tissues and organs must similarly be employed in integrative functions. Klevecz (Chap. 3) suggests that the fundamental macromolecular ultradian oscillator is a chaotic attractor, as the precision of the circadian rhythms produced by coupling cannot be explained on a limit cycle basis. Gilbert and MacKinnon (Chap. 4) discuss control of cell division cycles on the basis of oscillating coenzyme levels and show that

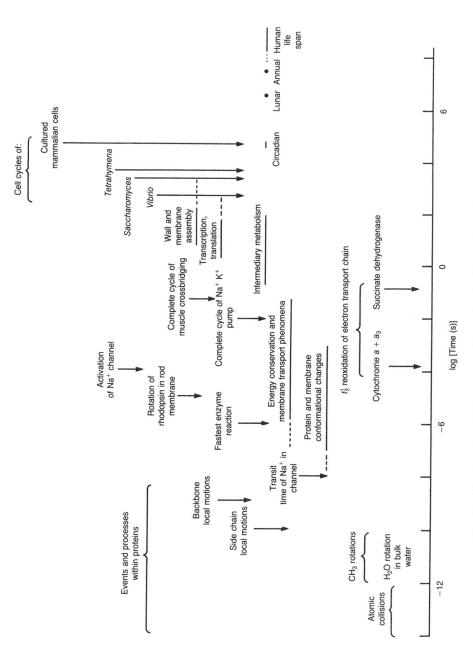

Fig. I.1. The dynamics of living systems. (Reproduced, with permission, from Lloyd 1987.)

disturbances of short-period rhythms alter replication frequency and may thereby cause cancer. Kyriacou et al. (Chap. 5) suggest that maybe we should begin with the circadian cycle and attempt to scale down to the ultradian (fast metabolic) domain of the *Drosophila* courtship song. On the other hand, the widely conserved *per* gene may be essential for coupling high frequency rhythms to generate circadian periodicities, but there are some objections to this. Dowse and Ringo (Chap. 6) stress that the strength of ultradian rhythms of individual flies varies inversely with the strength of their circadian rhythms. They go on to suggest that circadian rhythms are built from coupled ultradian rhythms, which in turn are constructed from even higher frequency rhythms.

Why stop there? A glance again at Fig. I.1 leads to a further hypothesis, implicit in the recurrent theme of these six chapters. The living organism shows self-similarities on *every* level of the temporal hierarchy. The oscillating ensemble has a fractal structure: the slow rhythms readily perceived by our senses are merely the emergent outputs of a system that, at its molecular core, has relaxation times measured in picoseconds. The dynamics of the living machine reveals and reflects an amplified version of its component molecular machines.

This coherent ensemble of oscillators is bombarded continually by a plethora of small molecules in aqueous solution. Intramolecular rotations and vibrations are an essential part of molecular recognition; thermal agitation is necessary for information flow, catalytic transport and transformation. Chaotic fluxes present advantages for new exploratory behaviour; these characteristics will not have been overlooked by the forces of evolutionary selection. All kinds of periodic behaviour (limit cycle, relaxation oscillatory, aperiodic, quasi-periodic, chaotic and hyperchaotic) will eventually be observed within and between every level of the temporal hierarchy. We are but beginning!

Reference

Lloyd D (1987) The cell division cycle. Biochem J 242: 313–321

Intracellular Time Keeping: Epigenetic Oscillations Reveal the Functions of an Ultradian Clock

D. Lloyd

Introduction

Unless special constraining elements are built in, any complex system will have a natural tendency to exhibit oscillatory behaviour; the more complex the system, the greater the possibilities for outputs of various frequencies and amplitudes. We see this for atomic motions within molecules, current flow in electronic circuits, concentrations of chemical reactants in the test tube or in the universe, and in mechanical structures (e.g. bridges and boats). Living organisms are extremely complex systems ($>10^9$ bits of information for a bacterium, $>10^{28}$ bits for a human being), and the dynamics of their subsystems span a wide range of time scales (Lloyd 1987; Fig. 1.1). The potential usefulness of periodic behaviour has not been overlooked in biological evolution (Lloyd and Volkov 1991; Table 1.1), and has resulted in an astonishing richness of dynamic behaviour. Detailed description of this panoply is only in its infancy, and our appreciation of most of its functional significance is even more rudimentary. Compared with the intensely studied and elaborately elucidated *structural* hierarchy of cellular components and constituents, the documentation of time structure is rather meagre; some parts of the frequency range are much better documented than others, but none is completely understood (Lloyd et al. 1982b).

Two of the time domains have received much attention:

1. The conspicuous rhymicity of organisms on a *circadian* ($\tau \approx 24\,\text{h}$) time scale led to the accurate description of the existence of a biological rhythm more than a quarter of a millenium ago. Since then, intensive investigations at all levels from those with animal population studies to those using cloned gene sequences have provided a detailed list of properties and biological functions. Despite a plethora of models, we still await the definitive mechanistic molecular explanation.

2. Periodic behaviour of biochemical reactions in the "*metabolic* domain" ($\tau < 5\,\text{min}$) have excited much interest since detailed studies on the "glycolytic oscillator" were initiated in Chance's laboratory in 1964 (Ghosh and Chance 1964). For many years the usefulness of this system for gaining an understanding of metabolic control principles has been evident. It is still not clear

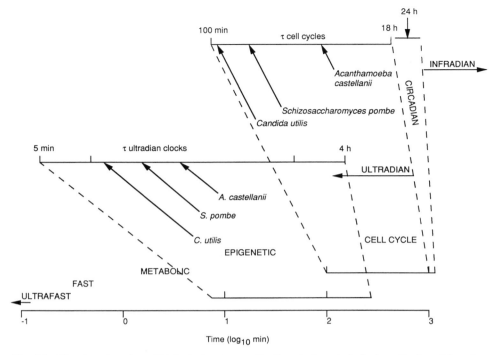

Fig. 1.1. The time domains of biological systems: *ultradian* clock periods (epigenetic domain) and cell division times (*cell cycle domain*) are indicated for different species of lower eukaryotes.

whether oscillations in concentrations of glycolytic intermediates do occur in vivo in natural populations, or whether they serve any biological function. Much more recently the key role of cyclic AMP-dependent oscillations in the cellular differentiation processes of the slime mould *Dictyostelium discoideum* provides a splendid example of *functional* periodicity (Goldbeter 1990). Inter- and intracellular signalling systems relying on generation of specific time-dependent metabolite concentration changes include the Ca^{2+} oscillator (Berridge 1989) and the many different hormones which show pulsatile release kinetics (Goldbeter 1989; Brandenberger, Chap. 7; Veldhuis, Chap. 8). Thus, studies of metabolic oscillations open up new chapters in our appreciation of biological systems control theory.

Between the metabolic and circadian time domains lies the "*epigenetic domain*". The upper boundary of this domain is determined by the cell cycle time (i.e. the time interval between successive cell division events, as short as 10 min for some species of bacteria, or as long as many days in some

Table 1.1. Some functions of oscillations in living systems

Process	Reference
1. Temporal organization	
(a) Coordination of processes in separate subcellular compartments	Turner and Lloyd 1971
(b) Separation of incompatible processes	Edwards and Lloyd 1980
(c) Phase entrainment	Edmunds 1984
(d) Frequency entrainment	Noble 1975
(e) Prediction (e.g. of dawn or dusk)	Bünning 1973
2. Spatial organization	
(a) Positional information specification	Goodwin and Cohen 1969
3. Signalling	
(a) Increased range	Nanjundiah 1973
(b) Frequency encoding	Rapp et al. 1981
4. Increased energy efficiency	Richter and Ross 1981

eukaryotes). Events occurring in the epigenetic domain include those of transcription and translation. The ultradian clock operates in this domain. For several reasons, and with few exceptions (Brodsky 1975; Brodsky and Nechaeva 1988), its pervasive influence has been largely overlooked.

Methodological Problems

Why have biologists been slow to recognize the intricacies of temporal organization, even though currently available techniques make accessible many of the time scales involved? Briefly the reasons are as follows:

1. The enormous successes of the reductionist approach; firstly with crude cell-free extracts, then with isolated organelles and membrane preparations, and finally with purified constituents have distracted the efforts of investigators away from studies of integrated cellular function.
2. The comparatively recent refinement of physical techniques suitable for non-invasive studies of intact biological entities (organisms or cells).
3. The mistaken philosophy that a "representatively sampled" population can give an understanding of the temporal organization of its component individuals still predominates.
4. Unswerving adherence to the principles of Claude Bernard's ideas about the homeostatic control of physiological processes, whereby steady-state operation is rigorously defended against the perturbations and potentially destructive influences of the noisy environment (Lloyd and Stupfel 1991).

Thus, most of the biochemical information currently available has been obtained by the use of samples taken from heterogeneous populations of cells or microorganisms under conditions which favour exponential growth, and much of the "steady-state" behaviour described represents a time average.

Methodology

Microbial cultures are usually grown without steps being taken to produce any synchronization of events occurring within individual organisms: a variety of methods are available for producing cultures in which growth and cell division processes become synchronous (Lloyd et al. 1982b). Favoured methods rely on the *selection* of a defined size class (usually the smallest and therefore "youngest" organisms) for the initiation of synchronous growth. Care must be taken to minimize the effects of possible sources of disturbance (nutrient deprivation, temperature changes, anaerobiosis, osmotic pressure changes, light–dark transitions, liquid shear forces, centrifugal forces etc.). A second approach to the production of synchronously dividing populations uses pro-cedures known to be reversibly inhibitory to growth processes. These methods *induce* synchronous division of the entire population of individuals, but usually give a distorted picture of cell cycle events due to the direct effects of the procedure, at least in the first cycle following synchronization.

Because of the intimate association of the ultradian clock with the cell division cycle (phase-locking, as revealed by quantization of cell cycle times, see below), any procedure giving synchronously dividing cultures also syn-chronizes many epigenetic oscillations, as these are closely coupled to the ultradian clocks of individual organisms of the population.

The experimental results that fit the models described in this chapter were obtained with populations of lower eukaryotes growing and dividing syn-chronously after a gentle procedure for the selection of small cells (Edwards and Lloyd 1978). To obtain these suspensions, organisms (e.g. the soil amoeba *Acanthamoeba castellanii*) were sedimented by low-speed ($10g$) centrifugation for a short time (2 min) while still in their growth media. The most slowly sedimenting subpopulations (the small, recently divided organisms) were decanted and cultured separately. Control (asynchronous) cultures were also simultaneously studied: these were obtained by mixing sedimented and non-sedimented organisms after centrifugation. The asynchronous cultures enable assessment of possible perturbative influences of the experimental procedures. Perfect synchrony is never obtained in practice, and the highest resolution of temporal organization requires observation of single organisms (e.g. *Paramecium tetraurelia*; Kippert 1985).

Experimental Results

In synchronously dividing cultures of lower eukaryotes, we have observed over the past 20 years (Table 1.2) oscillations in many different biochemical parameters; periods are of the order of 1 h and are a characteristic (Lloyd and Edwards 1987a) of the species studied. The most easily monitored is cellular

Table 1.2. Ultradian rhythms in synchronous cultures of lower eukaryotes

Species	Synchrony method	Temperature (°C)	Cell cycle τ (h)	Ultradian τ (min)	Measured variable	References
Schizosaccharomyces pombe 972h⁻ (glucose grown)	GSS CFSS	30	2.65	40	Respiration Adenine nucleotides	Poole et al. 1973 S. J. Jenkins and D. Lloyd, unpublished results
	CFSS DAIS				Mitochondrial ATPase Respiration	Edwards and Lloyd 1977 Poole 1977
	DAIS				Adenine nucleotides	Poole and Salmon 1978
	CFSS				Respiration, adenylates	El-Khayat and Lloyd 1978
	DAIS				NADH and flavin fluorescence	Bashford et al. 1980
	CHT				Septum formation	Kippert and Lloyd 1991
Crithidia fasciculata	GSS	30	5.5	66	Respiration Adenine nucleotides	Edwards et al. 1975
Tetrahymena pyriformis, ST	CFSS	30	2.5	50	Respiration, adenylates, heat production	Lloyd et al. 1978
T. thermophila *T. pyriformis* AII	CHT SR	20 29		240 31	Tyrosine aminotransferase Respiration	Michel and Hardeland 1985 Kippert 1987
Acanthamoeba castellanii, Neff	CFSS DCSS DCSS DCSS DCSS DCSS DCSS	30	7.8	69	Respiration Respiration, adenylates Total protein, RNA Flavin fluorescence Enzyme activities and amounts Cytochromes	Chagla 1978 Edwards and Lloyd 1978 Edwards and Lloyd 1980 Bashford et al. 1980 Edwards et al. 1981 Edwards et al. 1982 Lloyd et al. 1983

Table 1.2. *Continued*

Species	Synchrony method	Temperature (°C)	Cell cycle τ (h)	Ultradian τ (min)	Measured variable	References
Candida utilis, NCYC 193 (glucose grown)	CFSS	30	1.7	33	Respiration	Kader and Lloyd 1979
Dictyostelium discoideum	DCSS	22	6.0	60	Respiration, total protein, RNA	Woffendin 1982 Woffendin and Griffiths 1985
Paramecium tetraurelia	SC SC	27	5.7	69 69	Motility Cell division	Kippert 1987 Lloyd and Kippert 1987
Chlamydomonas reinhardii, CCAPII/32b mating type minus	LD DCSS	28	24 20	55 55	Chlorophyll a Chlorophyll a	Jenkins et al. 1989 Jenkins et al. 1990
Euglena gracilis, strain Z (no. 1224–5/25)	LD	23	24	240	Tyrosine aminotransferase	Neuhaus-Steinmetz et al. 1990

CHT, cyclic heat treatment (entrainment); (Kippert and Lloyd 1991); DCSS, centrifugal size selection, directly from culture; (Chagla and Griffiths 1978); CFSS, continuous flow size selection, directly from culture; (Lloyd et al. 1975); DAIS, 2'-deoxyadenosine-induced synchrony; (Poole 1977); GSS, gradient size selection procedures; (see original papers); LD, light–dark cycles (LD, 12:12); (Jenkins et al. 1989); SC, single cell observations; (Kippert 1987); SR, starvation and refeeding; (Lloyd and Kippert 1987).

respiration, as measured by oxygen consumption (e.g. Poole et al. 1973; Edwards and Lloyd 1978). Other observables include total cellular protein (Edwards and Lloyd 1980), heat production using microcalorimetry (Lloyd et al. 1978), the activities of various enzymes (Michel and Hardeland 1985), spectrophotometrically or immunologically detectable enzyme protein (e.g. for cytochrome c oxidase (Lloyd et al. 1983) or catalase (Edwards et al. 1981), respectively), and a natural inhibitor of mitochondrial ATPase (Edwards et al. 1982). These parameters show no oscillatory behaviour in asynchronous controls, and the oscillations are therefore not the consequences of external perturbation: they are rather the expression of an endogenous rhythm. Their characteristic periods place them in the epigenetic time domain, i.e. the observed frequencies suggest the operation of transcriptional and translational control processes, rather than that of the faster processes involving biochemical transformations of small molecules. These "epigenetic oscillations" are observable outputs of an ultradian clock, and several typical examples are different from all the other high frequency oscillatory phenomena hitherto described in biology, in that they exhibit a temperature-compensated period. Thus, measurement of periods of the oscillations of respiration or of the total protein of amoebae at different temperatures within the physiological range (20–30 °C)

Table 1.3. Effects of temperature upon period of ultradian rhythms

Organism	Rhythm	Temperature (°C)	Period (min)	Q_{10}	Reference
Acanthamoeba castellanii	Respiration	30	65	ND[a]	Lloyd et al. 1982a
		27	66		Lloyd and Edwards
		25	81		1987a
		22.5	80		Marques et al. 1987
		20	69		
	Protein	30	66	ND[a]	
		27	71		
		25	86		
		22.5	90		
		20	73		
Tetrahymena thermophila	Tyrosine aminotransferase activity	10–30	240–300	ND[a]	Michel and Hardeland 1985
Euglena gracilis	Tyrosine aminotransferase activity	16–31.5	240–300	<1.0	Balzer et al. 1989
Tetrahymena pyriformis	Respiration	33	30	1.12	Kippert 1987
		19	35		
Paramecium tetraurelia	Motility	33	70	1.04	Lloyd and Kippert 1987
		27	69		
		18	74		
	Cell division (quantal increments)	33	70	1.04	
		27	69		
		21	72		

[a] Q_{10} values not cited, as temperature compensation was not uniform over the range tested.

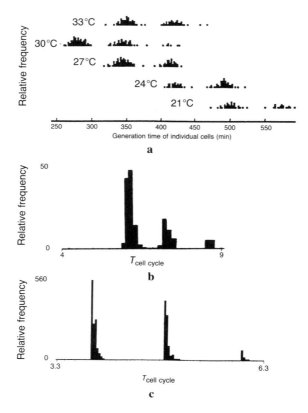

Fig. 1.2. Quantized cell cycle times. **a** Experimentally observed distributions of generation times of individual *Paramecium tetraurelia* cells at various steady-state growth temperatures. Each square represents timing of division of an isolated organism. (Reproduced, with permission from the Company of Biologists, from Lloyd and Kippert 1987.) **b** Frequency distributions of cell cycle times ($T_{cell\ cycle}$) from the model system incorporating noise (equation (1.3)) when $\tau_L = 1$; $\tau_R = 0.1 \mathcal{H} = 0.15$; $D = 0.4$; $\gamma = 1.5$; $\delta = 0.15$; $\Delta\eta = 0.3$; ς = random numbers; $\varepsilon[0,1]$; $C = 0.45$; $T_{UR} = 1.25$ and $\eta = 5.55$. (Reproduced, with permission, from Lloyd and Volkov 1990.) **c** $T_{cell\ cycle}$ for parameter values as follows: $\eta = 4.95$; $D = 0.4$; $\mathcal{H} = 0.15$; $\gamma = 1.5$; $\delta = 0.15$; $C = 0.45$; $\Omega = 5.0$; $\tau_L = 1$ and $\tau_R = 0.1$ for equation (1.2) (in the absence of a noise term). (D. Lloyd, A. L. Lloyd and L. F. Olsen, unpublished results.)

gave only a small decrease in the period (69–65 min). This property (Lloyd et al. 1982a; Table 1.3) resembles that extensively described for circadian systems and, in those systems, temperature compensation is the clearest indicator of a time-keeping function (Sweeney and Hastings 1960; Balzer and Hardeland 1988).

We therefore proposed that the ultradian clock, composed of (or coupled to) the observed epigenetic oscillations, has timing functions, and it was predicted that an important example of a time-controlled process would be that of cell division itself (Lloyd et al. 1982a). As the interdivision time of organisms (i.e. the time interval between successive cell division events) is strongly temperature dependent we suggested that control by a temperature-independent timer would give rise to an increase in cell cycle times by discrete quantal increments: hints of this process had been noted for *A. castellanii*. This model could also explain the previously observed quantal differences in cell cycle times in cultured mammalian cell lines (Klevecz 1976), and also the discrete ranges for photoentrainment of the ultradian mode of cell division in *Tetrahymena pyriformis* (Readey 1987).

Work with single cells of the large ciliate protozoon *Paramecium tetraurelia* further confirmed the operation of an ultradian clock and added a behavioural rhythm, that of motility, to the list of coupled outputs (Lloyd and Kippert 1987). Furthermore, measurement of the generation times of more than 600 individuals at different growth temperatures indicated clustering rather than random division times. The clusters are separated by a time interval which corresponds to the period of the motility rhythm. Again temperature compensation occurs (Fig. 1.2a), and values were 70.0 min at 33 °C, 69.2 min at 27 °C and 71.8 min at 21 °C.

Epigenetic oscillations in the respiration of growing cells were first observed during investigations of mitochondrial biogenesis in synchronously dividing cultures of *Schizosaccharomyces pombe* (Poole et al. 1973). Symmetrical cell division is a characteristic which makes this fission yeast preferable to budding yeast for many cell cycle studies: the attractions of its well-defined genetic system have led us to recent further studies of this organism (Kippert and Lloyd 1991). In the wild-type strain 972 h$^-$, the period length of the ultradian clock is between 40 and 45 min. Its influence on the cell division cycle is observed as a periodic change in the percentage of cells with a septum. Its period is independent of growth conditions (i.e. it is similar irrespective of whether organisms are grown on complex or defined media, or under glucose-repressed or glucose-derepressed conditions), and is almost independent of temperature over the range studied (between 26 and 34 °C). Persistence of the rhythm with little damping over 25 cycles suggests that cell–cell communication gives mutual synchronization. Measurements of the timings of DNA replication and mitosis (e.g. the production of binucleate cells which have not yet started to develop a septum) indicate that these are highly precise and under the control of the ultradian clock. Further work proceeds with cell division cycle mutants. Studies of the rhythm of septation in several different nuclear and mitochondrial mutants show that different strains have different periods (range 28–52 min). This is in contrast to the constancy of the wild-type period, and suggests that within a certain range no selective pressure is exerted on a particular period length. A similar conclusion was previously reached for different *species* of yeast and Protozoa (and also for different *strains* of *T. pyriformis*); each genotype has its own characteristic period (Table 1.2; Lloyd and Edwards 1987b).

It has been suggested repeatedly that the period of the circadian clock might be generated by frequency demultiplication (i.e. counting every *n*th output) of a higher frequency oscillator. The circadian clock would thereby be composed of ultradian rhythms. Recent evidence has indicated the simultaneous operation of ultradian oscillations and the circadian clock (e.g. for chlorophyll accumulation in *Chlamydomonas reinhardii* (Jenkins et al. 1989, 1990), motility and phototaxis in *Euglena gracilis* (Adams 1988, 1989; Balzer et al. 1989) and tyrosine aminotransferase activity (Neuhaus-Steinmetz et al. 1990).

Mathematical Models

I. E. Volkov, noting the remarkable quantization of cell cycle times in *P. tetraurelia* (Lloyd and Kippert 1987), demonstrated how such behaviour could be simulated by the interaction of the ultradian clock with a putative cell cycle (mitotic) oscillator (Lloyd and Volkov 1990, 1991).

The model is based on an earlier one of Chernavskii et al. (1977) in which a mitotic or cell division cycle oscillator with one slow variable (τ_L, of the order of hours) and one fast variable (τ_R, of the order of minutes) may be described by the following system of equations:

$$\tau_L \frac{dL}{dt} = \eta - 2LR - DL$$

$$\tau_R \frac{dR}{dt} = \mathscr{H} + LR - R^2 - \frac{\gamma R}{(R + \delta)} \qquad (1.1)$$

where L and R are concentration terms, τ_L and τ_R are their characteristic times, and η, D, H, γ and δ are velocity constants. Both the slow and fast components oscillate with the same period but with very distinctive waveforms. The non-symmetrical time dependence of the slow variable becomes more pronounced as the system approaches a bifurcation point (i.e. the non-proliferating state).

Modulation by an output of the endogenous temperature-compensated ultradian clock (period T_{UR}) is simulated (Lloyd and Volkov 1990) by the introduction of a harmonic term as a forcing function into the slow equation, which then becomes:

$$\tau_L \frac{dL}{dt} = \eta - 2LR - DL + C \sin \Omega t \qquad (1.2)$$

where

$$T_{UR} = \frac{2\pi}{\Omega} \ll T_{\text{cell cycle}}$$

The autooscillating solution considers that cells divide when $L(t)$ reaches a threshold to initiate a rapid phase of the limit cycle. To simulate natural variability of cell division cycle timing (Mustafin and Volkov 1982) a noise term ($\Delta\eta \times \zeta_i$) was included:

$$\tau_L \frac{dL}{dt} = \eta - 2LR - DL + \Delta\eta\zeta_i + C \sin \Omega t \qquad (1.3)$$

When $(\eta - \eta_{\text{BIF}}) \ll 1$ (BIF indicating bifurcation), $T_{\text{cell cycle}}$ has either a bimodal solution or a polymodal one, depending on the value of η.

Thus, the ultradian rhythm modulates the noisy limit cycle trajectory, which increases the probability that some parts of it reach a threshold for cell division. Both variability and quantization of cell cycle times (Fig. 1.2b) are therefore accounted for. Conditions necessary for quantization are: (a) operation in the neighbourhood of a bifurcation, (b) that the period of the cell cycle depends strongly on η, and (c) that the cell cycle oscillator be of the relaxation type rather than saw-toothed.

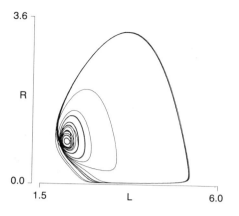

Fig. 1.3. Phase portrait of the system described in equation (1.2) with parameter values as in Fig. 1.2c. (D. Lloyd, A.L. Lloyd and L. F. Olsen, unpublished results.)

Further exploration of the dynamic structure of the equations (D. Lloyd, A. L. Lloyd and L. F. Olsen, unpublished results) in the *absence* of the noise term showed that solutions for η of between 3.00 and 4.50 in equation (1.2) show quasiperiodicity (circular or toroidal next-amplitude maxima plots). Higher values of η (e.g. $\eta = 4.95$) indicated chaotic dynamics. The phase portrait of R versus L (Fig. 1.3) and a time-one map (a plot of R versus L at time intervals of the ultradian period T_{UR}) indicate an intricate structure indicative of either (a) a complex torus with an integer dimension, or (b) a chaotic attractor with an almost integer fractal dimension (correlation dimension of Grassberger and Procaccia (1983) of 1.95). Calculation of the Lyapunov dimension D_L indicates a value of 2.03, and thus strongly suggests chaotic dynamics. Remarkably, a relative frequency plot of cell cycle times for $\eta = 4.95$ shows a trimodal distribution (Fig. 1.2c) with negatively skewed dimensions. Thus, dispersion of cell cycle times, multimodality (quantized cell cycle times) and chaotic dynamics are all phenomena that can result from the interaction of two oscillators (the ultradian clock and the cell cycle oscillator). This model differs from an earlier *stochastic* one (Lloyd and Volkov 1990, 1991) in that the complex dynamic structure is generated in the absence of external noise. The variable and quantized cell cycle times are consequences of a chaotic trajectory and thus have a purely *deterministic* basis. Chaos is common in multioscillator systems of weakly coupled units, and chaotic systems are generally more robust to perturbations than their periodic counterparts (Schaffer et al. 1986).

Continuing investigations (M. Stolyarov, personal communication) provide evidence that a purely deterministic model provides a good fit with experimental observations only for a narrow range of values of η and for limited relaxation characteristics of the cell cycle oscillator. Thus Stolyarov prefers to think of variability and quantization as arising from the effects of a white noise term.

Discussion

In lower eukaryotes, the outputs of the ultradian clock may have many functions: the clock provides a time base for intracellular coordination. We have suggested that cycles of activity of energy-yielding processes (Edwards and Lloyd 1978) are the consequences of timer-controlled alternating phases of high and low biosynthetic energy need (Edwards and Lloyd 1980; Edwards et al. 1981, 1982; Lloyd and Edwards 1984, 1986, 1987a,b). Studies on the effects of uncoupling energy conservation reactions from those of electron transport indicate that ADP levels limit respiratory rates, and the phase relationships between these two measured variables confirm this mechanism, both in *A. castellanii* (Edwards and Lloyd 1978) and in *Schiz. pombe* (Poole and Salmon 1978). Thus mitochondrial activities are determined by energy requirement on an epigenetic time scale rather than on a faster metabolic dynamic. Rhythms of motility in protozoa (Kippert 1985, 1987) provide an easily measured behavioural ultradian output. The cell division cycle requires precise temporal coordination and integration of both concurrent and sequential intracellular processes, and for these the ultradian clock provides the necessary time base.

There have been many models of the mitotic oscillator control of cell division (Gilbert 1968; Sel'kov 1970; Gilbert 1974, 1981; Kauffman and Wille 1975; Zeuthen 1978), and several suggestions for the identity of the putative oscillator. Experiments with cell cycle mutants of various lower eukaryotes (yeast, *C. reinhardii*, *Physarum*) in the late 1970s led to the conclusion that two major types of control were both essential (for a review, see Lloyd et al. 1982b). These were: (a) a sizer (organisms only divide when they become large enough), and (b) a timer (a certain time span ensures completion of all necessary processes). The past decade has seen the identification of various cyclin complexes; molecular oscillators common to yeasts (both *Sac. cerevisiae* and *Schiz. pombe*), starfishes, frogs and mice (Pelech 1990; Nurse 1990). The human homologue of the yeast tyrosine/threonine phosphatase (*cdc25* gene product) is probably a positive regulator of mitosis (Strausfeld et al. 1991). This "maturating promoting factor" system is thus highly conserved and has been termed a cellular clock; indeed the cell cycle itself has often been called a clock (Edmunds 1984). But an oscillator is only one component of a clock; and there is no evidence that cyclin/kinase systems constitute the *fundamental* mitotic oscillator (see Gilbert and MacKinnon, Chap. 4).

The two mathematical models presented here have both sizer (threshold) and timer (ultradian clock) elements, and explain how a temperature-dependent oscillator (the mitotic oscillator) is timed. Although originally conceived for a lipid-free radical mechanism, the state variables could equally well be thiol/disulphide (Sel'kov 1970) or phosphorylated/dephosphorylated proteins (Gilbert 1974). It is thus really a general concentration-state-dependent model, in which not only does a critical concentration (threshold) have to be attained but also a set value of the phase angle of the ultradian clock has to be satisfied. The novel feature of the present models is the participation of the ultradian clock as a forcing function in the differential equation for the slow variable. Interaction of the ultradian clock pulses with the cell cycle oscillator generates quantized cell cycle times similar to those observed experimentally. Whether or not variability of cell cycle timing arises purely from stochastic events or deterministically from the potentially chaotic behaviour of a relaxation oscillator interacting with ultradian clock pulses remains to be elucidated. Perhaps some systems operate by utilizing noisy limit cycles, others a strange attractor. For yet others, a combination of noise and chaos may provide. We

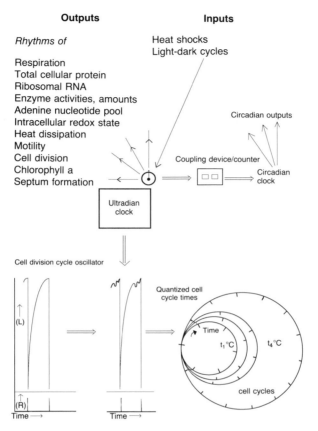

Fig. 1.4. The ultradian clock has multiple outputs (rhythms of respiration, adenine nucleotides, accumulating protein, enzyme activities and amounts, etc.): it also interacts with the cell division cycle oscillator to give quantal increments in cell cycle time as temperature is decreased. Ultradian clock pulses may also be counted to give circadian time keeping of diverse cell functions, including cell division time.

may speculate that if benefits accrue, any useful mechanism will be conserved. Benefits of dispersed and quantized cell cycle times may include the generation of diversity and the maintenance of functional independence of the individual (e.g. avoidance of entrainment). Too narrow an age structure may be a severe limitation for evolutionary survival (Conrad 1986).

In the natural environment, conditions of light intensity, temperature and nutrient supply do not often rival those provided in the laboratory for optimal growth rates. Rapid growth and multiplication under constant conditions does not make use of the resettable, long-period circadian timer, and instead relies on the ultradian clock. But natural populations, almost always multiplying slowly under less than optimum conditions (and often under conditions of nutrient deprivation), show circadian-controlled ("gated") cell cycle events (Sweeney 1982).

Indeed, evidence from diverse systems suggests that the circadian clock may be composed from ultradian rhythms (Fig. 1.4). Thus quantized departures from the wild-type period have been observed in clock mutants of *Neurospora crassa* (Gardner and Feldman 1980). The so-called arrhythmic *per*[0] mutant of *Drosophila melanogaster* shows ultradian rhythmicity when held in constant low-level red illumination (Helfrich 1985); further work suggests that the function of the *per*[+] gene product is to mediate the coupling of multiple ultradian oscillators to produce wild-type circadian rhythms (Dowse and Ringo 1987, and Chap. 6; Dowse et al. 1987). Disruption of normal rhythmicity of *Calliphora vicina* by constant bright light (Gibbs 1983), or of rats and hamsters by exposure to low temperatures (von Grosse 1985) or by mutation (Büttner and Wollnick 1984) reveals the composite ultradian rhythms. The damping of circadian rhythms at high unvarying levels of illumination (Bünning 1973) may be due to desynchronization of coupled oscillators.

Finally, we may speculate about timekeepers with frequencies even higher than those of ultradian clocks. A fascinating glimpse of possibly fractal (Brodsky, Chap. 2) self-similarities ("worlds within worlds") comes from the work on *Drosophila* courtship song, where a very high frequency clock operates (Kyriacou and Hall 1980, 1989; Kyriacou et al., Chap. 5). That changes in the period of this clock often go hand in hand with changes in circadian periods of various mutant flies suggests deep, but as yet unfathomed, interrelationships between very disparate time domains.

Summary

The time structure of a biological system is at least as intricate as its spatial structure. Although we have detailed information about the latter, our understanding of the former is still rudimentary. As techniques for monitoring intracellular processes continuously in single cells become more refined, it becomes increasingly evident that periodic behaviour abounds in all time domains.

Intracellular time keeping is essential. The presence of a temperature-compensated oscillator provides such a timer. The coupled outputs (epigenetic oscillations) of this ultradian clock constitute a special class of ultradian rhythm. These are undamped and endogenously driven by a device which shows biochemical properties characteristic of transcriptional and translational elements. Energy-yielding processes, protein turnover, motility and the timing of the cell division cycle processes are all controlled by the ultradian clock. Different periods characterize different species, and this indicates a genetic determinant. Suitably coupled, these ultradian rhythms may be used to construct the circadian clock.

Acknowledgements

Work with Dr E. I. Volkov and Dr M. N. Stolyarov was made possible by British Council and Royal Society support.

References

Adams KJ (1988) Circadian clock control of an ultradian rhythm in *Euglena gracilis*. J Interdiscipl Cycle Res 19:153–154

Adams KJ (1989) Circadian clock control of an ultradian rhythm in *Euglena gracilis*. In: Morgan E (ed) Chronobiology and chronomedicine: basic research and applications. Peter Lang, Frankfurt-am-Main, pp 13–22

Balzer I, Hardeland R (1988) Influence of temperature on biological rhythms. Int J Biometeorol 32:231–241

Balzer I, Neuhaus-Steinmetz V, Hardeland R (1989) Temperature compensation in an ultradian rhythm of tyrosine aminotransferase activity in *Euglena gracilis* Klebs. Experientia 45:476–477

Bashford CL, Chance B, Lloyd D, Poole RK (1980) Oscillations of redox state in synchronously dividing cultures of *Acanthamoeba castellanii* and *Schizosaccharomyces pombe*. Biophys J 29:1–12

Berridge MJ (1989) Cell signalling through cytosolic calcium oscillations. In: Goldbeter A (ed) Cell to cell signalling: from experiments to theoretical models. Academic Press, London, pp 449–460

Brodsky VY (1975) Protein synthesis rhythm. J Theor Biol 55:167–200

Brodsky VY, Nechaeva NV (1988) Protein synthesis rhythm (in Russian). Nauka, Moscow

Bünning E (1973) The physiological clock, circadian rhythms and biological chronometry, 3rd edn. English Universities Press, London

Büttner D, Wollnick F (1984) Strain-differentiated circadian and ultradian rhythms in locomotory activity of laboratory rat. Behav Genet 14:137–152

Chagla AH (1978) Growth and encystation of *Acanthamoeba*. PhD thesis, University of Wales

Chagla AH, Griffiths AJ (1978) Synchronous cultures of *Acanthamoeba castellanii* and their use in the study of encystation. J Gen Microbiol 108:39–43

Chernavskii DJ, Palamarchuk EK, Polezhaev AA, Solyanik GI, Burlakova EG (1977) Mathematical model of periodic processes in membranes with application to cell-cycle regulation. BioSystems 9:187–193

Conrad M (1986) What is the use of chaos? In: Holden AV (ed) Chaos. University Press, Manchester, pp 3–14

Dowse HB, Ringo JM (1987) Further evidence that the circadian clock in *Drosophila* is a population of coupled ultradian oscillators. J Biol Rhythm 2:65–76

Dowse HB, Hall JC, Ringo JM (1987) Circadian and ultradian rhythms in period mutants of *Drosophila melanogaster*. Behav Genet 17:19–35

Edmunds LN Jr (1984) Physiology of circadian rhythms in microorganisms. Adv Microb Physiol 25:61–148

Edwards C, Statham M, Lloyd D (1975) The preparation of large-scale synchronous cultures of the trypanosomatid *Crithidia fasciculata* by cell-size selection: changes in respiration and adenylate charge through the cell-cycle. J Gen Microbiol 88:141–152

Edwards SW, Lloyd D (1977) Mitochondrial ATPase of the fission yeast *Schizosaccharomyces pombe*: changes in activity and oligomycin-sensitivity during the cell cycle of catabolite-repressed and -derepressed cells. Biochem J 162:39–46

Edwards SW, Lloyd D (1978) Oscillations of respiration and adenine nucleotides in synchronous cultures of *Acanthamoeba castellanii*: mitochondrial respiratory control in vivo. J Gen Microbiol 108:197–204

Edwards SW, Lloyd D (1980) Oscillations in protein and RNA content during synchronous growth of *Acanthamoeba castellanii*: evidence for periodic turnover of macromolecules during the cell cycle. FEBS Lett 109:21–26

Edwards SW, Evans JB, Lloyd D (1981) Oscillatory accumulation of catalase during the cell cycle of *Acanthamoeba castellanii*. J Gen Microbiol 125:459–462

Edwards SW, Evans JB, Williams JL, Lloyd D (1982) Mitochondrial ATPase of *Acanthamoeba castellanii*: oscillating accumulation of enzyme activity, enzyme protein and F_1-inhibitor during the cell cycle. Biochem J 202:453–458

El-Khayat G, Lloyd D (1978) Changes in pool sizes of adenine nucleotides and ATPase in starved and synchronous cultures of *Schizosaccharomyces pombe* 972h$^-$. Proceedings of the Sixth International Symposium on Yeast, Montpellier, Abstract SII2

Gardner GF, Feldman JF (1980) The frq locus in *Neurospora crassa*: a key element in circadian clock organization. Genetics 96:877–886

Ghosh A, Chance B (1964) NADH oscillations in yeast. Biochem Biophys Res Commun 16:174–179

Gibbs FP (1983) Temperature dependence of the hamster circadian pacemaker. Am J Physiol 244:R607

Gilbert DA (1968) Differentiation, oncogenesis and cellular periodicities. J Theor Biol 21:113–122

Gilbert DA (1974) The nature of the cell cycle and the control of cell replication. BioSystems 5:197–204

Gilbert DA (1981) The cell cycle 1981. One or more limit cycle oscillations? S Afr J Sci 77: 541–546

Goldbeter A (ed) (1989) Cell to cell signalling: from experiments to theoretical models. Academic Press, London

Goldbeter A (1990) Rhythmes et chaos dans les systèmes biochimiques et cellulaires. Masson, Paris

Goodwin BC, Cohen MH (1969) A phase shift model for spatial and temporal organization in developing systems. J Theor Biol 25:49–107

Grassberger P, Procaccia I (1983) Measuring the strangeness of strange attractors. Physica 9D: 189–208

Helfrich C (1985) Untersuchungen über das circadiane System von Fliegen. Dissertation, University of Tübingen

Jenkins H, Griffiths AJ, Lloyd D (1989) Simultaneous operation of ultradian and circadian rhythms in *Chlamydomonas reinhardii*. J Interdiscipl Cycle Res 20:257–264

Jenkins H, Griffiths AJ, Lloyd D (1990) Selection-synchronized *Chlamydomonas reinhardii* display ultradian but not circadian rhythms. J Interdiscipl Cycle Res 21:75–80

Kader J, Lloyd D (1979) Respiratory oscillations and heat evolution in synchronous cultures of *Candida utilis*. J Gen Microbiol 114:455–461

Kauffman SA, Wille JJ (1975) The mitotic oscillator in *Physarum polycephalum*. J Theor Biol 55:47–93

Kippert F (1985) Evidence for the concept of quantized cell cycles: generation times of individual *Paramecium* cells is a discrete multiple of ultradian subcycles. Eur J Cell Biol 38 [Suppl] 9:6

Kippert F (1987) Temperature compensation of ultradian rhythms in ciliates. In: Hildebrandt G, Moog R, Rashke F (eds) Chronobiology and chronomedicine: basic research and applications. Peter Lang, Frankfurt-am-Main, pp 60–68

Kippert F, Lloyd D (1991) The ultradian clock of *Schizosaccharomyces pombe*: environmental stability but genetic variability. J Interdiscipl Cycle Res 22:138–139

Klevecz RR (1976) Quantized generation times in mammalian cells as an expression of the cellular clock. Proc Natl Acad Sci USA 73:4012–4016

Kyriacou CP, Hall JC (1980) Circadian rhythm mutations in *Drosophila melanogaster* affect a short-term fluctuation in the male courtship song. Proc Natl Acad Sci USA 77:6729–6733

Kyriacou CP, Hall JC (1989) Spectral analysis of *Drosophila* courtship song rhythms. Animal Behav 37:850–859

Lloyd D (1987) The cell division cycle. Biochem J 242:313–321

Lloyd D, Edwards SW (1984) Epigenetic oscillations during the cell cycles of lower eukaryotes are coupled to a clock: life's slow dance to the music of time. In: Edmunds LN (ed) Cell cycle clocks. Marcel Dekker, New York, pp 27–46

Lloyd D, Edwards SW (1986) Temperature-compensated ultradian rhythms in lower eukaryotes: periodic turnover coupled to a timer for cell division. J Interdiscipl Cycle Res 77:321–326

Lloyd D, Edwards SW (1987a) Temperature-compensated ultradian rhythms in lower eukaryotes: timers for cell cycle and circadian events? In: Pauly JE, Scheving LE (eds) Advances in chronobiology, pt A. Alan R. Liss, New York, pp 131–152

Lloyd D, Edwards SW (1987b) Epigenetic oscillations in synchronous cultures of lower eukaryotes. In: Hildebrandt G, Moog R, Rashke F (eds) Chronobiology and chronomedicine: basic research and applications. Peter Lang, Frankfurt-am-Main, pp 69–73

Lloyd D, Kippert F (1987) A temperature-compensated ultradian clock explains temperature-dependent quantal cell cycle times. In: Bowler K, Fuller BJ (eds) Temperature and animal cells. (Symp Soc Exp Biol 41) Cambridge University Press, Cambridge, pp 135–155

Lloyd D, Stupfel M (1991) The occurrence and functions of ultradian rhythms. Biol Rev 66: 275–299

Lloyd D, Volkov EI (1990) Quantized cell cycle times: interaction between a relaxation oscillator and ultradian clock pulses. BioSystems 23:305–310

Lloyd D, Volkov EI (1991) The ultradian clock: timekeeping for intracellular dynamics. In: Mosekilde E (ed) Complexity, chaos, and biological evolution. Plenum Press, New York, pp 51–60

Lloyd D, John L, Edwards C, Chagla AH (1975) Synchronous cultures of microorganisms: large-scale preparation by continuous-flow size selection. J Gen Microbiol 88:153–158

Lloyd D, Phillips CA, Statham M (1978) Oscillations of respiration adenine nucleotides and heat evolution in synchronous cultures of *Tetrahymena pyriformis* ST prepared by continuous-flow selection. J Gen Microbiol 106:19–26

Lloyd D, Edwards SW, Fry JC (1982a) Temperature-compensated oscillations in respiration and cellular protein content in synchronous cultures of *Acanthamoeba castellanii*. Proc Natl Acad Sci USA 79:3785–3788

Lloyd D, Poole RK, Edwards SW (1982b) The cell division cycle: temporal organization and control of cellular growth and reproduction. Academic Press, London

Lloyd D, Edwards SW, Williams JL, Evans JB (1983) Mitochondrial cytochromes of *Acanthamoeba castellanii*: oscillating accumulation of haemoproteins, immunological determinants and activity during the cell cycle. FEMS Microbiol Lett 16:307–312

Marques N, Edwards SW, Fry JC, Halberg F, Lloyd D (1987) Temperature-compensated ultradian variation in cellular protein content of *Acanthamoeba castellanii*. In: Pauly JE, Sheving LE (eds) Advances in chronobiology, pt A. Alan R. Liss, New York, pp 131–152

Michel V, Hardeland R (1985) On the chronobiology of *Tetrahymena*. III. Temperature compensation and temperature dependence in the ultradian oscillator of tyrosine aminotransferase. J Interdiscipl Cycle Res 16:17–23

Mustafin AT, Volkov EI (1982) On the distribution of cell cycle generation times. BioSystems 15:111–126

Nanjundiah V (1973) Chemotaxis, signal relaying and aggregation morphology. J Theor Biol 42:63–105

Neuhaus-Steinmetz U, Balzer I, Hardeland R (1990) Ultradian rhythmicity of tyrosine aminotransferase in *Euglena gracilis*: analysis by cosine and non-sinusoidal fitting procedures. Int J Biometeorol 34:28–34

Noble D (1975) The initiation of the heartbeat. Clarendon Press, Oxford

Nurse, P (1990) A universal control mechanism regulating onset of M-phase. Nature 344:503–508

Pelech S (1990) When cells divide. The Sciences 1:23–28

Poole RK (1977) Development of respiratory activity during the cell cycle of *Schizosaccharomyces pombe* 972 h⁻: respiratory oscillations and heat dissipation in cultures synchronized with 2′-deoxyadenosine. J Gen Microbiol 103:19–27

Poole RK, Salmon I (1978) The pool sizes of adenine nucleotides in exponentially growing, stationary phase and 2′-deoxyadenosine-synchronized cultures of *Schizosaccharomyces pombe* 972 h⁻. J Gen Microbiol 106:153–164

Poole RK, Lloyd D, Kemp RB (1973) Respiratory oscillations and heat evolution in synchronously dividing cultures of the fission yeast *Schizosaccharomyces pombe* 972 h⁻. J Gen Microbiol 77:209–220

Rapp PE, Mees AI, Sparrow CT (1981) Frequency dependent biochemical regulation is more accurate than amplitude dependent control. J Theor Biol 90:531–544

Readey M (1987) Ultradian photosynchronization in *Tetrahymena pyriformis* GLC is related to model cell generation time: further evidence for a common timer model. Chronobiol Int 4: 195–208

Richter PH, Ross J (1981) Concentration oscillations and efficiency: glycolysis. Science 211:715

Schaffer WM, Ellner S, Kot M (1986) Effects of noise on some dynamical models in ecology. J Math Biol 24:479–523

Sel'kov EE (1970) Two alternative self-oscillating stationary states in thiol metabolism – two alternative types of cell division, normal and malignant ones. Biophysika 15:1065–1073

Strausfeld U, Labbé JC, Fesquet D, Cavadore JC, Picard A, Sadhu K, Russell P, Dorée M (1991) Dephosphorylation and activation of a p34^{cdc2}/cyclin B complex in vitro by human CDC25 protein. Nature 351:242–245

Sweeney BM (1982) Interaction of the circadian cycle with the cell cycle in *Pyrocystis fusiformis*. Plant Physiol 70:272–276

Sweeney BM, Hastings JW (1960) Effects of temperature upon diurnal rhythms. Cold Spring Harbor Symp Quant Biol 25:87–104

Turner G, Lloyd D (1971) Effects of chloramphenicol on growth and mitochondrial function of the ciliate protozoon *Tetrahymena pyriformis* strain ST. J Gen Microbiol 67:175–188

von Grosse W-R (1985) Zur endogenen Grundlage der circadianen Activität bei *Calliphora vicina* R.D. Zool Jb Physiol 89:491–507

Woffendin C (1982) The growth and differentiation of *Dictyostelium discoideum*. PhD thesis, University of Wales

Woffendin C, Griffiths AJ (1985) Changes in respiratory activity and total protein during syn-
  chronous growth of *Dictyostelium discoideum*. FEMS Microbiol Lett 29:203–207
Zeuthen E (1978) Induced reversal of order of cell division and DNA replication in *Tetrahymena*.
  Exp Cell Res 116:39–46

Rhythms of Protein Synthesis and Other Circahoralian Oscillations: The Possible Involvement of Fractals

V. Y. Brodsky

Protein Synthesis Rhythms: History, Terminology and Phenomenology

The phenomenon of periodicity of protein synthesis was discovered only 30 years ago. Initially fluctuations in cell dry weight and total protein content in the cell were described (Brodsky 1959, 1961). The period of these oscillations (close to 1 h) strongly differed from the period of other biological rhythms known at that time. The periodicity was not necessarily associated with the rhythm of protein synthesis, but could have been due to oscillations of protein secretion from the cells. This second hypothesis was later ruled out on the basis of appropriate experiments. Still later it was reported that amino acid incorporation into proteins fluctuates with a period of approximately 1 h (Brodsky et al. 1967; Mano 1968). Corrections for the pool of labelled amino acids did not abolish the periodicity of the incorporation rate; in one case the constancy of the endogenous pool was established (Brodsky 1975), and thus the amino acid pool cannot be a source of rhythmicity in the incorporation rate. Moreover the rhythm of leucine incorporation into proteins has also been shown in a cell-free system (Boikov et al. 1990), i.e. in the presence of excess amino acids.

At present, literature about periodicity of protein synthesis contains more than 100 references (see Tables 2.1–2.3 for some of them). Whereas it has become clear that rhythmic protein synthesis is a general phenomenon, weight fluctuations have been observed in only some of the cells studied.

The observed periods of oscillations in protein synthesis rates varies from 20 to 120 min, thus being different from circadian rhythms, not only in the duration, but also because there is no external pacemaker. The intrinsically cellular nature of these rhythms is emphasized many times in this review. The term "ultradian" in this case is not quite correct because similarity to "circadian" implies a genealogical similarity, which appears not to be the case for many rhythms with a period of approximately 1 h. The fundamentally different nature of such rhythms is reflected in the term which was proposed

Table 2.1. Circahoralian rhythms of protein levels and incorporation rates

	Period (min)	References
A. Dry weight and protein concentration		
Ganglial cells of the retina (frog, chicken, mouse, cat)	30–70	Brodsky 1959, 1961, 1966; Brodsky and Arefieva 1968; Brodsky and Neverova 1968; Beckchanov 1978
Acinary cells of parotid and exorbital glands of the rat	20–80	Brodsky 1964; Brodsky et al. 1967, 1973; Nechaeva and Fateeva 1973, 1974
Rat and mouse hepatocytes	40–60	Novikova and Nechaeva 1979
Cells in vitro (Hep-2, glioma C-6, monkey kidney)	20–60	Veksler et al. 1973; Gilbert and Tsilimigras 1981; Tsilimigras and Gilbert 1977
Acanthamoeba in culture	30–60	Edwards and Lloyd 1978; Lloyd et al. 1982
Living cells (Ehrlich ascites, gliocyte, neuron)	20–90	Kostenko et al. 1973; Svanidze and Didimova 1974; Brodsky et al. 1977; Zaguskin et al. 1980; Nechaeva et al. 1985
B. Incorporation of amino acids into proteins		
Parotid gland, slices in vitro	20–90	Brodsky et al. 1967, 1979; Brodsky, 1975; Nechaeva et al. 1980
Mouse retina in vitro	30–80	Brodsky et al. 1978
Pancreas, slices in vitro	30–60	Brodsky et al. 1979
Hepatocytes, slices and monolayer in vitro	30–80	Novikova and Nechaeva 1979
Sea urchin embryos	50–60	Mano 1968, 1970, 1975
Mitylus gill, pieces in vitro	20–70	Harazova and Nechaeva 1988
Human stomach, biopsy samples	30–70	Brodsky et al. 1984; Fateeva et al. 1987
Cell-free system from rat hepatocytes	30–60	Boikov et al. 1990

quite some time ago: "circahoralian rhythms". It is demonstrated below that oscillations with a period of approximately 1 h are characteristic not only of certain intracellular processes but also of some organ functions. The latter oscillations may indeed resemble ultradian rhythms as defined by Halberg (Aschoff 1981; Schultz and Lavie 1985).

Circahoralian oscillations of enzyme activities have been described in several cell types (Table 2.2). In certain bacteria, oscillations of the activities of several enzymes were attributed to rhythms in the rates of their syntheses (Kuempel et al. 1965; Masters and Donachie 1966). Of course, any rhythmicity in the activity of an enzyme need not necessarily be due to the rhythmic rate of its synthesis, but could be attributed to changes in its structure and/or changes in the intracellular medium. A good example is circahoralian periodicity of lactate dehydrogenase activity in mammalian red blood cells, which do not synthesize protein. Rhythm of the lactate dehydrogenase activity in fibroblasts in vitro is retained in the presence of puromycin (Duffy 1971). Interesting examples of circahoralian oscillations of enzyme activities which are not

Table 2.2. Circahoralian rhythms of enzyme activities

Enzyme and species	Period (min)	References
Galactosidase, tryptophanase, histidinol dehydrogenase, alkaline phosphatase, *Escherichia coli*	30–70	Kuempel et al. 1965; Knorre 1968, 1973
Ornithine transcarbamylase, saccharase, *Bacillus subtilis*	50–80	Masters and Donachie 1966
Amidase, *Pseudomonas aeruginosa*	30–40	Boddy et al. 1967
Yeast glucosidase	90	Halvorson et al. 1966
Catalase, *Acanthamoeba*	50–60	Lloyd et al. 1982
Proteases and ATPases in sea urchin embryos	40–60	Mano 1970; Petzelt 1972
Glucose-6-phosphate dehydrogenase, lactate dehydrogenase and aldolase in hamster and BHK cells in vitro	20–60	Gilbert 1974; Gilbert and Tsilimigras, 1981
Lactate dehydrogenase in human erythrocytes	30–60	Duffy 1971; Duffy and Sanderson 1971
Ornithine decarboxylase in rat hepatocytes in vitro	30–60	Jarigin et al. 1978

Table 2.3. Circahoralian oscillations of hormone levels in blood plasma

Hormone, species	Periods (min)	References
Rat LH	50–60	Kinosita et al. 1982
Monkey LH	40–90	Dierschke et al. 1970
Human LH	120	Kapen et al. 1975
Human thyroid hormone	80	Van Cauter and Honinsky 1985
Rat GH	180	Tannenbaum and Martin 1976
Human GH	150	Finkelstein et al. 1972
Human follicle-stimulating hormone	120	Kapen et al. 1975
Human adrenocorticotrophic hormone	90	Van Cauter and Honinsky 1985

LH, luteinizing hormone; GH, growth hormone.

due to any changes in amounts of protein have been described by Gilbert and Tsilimigras (1981).

Circahoralian periodicities in the levels of protein hormones in mammalian blood plasma have been described (Table 2.3). This information of course does not provide any clues about the rhythm of synthesis but rather about fluctuations of secretion rate or the rate of hormone disappearance from blood. Detection of sex differences in the growth hormone patterns (Shapiro et al. 1989) is of interest. Circulating growth hormone oscillates with circahoralian frequencies in female rats but the hormone concentration has bursts with ultradian (4–5 h) periods in males.

Evidence for the circahoralian rhythms of secretion of liver proteins has been obtained (Brodsky et al. 1983). This oscillation occurs together with the protein synthesis rhythm in the same cells.

Circahoralian periodicity of the axoplasmic flow has been described (Weiss 1972, 1974). This phenomenon can be due either to the rhythm of protein

Table 2.4. Intracellular circahoralian rhythms other than those of protein synthesis

Rhythm and system	Period (min)	References
Permeability for amino acids in retina, liver and parotid in vivo and in vitro	30–80	Nechaeva and Fateeva 1974; Brodsky 1975; Brodsky et al. 1978, 1983
cAMP level in rat hepatocytes in vitro	20–60	Jarigin et al. 1979
cAMP level in sea urchin embryos	40–60	Ishida and Yasumasu 1982
Calcium concentration in sea urchin embryos	40–60	Clothier and Tumorian 1972
The total sum of adenylates in rat muscle	30–60	Tornheim and Lowenstein 1974
AMP, ADP and ATP levels in *Acanthamoeba castellanii*	60–70	Edwards and Lloyd 1978, 1980
ATP level in hepatocytes in vitro	30–70	Jurovitsky et al. 1989
Respiration in *Acanthamoeba castellanii*	70–90	Lloyd et al. 1982
Concentration of polyamines in sea urchin embryos	50–60	Kusunoki and Yasumasu 1976
Cytoplasmic pH in *Tetrahymena pyriformis*, sea urchin embryos, CHC cells and hepatocytes in vitro	30–70	Gillies and Deamer 1979; Litinskaya et al. 1982, 1987; Boshkova et al. 1987

synthesis in the perikaryon or to rhythmic transport of protein into the axon. It has been calculated that the presence of such a pulse flow of axoplasm requires rhythmicity of protein synthesis with a period of about 1 h (Nadelhaft 1976). Similar periodicities have also been demonstrated for other neurons (Brodsky et al. 1978).

Other Circahoralian Intracellular Rhythms: The Pacemaker Problem

Apart from protein synthesis rhythms, other intracellular circahoralian rhythms were described during the 1970s and 1980s (Table 2.4). At least some of these can be associated with the rhythm of protein synthesis. They can be attributed to changes in concentrations and activities of the corresponding enzymes. Alternatively metabolic rhythms themselves can generate oscillations in the rate of protein synthesis. Circahoralian periodicity of ATP levels in cells and oscillations of the pH have been considered as possible pacemakers for the rhythm of protein synthesis rate.

Lloyd and co-workers (Edwards and Lloyd 1978, 1980; Lloyd et al. 1982) were the first to demonstrate intracellular circahoralian periodicities of adenylate levels (and specifically of ATP). Several full periods of ATP oscillations have been observed in the course of one cell cycle in *Acanthamoeba*. This was also so for RNA and protein levels. In other words these metabolic oscillations were not just events of the mitotic cycle.

After this, it was suggested that the rhythm of ATP concentration might be a pacemaker for the rhythm of protein synthesis. Jurovitsky et al. (1989) found a circahoralian rhythm of ATP levels in rat hepatocytes in vitro; many peaks of this rhythm were in the phase opposite to that of the rhythm of protein synthesis rate in the same monolayer. A circahoralian rhythm of tRNA amino-acylation rate has also been found in the same culture, and often it is in the same phase as the rhythm of leucine incorporation into hepatocyte proteins (Boikov et al. 1990). Earlier, a similar mode of aminoacylation was found in eggs of the sea urchin *Pseudocentrotus depressus* during cleavage (Mano 1975).

The hypothesis that ATP plays a key role in the initiation of the protein synthesis rhythm has never been confirmed experimentally. The addition of 1 mM ADP to the medium bathing a monolayer of hepatocytes increased ATP level by a factor of 2–3, but periodicity of the protein synthesis rate was retained (Jurovitsky et al. 1990). The cell-free system used by Boikov et al. (Eisenstein and Harper 1984) contained 2 mM ATP and 1 mM GTP, i.e. these nucleotides were in a large excess; nevertheless, the rhythm of protein synthesis rate was quite distinct.

Another interesting idea put forward was that oscillations in intracellular pH might serve as a pacemaker for the rhythm of protein synthesis. Studies of hepatocytes in vitro (Litinskaja et al. 1987) have indeed demonstrated the dependence on pH of lysine incorporation into cytoplasmic proteins. However, changes of lysine pool size remained the same. When the incorporation was corrected for pool size, there was no significant correlation of protein synthesis with intracellular pH. In other words, pH changes do not initiate oscillations in the rate of protein synthesis.

We have been able to smooth oscillations of protein synthesis rate considerably by adding putrescine to the medium for the cultivation of hepatocytes (Nechaeva et al. 1980). Polyamines can affect protein synthesis: circahoralian rhythms of polyamine level were discovered in blastomeres of the sea urchin *Hemicentrotus pulcherrimus* (Kusunoki and Yasumasu 1976). However, the effects of polyamines are mediated by numerous factors and are temporally dissociated from protein synthesis.

Main Properties of Intracellular Circahoralian Rhythms: Their Possible Fractal Nature

There are several general characteristics of circahoralian intracellular rhythms which provide evidence for their endogenous intrinsic nature.

1. A pacemaker responsible for periodicity in protein synthesis remains to be identified. These studies have to be continued, but it cannot be ruled out that the pacemaker simply does not exist, or that there may be two (or more) pacemakers for compound oscillations. At least, it is difficult to imagine that there is an external geophysical pacemaker. Oscillations of the diameter of the sun with a mean period of about 2.5 h, have been described (Brookes et al. 1976; Severny et al. 1976). There are also mechanical oscillations of the earth with the period of 1–2 h which are different from earthquakes (Linkov et al. 1982). So far these geophysical oscillations cannot be linked to patterns of cell

Table 2.5. Cyclic processes with 40–60 min periodicities in developing embryos remaining after the suppression of cleavage divisions

Process	Treatment	Species	References
Thymidine uptake	Cycloheximide	*Hemicentrotus pulcherrimus*	Suzuki et al. 1977
DNA synthesis	Colchicine	*Hemicentrotus pulcherrimus*	Suzuki and Mano 1973
	Cold	*Lutehinus*	Brookbank 1976
Incorporation of amino acids	Colchicine	*Hemicentrotus*	Mano 1968, 1970
	Cold	*Anthocidaris*	Suzuki and Mano 1973
Activity of Ca^{2+}-dependent ATPase	Colchicine	*Paracentrotus lividus*	Petzelt 1972
The level of SH groups in proteins	Colchicine	*Clypeaster*	Dan and Ikeda 1971

metabolism by any reasonable considerations. It is most probable that periodic oscillations of metabolism belong to intrinsic cell properties.

2. Many circahoralian rhythms are fully uncoupled from the mitotic cycle. This is self-evident for neurons, hepatocytes and other glandular cells which do not divide. In cleaving sea urchin embryos, metabolic cycles have the same duration as the division cycle. Many of the metabolic cycles, however, continue even after divisions are suppressed or even after enucleation of the embryo (with the exception of the reports presented in Tables 2.1, 2.2 and 2.4; see Table 2.5).

3. The finding of circahoralian rhythms in blastomeres and various differentiated cells, in various eukaryotes including Protozoa, as well as in various prokaryotes, provides evidence for this property having a fundamental general nature.

4. Circahoralian rhythms are extremely irregular, and within 3–4 h of observation, periods with the duration of 60–100 min and 20–30 min can be observed in any one experiment. Therefore, conventional circadian concepts such as acrophase and so on, are devoid of any particular sense in this case. It is probable that circahoralian periodicity comprises a set of varied and irregular oscillations. The initial irregularity and inevitably low number of periods observed in any one experiment (3–4 h in studies of rhythms of protein synthesis) make spectral analysis inappropriate. However, circahoralian rhythms can be modified experimentally. Thus, in a rat taken directly from the animal house, periods of protein synthesis rate in the parotid vary from 20 to 100 min (Brodsky and Nechaeva 1971). If the rat is fed at 2-h intervals, a periodicity of 20–50 min is found; whereas if the interval between feedings equals 6 h, only long periods with a duration of 70–110 min can be observed. Irregularity is retained, but at the edges of the original spectrum.

Temperature affects the mean level but not the period of oscillations. Thus, respiration rhythms and rhythms of protein content in *Acanthamoeba* have been observed in the temperature range 25–30 °C, which provides an important argument supporting the endogenous nature of these rhythms (Lloyd et al. 1982). Decrease of temperature from 23 °C to 11 °C drastically reduces protein synthesis rate in cleaving sea urchin embryos but does not change the periods (Mano 1970). In gills of *Mitylus* studied during different seasons at water temperatures varying from 1.5 to 16 °C, the average incorporation rate differed enormously, but the set of periods in the rhythm of protein synthesis remained virtually unchanged (Harazova and Nechaeva 1988).

Circahoralian rhythms were observed in synchronized cultures (Lloyd et al. 1982) as well as in cultures which were not synchronized deliberately (see other examples in Tables 2.1 and 2.2). In the latter case, an overall rhythm may be the result of cellular interactions which lead to self-synchronization. The possibility of self-synchronization of interacting oscillators was demonstrated long ago with mathematical models (Gelfand and Zeitlin 1960), and was confirmed for some cellular systems (Robertson et al. 1972; Goshima 1973).

The biological significance of rhythmicity, including that of short rhythms, is usually attributed to homeostasis, maintenance of stability of populations of organisms, and functions of the individual organism or of its tissues and cells through feedback mechanisms. As early as 1963 Goodwin, having no information about the original publications on oscillations of protein content (Brodsky 1959, 1961), described the theory of feedback in protein synthesis and, moreover, was able to predict approximately circahoralian oscillatory kinetics. Such considerations are still important. However, homeostasis alone cannot account for the extreme irregularity of circahoralian periods and the absence of a known pacemaker. Theoretically, irregularity of oscillations can be a result of summation of practically regular sinusoids. Thus, if we take into account only the mean periods, the irregular rhythm of protein synthesis in the rat parotid can result from the circumstance that proteins change with a mean period of about 50 min in acinary cells, whereas periodicity in the intralobular ducts is 90 min, and in the interlobular ducts 120 min (Nechaeva and Fateeva 1973). However, summation of sinusoids does not appear to be an attractive concept because of the extreme irregularity of each of the relevant tissue rhythms i.e. acinary, intralobular or interlobular. Irregularity is very distinct in studies of individual cells. Thus, in a microinterferometric study of an isolated crayfish (*Astacus astacus*) neuron, oscillations in the dry weight occurred with periodicity of 20–70 min over 2–4 h of a single cell observation (Brodsky et al. 1977; Zaguskin et al. 1980). To explain the marked irregularity of a compound rhythm a hypothetical discussion of two or even more pacemakers for each variable may be advanced (for different proteins, for instance). In such a case, summation of sinusoids results in variation of frequencies and amplitudes to give a smooth compound curve.

In addition to the homeostatic model, another interpretation of circahoralian rhythms is now possible. These oscillations may reflect a certain predetermined chaotic organization similar to that of fractals, well-known in physics (Mandelbrot 1983), and discussed more recently in biology (Goldberger and West 1987). In this case, oscillations appear to be an intrinsic property of the process.

Fractals vary in their periodicity and may be found on different time scales; they do not depend on a pacemaker. The possible biological significance of these apparently unstable oscillations is that they can be involved in cellular adaptation to the changing environment.

Circahoralian periodicities of various processes (such as energy metabolism, permeability, protein synthesis and others) mutually affect one another, but it may well be that they are not organized hierarchically.

Short-Period Rhythms in Organ and Body Functions: Fractals and Ultradians

Along with intracellular circahoralian rhythms, similar periodicity of organ functions and organismic behaviour has been described. A few communications on this subject were published during the first decades of this century but the main data have come from studies performed during last 20 years. Some organ rhythms with short periods may be similar in their nature to ultradian rhythms, historically due to diurnal effects. But it cannot be ruled out that they are associated with endogenous intracellular rhythms and, primarily, with the rhythm of protein synthesis rate.

Rhythms of Respiration

The circahoralian rhythm of respiration in amoebae (Table 2.4) was mentioned above. Similar rhythms of respiration have been described in the yeast (Mochan and Pye 1972; Meyenburg 1973). Circahoralian components have been found in studies of the daily rhythm of respiration in rats and mice (Kayser and Hildwein 1974a; Sacher and Duffy 1978). In rats, the daily curve is formed by circahoralian peaks of approximately similar amplitude: at night the mean level of respiration was increased. In mice, minimal peak values do not differ in the daytime and at night; however, the maxima of circahoralian peaks differ strongly. In animals active in the daytime, e.g. quails or guinea pigs, the circadian rhythms of respiration appear to have their source in circahoralian oscillations which are more distinct in the daytime than at night (Kayser and Hildwein 1974a,b; Stupfel et al. 1979, 1981). Respiration rate measured for 1 min shows six to eight peaks over 6 h of measurements in humans (Horne and Whitehead 1976).

All these circahoralian periodicities of the respiration rate are extremely irregular, just as are intracellular rhythms. Their fractal nature is apparent: periods with a duration of about 1 min give rise to circahoralian periodicity, which may in turn be a source of circadian rhythms.

Variation of Heart Rate

Heart rate shows circahoralian oscillations with a period of 20–120 min (Orr and Hoffman 1974; Orr et al. 1974; Hoppenbrouwers et al. 1978; Livnat et al. 1984). Computations of heart rhythms provided an example of fractals in the important study of Goldberger and West (1987). It is interesting that the effect of frequency selection has been described for the heart rate (as well as for the rhythm of protein synthesis rate in the parotid, see above). Under various conditions of physical and mental load in humans, mainly fast (20–30 min period) oscillations, were found; alternatively there were slow oscillations but also circahoralian, i.e. with a period of 90–130 min (Smirnov et al. 1980).

Rhythms in the Digestive System

Circahoralian rhythms of protein synthesis rate in parotid and liver, some of which are associated with digestion, have already been noted. Similar rhythms of protein synthesis rate have been described in the pancreas (Brodsky et al. 1979). At the beginning of the century I. P. Pavlov and his associates found short "hunger" rhythms of stomach and intestinal contractions, with periods of about 1 h. High frequency spikes and their integral derivatives with periods of minutes and hours duration have been observed in rat intestines using modern electrophysiological methods (Golenhofen 1970; Rukebush and Weekes 1976; Shemerovsky 1981). Circahoralian oscillations of stomach muscle contractions correlate with the activation, shown by electroencephalography (EEG), of the motor zones of the cerebral hemispheres (Lebedev 1983).

Rhythms of Brain Activity and Rhythms of Behaviour

The primary signals of brain electrical activity are known to have shorter periods than does respiration or heart rate. However, a long time ago, summation of nerve pulses demonstrated circahoralian periods. The first observations were made concerning sleep phases. Rapid eye movement (REM) sleep in humans has a periodicity of about 50–90 min, whereas in the dog, cat and rabbit its periodicity is 20–30 min (Kripke et al. 1968; Webb 1971; Sterman et al. 1972). Circahoralian periodicity in brain activity has been also described using EEG in the state of wakefulness (Kripke 1972, 1974).

Circahoralian periodicity of brain activity may be similar to oscillations of vegetative functions. Both groups of cycles can determine certain periodicities of mobility and behaviour of animals. The first observation of circahoralian oscillations of mobility in short-tailed voles (*Microtus*) was made as early as 1933 by Davis. Later it became possible to distinguish food motivation and intrinsic brain signalling (Lehman 1976; Daan and Slopsema 1978). Periods of mobility were circahoralian but with a distinct circadian maximum. In shrews and guinea pigs, in contrast to voles, circahoralian peaks of mobility have been found that do not have a daily maximum (Kayser and Hildwein 1974b). The motor activity of rhesus monkeys can be seen only in the daytime: eight to ten maxima, each with approximately similar amplitudes, were observed over a 12-h period (Delgado-Garcia et al. 1975, 1976; Maxim et al. 1976; Lewis and Kripke 1977). Circahoralian rhythms with periods of 30–60 min have been found in non-motivated ("hand–mouth") behaviour in resting humans (Oswald et al. 1970) and a similar rhythm has been described in the productivity of monotonous work (Murell 1971).

Analysis of the nature of mobility cycles and behaviour is complex. In addition to spontaneous brain activity, we have to take into account the food motivation, and in humans also those characteristics associated with habits or culture. Still, the similarity between periodicity of vegetative functions and mobility on one hand and the cycle of brain activity on the other, and the relation of these rhythms to the periodicity of protein synthesis rate and other metabolic rhythms, deserve special analysis.

Since circadian rhythm can be produced by summation of circahoralian oscillations, it is apparent that any changes of circahoralian rhythm parameters may affect the resulting circadian rhythm. Therefore a study of the mutual effects of circahoralian and circadian rhythms is of considerable interest.

A Model for Interaction of the Circahoralian and Circadian Rhythms: Adaptivity of Circahoralian Rhythms

The ideal circadian rhythm of the type

$$S(t) = B + A[\sin(Wt + P)]$$

is compared with the integral circadian rhythm produced by summation of circahoralian oscillations of the type

$$f^{(i)}(t) = b^{(i)} + (a^{(i)} + k_1^{(i)}z) \sin[(w^{(i)} + k_2^{(i)}z)t + (p^{(i)} + k_3^{(i)}z)]$$

where B and b are mean levels, A and a are amplitudes, W and w are angular frequencies, P and p are phases of oscillation, t is time, i is the curve number ($1 < i < N$) and z is a random number from 0 to 1 (its inclusion simulates the uncontrolled fluctuations of corresponding variables); k_{1-3} are arbitrary coefficients (Zaguskin et al. 1991).

The integral circadian rhythm is the sum of three groups of curves with characteristic periods of 20 min, 1 h and 2 h. Their mean levels, amplitudes and phases of oscillation were chosen in the three variants with differences of 6%–8%. For every time t, the sum of the three groups of rhythms was calculated:

$$F(t) = \Sigma f^{(i)}(t)$$

The resultant rhythm was optimized by checking the approximation of its parameters to the ideal circadian rhythm.

At each time point we optimized the goal function $Q(t)$ (defined as the absolute value of difference between $F(t)$ and $S(t)$) by variation of the vector (**V**) of one of the parameters (b, a or p) common for the whole set of initial frequencies. Minimization of the Q function was performed using a matrix random search algorithm

$$Q_t = |F_t(\mathbf{v}) - S_t| \xrightarrow[(\mathbf{V})]{} \text{minimum}$$

$$[\mathbf{V}_\varepsilon \langle b, a, w, p \rangle]$$

At a certain time, the phase of the "ideal" sinusoid $S(t)$ was changed by half of the period. This allowed us to observe the tuning of the optimized rhythm to drastic changes of the original kinetics.

Fig. 2.1. Changes of the ideal circadian sinusoid (s) and a circadian rhythm composed of circahoralian oscillations (f) and after its optimization (o) (see corresponding equations in the text). **a** and **b** Two periods of time within the 24-h cycle. Step values are equal to 2 min and, consequently, the beginning of the curve corresponds to 116×2 min from the start of computer simulation, whereas the beginning of the curve in **b** refers to 335×2 min in the same cycle. After 20 min on the curve in **a**, a phase shift of the ideal curve (s) was performed for half a period. The double-dashed line (-- --) is ordinate (time); figures on the left and corresponding marks on single-dashed lines (---) (I) is abscissa (numerical values of model curves). Gaps between the single-dashed lines correspond to 10 steps, i.e. 20 min. q characterizes the difference between (s) and (o) at each time point.

```
   Step 116:
-14.5     -4.542       5.430       15.403       25.375       35.348       45.3
                     ¦ q                                          f        o    s
                     ¦ q                                          f       o     s
                     ¦ q                                           f     o      s
                     ¦  q                                           f  o        s
                     ¦ q                                             f  o       s
                     ¦ q                                             f   o      s
                     ¦ q                                             f    o     s
 q                   ¦                                    s          f    o
 ---------Iq---¦----I---------I---------Is---o----I---f-----
           q   ¦                             s     o          f
            q  ¦                             s     o             f
           q   ¦                             s      o           f
             q ¦                             s      o           f
             q ¦                             s      o           f
             q ¦                             s      o          f
             q ¦                             s      o         f
            q  ¦                             s       o       f
             q ¦                             s       o      f
 ---------q----¦----I---------I---------Is---o----If--------
           q   ¦                             s      o      f
            q  ¦                             s      o      f
           q   ¦                             s       o     f
            q  ¦                             s       o     f
            q  ¦                             s       o     f
             q ¦                             s       o    f
            q  ¦                             s        o  f
            q  ¦                             s       o   f
             q ¦                             s        o f
 ---------q----¦----I---------I---------Is----o-f-I---------
           q   ¦                             s        o f
            q  ¦                             s        o   f
             q ¦                             s        o    f
            q  ¦                             s         o   f
            q  ¦                             s         o    f
            q  ¦                             s         o    f
             q ¦                             s          o   f
           q   ¦                             s          o   f
              q¦                             s        o    f
 ---------Iq---¦----I---------I---------Is--o-----f---------
           q   ¦                             s        o    f
           q   ¦                             s         o   f
             q ¦                             s         o   f
            q  ¦                             s          o  f
            q  ¦                             s          o      f
              q¦                             s        o       f
             q ¦                             s        o       f
           q   ¦                             s         o    f
            q  ¦                             s         o  f
 ---------q----¦----I---------I---------Is---o--f-I---------
           q   ¦                             s         o f
          q    ¦                             s        of
            q  ¦                             s        o  f
           q   ¦                             s        o  f
           q   ¦                             s         o   f
             q ¦                             s        o   `f
             q ¦                             s        o    f

                              a
```

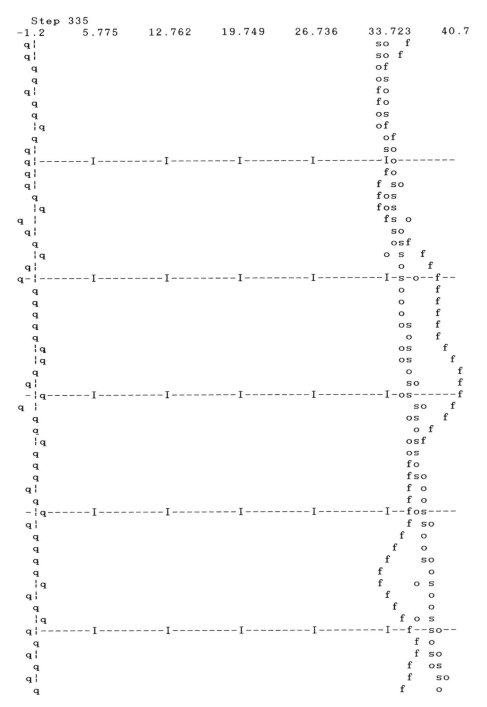

Fig. 2.1. *Continued*   **b**

Computer simulation at $N = 9$ has demonstrated a low efficiency of search for the oscillation amplitude. Search according to the phase or frequency is effective over a narrow range of changes and after a certain time. The circadian rhythm was significantly affected by the mean levels of the circahoralian oscillations and over a wide range of changes of the tuning coefficients of the model.

The efficiency of search for the middle level is so great that there is no need to present the corresponding plot: the ideal curve and the optimized curve coincide virtually completely. Search for the optimal phase is extended in time (Fig. 2.1a and b). In the first half of this plot (a) the ideal sinusoid looks like the straight line, both before the half-period shift was made (for the first 20 min), and after it, during the first hour. It should be remembered that 20-min intervals are too small for comparison with the 24-h cycle. Especially at the maximum or at the minimum of the cycle, 1-h intervals should look like straight lines. However, in the ascending or descending part (Fig. 2.1b), certain deviations from the straight line are apparent. The optimized curve (o) approaches the ideal curve (s) immediately after the shift (of the ideal), although the differences are observable within the first few hours (see also q symbols characterizing absolute deviations, s − o). After a few hours, phases of the ideal circadian rhythm on the optimized curve constructed from the circahoralian oscillations begin to coincide and then merge completely.

These calculations have been confirmed by observations of real biological systems. Above, we referred to results of continuous or more often discrete measurements of respiration intensity, heart rate or animal mobility. The circadian sinusoid (when present) consisted of circahoralian peaks. At the maximum of the circadian curve, the mean level of the circahoralian rhythms was also higher, but their amplitudes did not necessarily change.

Phase shifting can be confirmed by simple observations from life. After flying over several time zones, the phases of a person's inner rhythms become adapted to local conditions at best only after a few hours ("at best" refers to optimization in the model described here). The mean level of processes changes immediately after the journey.

If circadian periodicity is interpreted as the effect of environmental factors on a certain process, the search for optimal tuning of this process can naturally be interpreted as selection of circahoralian (and other) oscillations, i.e. optimal under specific conditions of the environment. With this model, circahoralian oscillations are more akin to fractals than to the sinusoidal harmonics of the homeostatic hypothesis. Circadian oscillations of environmental factors in turn affect circahoralian rhythms by canalizing these stochastic changes.

As pointed out for experiments with rat feeding at different intervals, circahoralian oscillations can also adapt to periods shorter than circadian ones, i.e. to 2-, or 6-h periodic external stimuli. In this case periodic nerve stimulation of the salivary gland resulted in the adaptive change of the average period and mean level of protein synthesis rate oscillations. The effect of the functional load on parameters of the circahoralian rhythm was also clearly demonstrated in studies of a single living neuron of the crayfish mechano-receptor (Zaguskin et al. 1980). When the neuron was stimulated periodically, the mean level and phase of the circahoralian rhythm of the dry weight equivalent of a single cell was changed.

Summary

That many cell functions have an intrinsic instability has been emphasized in studies of biological rhythms over the last 30 years. Rhythms of protein synthesis and of many other functions are characterized by periods of about 1 h, and have become known as circahoralian rhythms. Some of these endogenous rhythms may, like circadian rhythms, have a diurnal origin, whereas others are more irregular. They have been demonstrated in cell-free systems, as well as in intact cells, and in tissues and organs. Some of the rhythms monitored in organs may be causally related to those with similar periods observed intracellularly. Their endogenous pacemaker remains uncharacterized and there is no known regular external pacemaker. At all levels of organization (intracellular to organs) these rhythms are sometimes irregular. Circahoralian rhythms can be entrained by various environmental stimuli (e.g. food, physical or mental loads), and entrainment provides a means of adaptation to facilitate homeostasis. Interactions between circahoralian and circadian rhythms play essential roles in these functions, and summation of higher frequency rhythms can provide circadian rhythmicities. A model for these interactions is presented together with a hypothesis of the fractal nature of biological rhythms.

References

Aschoff J (ed) (1981) Biological rhythms. Plenum Press, New York London
Beckchanov AN (1978) Maturation of the eye structures in ontogenesis (in Russian). J Evol Biochim Physiol 14:412–414
Boddy A, Clark PH, Houldsworth MA, Lilly MD (1967) Regulation of amidase synthesis by *Pseudomonas aeruginosa* 8602 in continuous culture. J Gen Microbiol 48:134–145.
Boikov PY, Novikova TE, Shevchenko NA, Brodsky VY (1990) Circahoralian rhythm of tRNA aminoacylation in hepatocytes in vitro (in Russian). Bull Exp Biol Med 109:393–396
Boshkova VP, Litinskaya LL, Sidorova VJ, Koslov DA (1987) Changes of intracellular pH by the sea urchin egg divisions (in Russian). Ontogenez 18:134–139
Brodsky VY (1959) Quantitative study of RNA and protein in the retinal cells (in Russian). Trans 1st Cytochem Conf Moscow, pp 7–9
Brodsky VY (1961) Protein synthesis in the retinal cells (in Russian). Tsitologia 3:312–326
Brodsky VY (1964) Oscillations of proteins in the rat exorbital gland (in Russian). Dokl Akad Nauk 157:171–174
Brodsky VY (1966) Protein synthesis and cell functions (in Russian). Nauka, Moscow
Brodsky VY (1975) Protein synthesis rhythm. J Theor Biol 55:167–200
Brodsky VY, Arefieva AM (1968) Protein changes in ganglionic cells of mouse retina (in Russian). Tsitologia 10:329–333
Brodsky VY, Nechaeva NV (1971) Protein rhythm in the parotid gland. Tsitologia 13:221–230
Brodsky VY, Neverova ME (1968) Dry weight kinetics in the chick ganglionic cells (in Russian). Dokl Akad Nauk SSSR 181:217–220
Brodsky VY, Nechaeva NV, Dmitrieva VI (1967) Protein mass and protein synthesis in the parotid gland (in Russian). J Obschei Biol 28:423–428
Brodsky VY, Nechaeva NV, Prilutsky VI (1973) Cell memory by protein content kinetics in the parotid gland (in Russian). Tsitologia 15:177–182
Brodsky VY, Zaguskin SL, Lebedev EA (1977) UV cytophotometry of a single living neuron (in Russian). Tsitologia 19:931–936
Brodsky VY, Arefieva AM, Kunina IM (1978) Oscillations of ³H-leucine incorporation into protein of the retinal cells (in Russian). Bull Exp Biol Med 80:351–354

Brodsky VY, Veksler AM, Litinskaya LL, Nechaeva NV (1979) Comparison of different circahoralian rhythms (in Russian). Tsitologia 21:976–979

Brodsky VY, Novikova TE, Fateeva VI, Nechaeva NV (1983) Protein synthesis and secretion rhythms in a monolayer culture of hepatocytes (in Russian). Tsitologia 25:156–161

Brodsky VY, Rapaport SI, Nechaeva NV, Fateeva VI, Rasulov MI (1984) The circahoralian protein synthesis rhythm in biopsies of the human stomach (in Russian). Bull Exp Biol Med 94:612–614

Brookbank JW (1976) DNA and RNA synthesis by fertilised cleavage arrested sea urchin eggs. Differentiation 1:33–39

Brookes JR, Isaak GR, van der Raay HB (1976) Observations of free oscillations of the sun. Nature 259:92–95

Clothier G, Tumorian H (1972) Calcium uptake and release by dividing sea urchin eggs. Exp Cell Res 75:105–110

Daan S, Slopsema S (1978) Short time rhythms in foraging behaviour of the common vole, *Microtus arvalis*. J Comp Physiol 127:215–227

Dan K, Ikeda M (1971) On the system controlling the time micromere formation in sea urchin embryos. Dev Growth Diff 13:285–301

Davis DHS (1933) Rhythmic activity in the short-tailed vole, *Microtus*. J Anim Ecol 2:232–238

Delgado-Garcia JM, Gran C, De Feudis P, Delgado JMR (1975) Ultradian rhythms in monkeys. Exp Brain Res 23:53–54

Delgado-Garcia JM, Gran C, De Feudis P, Del Pozo F, Jimenez JM, Delgado JMR (1976) Ultradian rhythms in the mobility and behavior of rhesus monkeys. Exp Brain Res 25:79–91

Dierschke DI, Bhattacharya AN, Atkinson LE, Knobil E (1970) Circahoral oscillations of plasma LH levels in the ovariectomized rhesus monkeys. Endocrinology 87:850–853

Duffy P (1971) Studies on the control of lactate dehydrogenase activity in mammalian cells. I. Biochim Biophys Acta 244:606–612

Duffy P, Sanderson JC (1971) Studies on the control of lactate dehydrogenase activity. II. Biochim Biophys Acta 244:613–617

Edwards SW, Lloyd D (1978) Oscillations of respiration and adenine nucleotides synchronous cultures of *Acanthamoeba castellanii*: mitochondrial respiratory control in vivo. J Gen Microbiol 108:197–294

Edwards SW, Lloyd D (1980) Oscillations in protein and RNA content during synchronous growth of *Acanthamoeba castellanii*. FEBS Lett 109:21–26

Eisenstein RS, Harper AE (1984) Characterization of a protein synthesis system from rat liver. J Biol Chem 259:9922–9928

Fateeva VI, Nechaeva NV, Komarov FI, Rapaport SI, Rasulov MI, Harajan LV, Brodsky VY (1987) Protein synthesis rhythm by ulcer aggravations and remissions (in Russian). Klin Med (Mosk) 65:68–71

Finkelstein JW, Roffwarg HP, Boyar RM, Kream J, Hellman L (1972) Age-related change in the twenty-four-hour spontaneous secretion of growth hormone. J Clin Endocrinol Met 35:665–670

Gelfand IM, Zeitlin ML (1960) Models of control systems (in Russian). Dokl Akad Nauk SSSR 131:1242–1245

Gilbert D (1974) The nature of the cell cycle and the control of cell proliferation. BioSystems 5:197–206

Gilbert D, Tsilimigras CWA (1981) Cellular oscillation: relative independence of enzyme activity rhythms and periodic variations in the amount of extractable protein. S Afr J Sci 77:66–83

Gillies RI, Deamer DW (1979) Intracellular pH changes during the cell cycle in *Tetrahymena*. J Cell Physiol 100:23–32

Goldberger AL, West BJ (1987) Fractals in physiology and medicine. Yale J Biol Med 60:421–435

Golenhofen K (1970) Slow rhythms in smooth muscle. In: Bulbring E (ed) Smooth muscle. Arnold, London, pp 316–342

Goodwin BC (1963) Temporal organization in cells. Academic Press, London New York

Goshima K (1973) A study of the prevservation of the beating rhythms of single myocardial cells in vitro. Exp Cell Res 80:432–438

Halvorson HO, Bock RM, Tauro P, Epstein R, Laberge M (1966) Periodic enzyme synthesis in synchronous cultures of yeast. In: Cameron IL, Padilla GM (eds) Cell synchrony. Academic Press, New York London, pp 102–116

Harazova AD, Nechaeva NV (1988) Protein synthesis in tissues of mollusc, *Mytilis edulis* (in Russian). J Obshei Biol 49:365–374

Hoppenbrouwers T, Ugartechea JC, Combs CD, Hodgman JE (1978) Studies of maternal–fetal interaction during the last trimester of pregnancy. Exp Neurol 61:136–153

Horne JA, Whitehead M (1976) Ultradian and other rhythms in human respiration rate. Experientia 32:1165–1167

Ishida K, Yasumasu I (1982) The periodic changes in adenosine-3', 5'-monophosphate concentration in sea urchin eggs. Biochim Biophys Acta 720:266–273

Jarigin KN, Nechaeva NV, Fateeva VI, Trushina ED, Brodsky VY (1978) Ornithine decarboxylase activity rhythm in slices of the rat parotid gland (in Russian). Bull Exp Biol Med 86:726–728

Jarigin KN, Nechaeva NV, Fateeva VI, Novikova TE, Brodsky VY (1979) Circahoralian rhythm of cAMP concentration (in Russian). Bull Exp Biol Med 88:711–713

Jurovitsky JG, Nechaeva NV, Novikova TE, Brodsky VY (1989) Oscillation of protein synthesis rate and ATP concentration in hepatocytes in vitro (in Russian). Dokl Akad Nauk SSSR 304:480–482

Jurovitsky JG, Nechaeva NV, Novikova TE, Fateeva VI (1990) Possible relation between oscillations of protein synthesis rate and nucleotide concentration (in Russian). Isvest Akad Nauk 2:165–171

Kapen S, Boyar R, Hellman L, Weitzman E (1975) Twenty-four-hour patterns of luteinizing hormone secretion in humans. Prog Brain Res 42:103–113

Kayser C, Hildwein G (1974a) Le rhythme circadien de la consommation d'oxygène et de l'activité locomotrice du rat. Arch Sci Physiol 28:81–111

Kayser C, Hildwein G (1974b) Evolution de la consommation d'oxygène et de l'activité du cobaye au cours du nycthémère. Arch Sci Physiol 28:1–23

Kinosita F, Nakai Y, Katakami H, Imura H (1982) Suppressive effect of dinorphin on luteinizing hormone release in conscious castrated rats. Life Sci 30:1915–1919

Knorre WA (1968) Oscillations of the rate of synthesis of galactosidase in Escherichia coli. Biochem Biophys Res Commun 31:812–817

Knorre WA (1973) Oscillations in the epigenetic system. In: Chance B, Pye EK, Ghosh AK (eds) Biological and biochemical oscillators. Academic Press, New York, pp 449–455

Kostenko GA, Litinskaja LL, Veksler AM, Eidus LH (1973) Synchronous oscillations of cell parameters in Ehrlich ascites carcinoma (in Russian). Dokl Akad Nauk SSSR 211:714–717

Kripke DF (1972) An ultradian biologic rhythm associated with perceptual deprivation and REM sleep. Psychosom Med 34:221–234

Kripke DF (1974) Ultradian rhythms and sleep. In: Scheving LE, Halberg F, Pauly JE (eds) Chronobiology. Igaku Shoin Ltd, Tokyo, pp 475–477

Kripke DF, Reite GV, Pegram GV (1968) Nocturnal sleep in rhesus monkeys. Electroencephalogr Clin Neurophysiol 24:582–586

Kuempel PL, Masters M, Pardee AB (1965) Bursts of enzyme synthesis in the bacterial duplication cycle. Biochem Biophys Res Commun 18:858–867

Kusunoki S, Yasumasu I (1976) Cyclic change in polyamine concentration in sea urchin eggs related with cleavage cycle. Biochem Biophys Res Commun 68:881–885

Lebedev NN (1983) Neurophysiology of intestinal functions (in Russian). Physiol J (Kiev) 29:526–542

Lehman U (1976) Short-term and circadian rhythms in the behaviour of the vole Microtus agrestis. Oecologia 23:185–199

Lewis BD, Kripke DF (1977) Ultradian rhythms in hand–mouth behavior of the rhesus monkey. Physiol Behav 18:283–286

Linkov EM, Petrova LN, Savina NG, Janovskaja TB (1982) Superlong oscillation of the earth (in Russian). Dokl Akad Nauk SSSR 262:321–324

Litinskaja LL, Ogloblina TA, Veksler AM, Agroskin LS (1982) pH kinetics in living cultured cells measuring using an indicator dye, neutral red (in Russian). Tsitologia 24:1215–1222

Litinskaja LL, Veksler AM, Nechaeva NV, Novikova TE, Fateeva VI, Brodsky VY (1987) The influence of pH change on protein synthesis rhythm in cultured cells (in Russian). Tsitologia 29:917–921

Livnat A, Zehr JE, Brotten TP (1984) Ultradian oscillations in blood pressure and heart rate in free-running dogs. Am J Physiol 246:R817–R824

Lloyd D, Poole RK, Edwards SW (1982) The cell division cycle. Temporal organization and control of cell growth and reproduction. Academic Press, New York London

Mandelbrot BB (1983) The fractal geometry of nature. WH Freeman, New York

Mano Y (1968) Regulation system of protein synthesis in early embryogenesis in the sea urchin. Biochem Biophys Res Commun 33:877–882

Mano Y (1970) Cytoplasmic regulation and cyclic variation in protein synthesis in the early cleavage stage of the sea urchin embryo. Dev Biol 22:433–460

Mano Y (1975) Systems constituting the metabolic sequence in the cell cycle. BioSystems 7:54–65

Masters M, Donachie WD (1966) Repression and the control of cyclic enzyme synthesis. Nature 209:476–479

Maxim PE, Bowden DM, Sackett GP (1976) Ultradian rhythms of solitary and social behaviour in rhesus monkeys. Physiol Behav 17:337–344

Meyenburg HK (1973) Stable synchrony oscillations in continuous cultures of *Saccharomyces cerevisiae* under glucose limitation. In: Chance B, Pye EK, Ghosh AK (eds) Biological and biochemical oscillators. Academic Press, New York, pp 411–417

Mochan E, Pye EK (1972) Rhythmic metabolism and cell–cell communication in yeast. In: Frank GM (ed) Thesis IV Int Biophys Congr Moscow, vol 4, pp 61–62

Murell KF (1971) Industrial work rhythms. In: Colquhoun WP (ed) Biological rhythms and human performance. Academic Press, New York London, pp 241–272

Nadelhaft I (1976) Dynamics of fast axonal transport. Biophys J 16:1125–1130

Nechaeva NV, Fateeva VI (1973) Dynamics of protein synthesis in intralobular ducts of the rat parotid gland (in Russian). Arch Anat Hist Embriol 65:60–66

Nechaeva NV, Fateeva VI (1974) Rhythm of protein content in the duct cells of the rat parotid gland (in Russian). Tsitologia 16:1033–1036

Nechaeva NV, Jarigin KN, Fateeva VI, Novikova TE, Brodsky VY (1980) Polyamines and the protein synthesis rhythm (in Russian). Bull Exp Biol Med 90:211–214

Nechaeva NV, Litinskaja LL, Hartman J, Lodin Z (1985) Circahoralian changes in the size and dry weight of glioma G6 cells in culture. Physiol Bohemosl 35:162–170

Novikova TE, Nechaeva NV (1979) Rhythms of protein concentration and ^{3}H-lysine incorporation in hepatocytes in vitro (in Russian). Arch Anat Hist Embriol 77:103–107

Orr WC, Hoffman HJ (1974) A 90 min cardiac biorythm. IEEE Trans Biomed Eng 21:130–143

Orr MC, Hoffman HJ, Hegge FW (1974) Ultradian rhythms in extended performance. Aerosp Med 45:995–1000

Oswald I, Merrington J, Lewis H (1970) Cyclical "on demand" oral intake by adults. Nature 225:959–960

Petzelt C (1972) Ca^{2+}-activated ATP-ase during the cell cycle of the sea urchin. Exp Cell Res 70:333–339

Robertson A, Drage D, Cohen M (1972) Control of aggregation in *Dictyostellium discoideum* by an external periodic pulse of cyclic adenosine monophosphate. Science 175:333–335

Ruckebush M, Weekes EC (1976) Insulin resistance and related electrical activity of the small intestine. Experientia 32:1163–1165

Sacher GA, Duffy PH (1978) Age changes in rhythms of energy metabolism, activity and body temperature in *Mus* and *Peromyscus*. In: Samis HV, Capobianco S (eds) Aging and biological rhythms. Plenum Press, New York, pp 105–125

Schultz H and Lavie P (eds) (1985) Ultradian rhythms in physiology and behavior. Springer, Berlin Heidelberg New York

Severny AB, Kotov VA, Tsap TT (1976) Observations of solar pulsation. Nature 259:87–89

Shapiro BH, MacLeod JN, Pampori NA, Morissey JJ, Larenson DP, Waxman DJ (1989) Signalling elements of the ultradian rhythm of circulating growth hormone. Endocrinology 125:2935–2944

Shemerovsky KA (1981) Temporal organization of an ultradian rhythm in duodenum. In: Komarow FI (ed) Chronobiology and chronomedicine (in Russian). Medicine, Moscow, pp 253–254

Smirnov KM, Navakatikjan AO, Gambashidze KM, Hovanov BV (1980) Biological rhythms and work (in Russian). Nauka, Moscow

Sterman MB, Lucas EA, Macdonald LR (1972) Periodicity within sleep and operant performance in the cat. Brain Res 38:327–341

Stupfel M, Davergne M, Perramon A, Lemercerre C, Gorlet V (1979) Rhythmes ultradiens respiratoires de quatre petits vertébrés. C R Acad Sci 289(D):675–678

Stupfel M, Perramon A, Gourlet V, Thierry H (1981) Light–dark and societal synchronization of respiratory and motor activities in mice, rats, quinea-pigs and quails. Comp Biochem Physiol 70A:265–274

Suzuki N, Mano Y (1973) Dependence of periodic DNA synthesis on protein synthesis. Dev Growth Diff 15:113–126

Suzuki N, Neki T, Mano Y (1977) Reaction step requiring protein synthesis in DNA synthesis in sea urchin embryos. Experientia 33:15–16

Svanidze IK, Didimova EV (1974) Dry weight dynamics in living glial cells (in Russian). Tsitologia 16:187–192

Tannenbaum GS, Martin JB (1976) Evidence for an endogenous ultradian rhythm governing growth hormone secretion in the rat. Endocrinology 98:562–570

Tornheim K, Lowenstein JM (1974) The purine nucleotide cycle. J Biol Chem 249:3241–3247

Tsilimigras CWA, Gilbert DA (1977) High frequency, high amplitude ocillations in the amount of "protein" extractable from cultured cells. S Afr J Sci 73:123–125

Van Cauter E, Honinsky E (1985) Pulsatility of pituitary hormones. In: Schultz H, Lavie P (eds) Ultradian rhythms in physiology and behavior. Springer, Berlin Heidelberg New York, pp 41–60

Veksler AM, Kostenko GA, Eidus LH, Litinskaja LL (1973) Study of living cells using scanning microscopes. In: Frank GM (ed) Studies of cell populations using scanning technique. Nauka, Pushchino, pp 101–112

Webb WB (1971) Sleep behavior as a biorhythm. In: Colquhoun WP (ed) Biological rhythms and human performance. Academic Press, New York London, pp 149–177

Weiss PA (1972) Neuronal dynamics and axonal peristaltics. Proc Natl Acad Sci USA 69: 1309–1312

Weiss PA (1974) Dynamics and mechanisms of neuroplasmic flow. In: Fuxe K et al. (eds) Dynamics of degeneration and growth in neurons. Pergamon Press, New York, pp 203–213

Zaguskin SL, Nemirowsky LE, Tshukovsky AB, Vachtel NM, Brodsky VY (1980) Nissl rhythm in a living neuron (in Russian). Tsitologia 22:982–988

Zaguskin SL, Grinchenko SN, Brodsky VY (1991) Interactions between circahoralian and circadian rhythms: a mathematical model. Izvest Akad Nauk (in press)

Chapter 3

A Precise Circadian Clock from Chaotic Cell Cycle Oscillations

R. R. Klevecz

In this chapter I attempt to set out in detail the background, the experimental results and computer simulations that have led to the idea that the timing of cell division, the scheduling of events within the cell cycle, as well as the many exquisitely precise timing functions of circadian rhythms, are all expressions of a fundamental macromolecular ultradian oscillator. A novel component of this argument is the idea that the underlying oscillator is not a periodic, limit cycle oscillator, as is commonly supposed, but is a strange, or chaotic attractor. This view is consistent with the long-recognized truth that oscillations in cellular constituents that are seen in the cell cycle of isolated eukaryotic cells are rarely impressive statistically. I show that, when large numbers of cells are brought together and caused to share some of one of the oscillating components, precise periodicity is a frequent result. The two outcomes of interest to biological self-organization, periodic spiral waves and chaotic synchronization, emerge from the interaction of large fields of chaotic cellular attractors coupled by diffusion. Self-organization of periodic structures from imprecise components appears to require the early establishment of an antipodal phase relationship between a few cells near the core of the spiral.

Background

The macroscopic properties of biological rhythmicity have been characterized in the course of the past 30 years (Pittendrigh 1960; Pittendrigh and Daan 1976), but, while it is generally agreed that circadian rhythms are engendered by a cellularly based oscillator or ensemble of oscillators that specify the phase and period of the biological clock, identifying these oscillators with any particular element of cellular chemistry has proved difficult. More problematic is the cell and tissue distribution of a functional clock. "Clockedness" often appears diminished in proportion to the degree that the system is simplified. Until recently it was thought that the clock was a limit cycle oscillator (see p. 47) of circadian duration (Winfree 1970). It now seems far more likely that the underlying chemistry is expressed in a higher frequency oscillator (Klevecz et al. 1984; Dowse and Ringo 1987; Edmunds 1988). The problem has been how to construct a plausible, higher frequency, ultradian oscillating system that could be coupled to achieve the required increase in period. Heretofore, analytical and numerical approaches to this question have not yielded a satisfactory answer (Goodwin 1963; Higgins 1967; Winfree 1967; Pavlidis 1973).

While the cell cycle is one of the more overt expressions of biological rhythmicity, and exhibits ultradian macromolecular oscillations (Klevecz 1969a; Klevecz and Kapp 1973), the nature of the relationship between circadian rhythms, higher frequency cellular oscillators and the cell replication cycle is only beginning to be defined (Edmunds 1984; Klevecz et al. 1984). The isolated metazoan cell in culture exhibits a pattern of time keeping that is ultradian and appears sometimes deterministic and sometimes probabilistic (Burns and Tannock 1970; Petrovic et al. 1984; Klevecz and Shymko 1985). Similar behaviour is seen in unicellular eukaryotes growing with generation times of less than 24 h (Edwards and Lloyd 1978; Chisholm and Costello 1980; Readey 1987). This pattern has been described as "quantized" and modelled as a limit cycle oscillator with added noise (Klevecz 1976; Klevecz et al. 1980a; Shymko and Klevecz 1981). It has also been noted that such behaviour is characteristic of deterministic chaos (Klevecz et al. 1991).

In an attempt to combine the formal and molecular approaches to dynamics, and to relate what was known about cellular time keeping to circadian rhythms (Klevecz 1969a; Klevecz et al. 1984), it was suggested that the "quantal" oscillator represented the chemical core of time keeping in both the cell cycle and circadian rhythms. In agreement with this suggestion, Kaufmann and Wille (1975) found that oscillations endured in the absence of cell division and argued that cell division was but one more downstream event, gated by this high frequency oscillator/clock. In further support of this idea it now appears that *Clock* mutants of *Drosophila* (Konopka and Benzer 1971), *Neurospora* (Feldman and Hoyle 1976), *Chlamydomonas* (Bruce 1972) and hamster (Ralph and Menaker 1988) show temporal clustering: at 2.5- and 5-h intervals in *Neurospora*, and at 4.5–5-h intervals in *Drosophila* (19.5, 24 and 28–29 h), while per^0 and per^- mutants originally described as arrhythmic, when subjected to power spectrum analysis are ultradian multiples of the fundamental rhythm (12, 8 and possibly 4 h) (Dowse and Ringo 1987). Fig. 3.1 shows a schematic summary of quantal periodicity. The range and temporal distribution, the "noisy and uncertain periodicity" seen particularly in the per^0 mutants, is what might be expected if one sampled the frequency of passage through a restricted portion of the surface of a chaotic attractor.

The behaviour of large arrays of chaotic Rössler attractors coupled diffusively has been studied. The purpose in doing this was to model the behaviour of a population of cells when they are brought together as an aggregate. Because spiral propagating waves and concentric banding patterns typically appear as part of the developmental sequence in aggregates of both prokaryotic and eukaryotic cells, it seems plausible to suggest that many of the characteristic patterns of organization seen in development, as well as the temporal features of biological clocks, emerge from the collective behaviour of populations of cells tightly coupled by diffusion in a tissue. This might perhaps be facilitated by proteoglycans in gap junctions (Bargiello et al. 1987), rather than as a unique mono- or oligogenic property of individual cells.

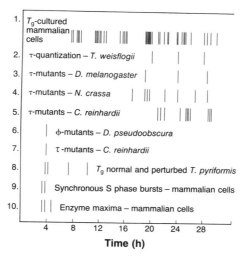

Time (h)

Fig. 3.1. Expression of a fundamental oscillator in circadian and ultradian rhythms. Each line represents a peak of occurrence. Line 1: quantized variation in generation times (T_g) of cultured mammalian cell lines. T_g values were determined from the published data on cells synchronized by mitotic selection or from intermitotic time lapse analyses of random cultures. The list is not exhaustive but represents a sampling of papers published between 1961 and 1991 in which the stated generation time could be directly confirmed in the data. Wherever possible modal generation times were obtained, and reports stating only population doubling time were excluded (Klevecz 1976). Line 2: polymodal distribution of generation times in the marine diatom *Thalassiosira weisflogii* growing in the circadian mode with normal τ also expressed (Chisholm and Costello 1980). Line 3: Long- and short-period (τ) mutants of *Drosophila melanogaster* isolated by Konopka and Benzer (1971). Line 4: τ-mutants of *Neurospora crassa* (Feldman and Hoyle 1976; Feldman et al. 1979; Feldman and Dunlap 1983). Line 5: phototaxic τ-mutant of *Chlamydomonas reinhardii* picked and isolated by Bruce (1972) from cultures treated with nitrosoguanidine. Line 6: phase angle (ϕ) early eclosion mutants selected by Pittendrigh (1965) by continuous selection through 50 generations for early emerging *Drosophila pseudoobscura*. Line 7: τ-mutants of *C. reinhardii* isolated by Bruce (1972) from cultures treated and selected for period (τ) changes. Line 8: oscillatory variations in T_g of *Tetrahymena pyriformis* perturbed by continuous incubation in low levels of actinomycin D. Normal T_g in these cultures is 4–4.5 h (Jauker and Cleffman 1970). Line 9: interval between synchronous bursts in DNA synthesis in the S phase of mammalian cells (Klevecz 1969b; Klevecz and Kapp 1973; Collins 1978; Kapp et al. 1979; Holmquist et al. 1982) scored from a maximum slope of [³H]thymidine incorporation rate between peaks. Line 10: intervals between peaks in maximum enzyme activity or levels in the cell cycle of synchronous hamster cells in culture (Klevecz and Ruddle 1968; Klevecz and Kapp 1973; Klevecz 1975; Forrest and Klevecz 1978).

Spiral propagating waves and concentric banding patterns appear as central features of spatial and temporal organization in prokaryotic (Hoeniger 1964; Shapiro 1988) and eukaryotic (Gerisch 1971; Devreotes 1982) cells growing as aggregates (Fig. 3.2a–c), and in heart muscle (Allessie et al. 1977; Winfree 1987), neurocortex (Winfree 1987) and retinal (Martins-Ferreira et al. 1974) tissues of higher eukaryotes. When cell-to-cell communication is reduced or ablated in the temporally *period per* mutant system of *Drosophila*, it results in altered clock period and in the case of per^0, high frequency, possibly chaotic rhythms (Dowse and Ringo 1987; Dowse et al. 1987). Here, the possibility that periodicity arises in multicellular biological systems, or in aggregates of unicells, from neighbourhood coupling of autonomous cellular oscillators that

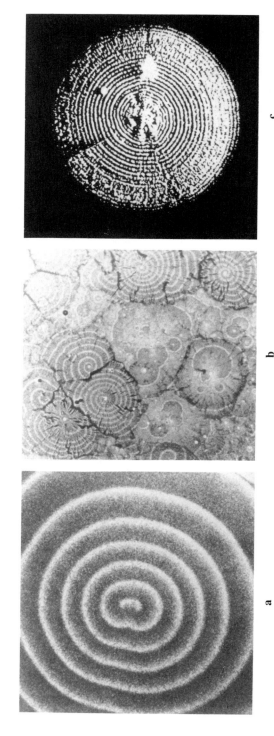

a   b   c

Fig. 3.2. Spatiotemporal organization seen in continuous chemical systems and unicellular eukaryotes. **a** Concentric banding seen in the Belusov–Zhabotinski reaction as a function of initial conditions (band interval approximately 1 cm). **b** Mixture of spirals and concentric bands seen during aggregation in *Dictyostelium discoideum* (band interval approximately 0.25 cm). **c** Archimedean spiral zonation pattern seen in *Nectria cinnibarina* (band interval approximately 0.25 cm).

are individually chaotic (Klevecz et al. 1991) has been further explored. Starting from a variety of initial conditions, with toroidal, no-flux, fixed or periodically driven boundaries, dynamic systems comprising a large array of diffusively coupled chaotic Rössler attractors were seen to organize spontaneously into spirals (compare simulations shown in Fig. 3.3a–c with 3.2a–c) or concentric bands in which wave propagation was periodic, or into synchronous fields in which the chaotic behaviour of the uncoupled attractors continued to be expressed.

Experimentally, the organization of spiral waves has been reported to show a strong dependence on initial conditions, yielding a few very specific patterns that are nevertheless difficult to replicate. In the ascomycete *Nectria*, concentric rings form in two-thirds of the colonies, while the remainder form archimedean spirals, right- and left-handed in approximately equal numbers, or twin intercalated spirals (Bourret et al. 1969; Winfree 1973; Fig. 3.2c). Similar patterns are seen in *Dictyostelium*, except that the spiral is formed by a periodic inward motion of cells (Gerisch 1971) toward the rotor, which may be viewed as a sink (Fig. 3.2b). Explicit descriptive equations of wave propagation in excitable biological tissues or chemical media have heretofore been considered essential to progress in this field and much attention has been focused on approximate representations (Keener 1986; Tyson and Keener 1988; Gerhardt et al. 1990) as well as on experimental quantification (Muller et al. 1987), using the Belousov–Zhabotinski reaction as a model (Fig. 3.2a).

The interaction of coupled attractors may play a role also in the organization of many physical and inorganic chemical systems. Recently, Pecora and Carroll (1990), for example, reported linking two chaotic Rössler or Lorenz attractors with a common driving signal and observed stable synchronization. These authors state that this sharing of variables is a necessary condition for synchronization. We have found that synchronization in large arrays of chaotic attractors occurs for a relaxed set of conditions and that synchronization is but one, and possibly the least interesting, of a set of outcomes in orthogonally coupled large arrays.

The Cell Cycle as a Noisy Clock

The question of the function of ultradian rhythms has two components, one of which has to do with the notion that ultradian rhythms are rhythmic. It seems possible on the basis of evidence in the literature and the work reported here that they are not, if we understand "rhythms" to mean having a precise period in the sense that a cosine function has a precise period. I think it is difficult to argue that the observed ultradian rhythms in most cases approach even the precision of circadian rhythms, and we must add that, in circadian rhythms, the vaunted high precision is often a case of selection by the investigator of the experimental systems. So, for example, *Drosophila pseudoobscura* is precise; *D. melanogaster* is not (Dowse and Ringo 1989). As to function, I argue that, paradoxically, the very imprecision of higher-frequency oscillations may make them more suitable as building blocks of circadian rhythms. Moreover, the complexity of biological systems makes it difficult to think that control systems would comprise just two variables, giving rise to limit cycle

c

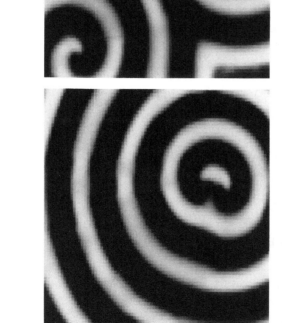

b

a

Fig. 3.3. Autogenous spatiotemporal organization seen in diffusively coupled cellular arrays of chaotic oscillators: a simulation of the patterns seen in Fig. 3.2. **a** and **b** Concentric bands and Archimedean spirals formed by a large (8100 cell) field of chaotic Rössler attractors. In these simulations the Rössler attractor: $dx/dt = -y - z$, $dy/dt = x + ay$, $dz/dt = b + z(x - c)$, was used with $a = b = 0.2$, $c = 5$ and random uniformly distributed initial conditions ($-3 < x < 3$; $-3 < y < 3$; $0 < z < 1$) and was coupled diffusively: $dx/dt = -d(4x -$ (sum of 4 neighbouring cells)) where $d = 0.25$, and with a time step of 0.01. No-flux boundary conditions were used. **c** All conditions are the same as in **a** and **b** except that the boundary conditions were varied sinusoidally with a period equal to that of the average fundamental period of the uncoupled attractor. All Figures are to approximately the same scale.

oscillations, though it is possible. With this in mind I would argue that it is manifestly inappropriate to model biological rhythms as a simple limit cycle oscillation of circadian duration expressed throughout the tissue or organism. This should be greeted as good news by the contributors to this volume, since such models, popular for the past 15 years, relegate ultradians to some sort of spurious noise that occurs aberrantly as a secondary or downstream event in clock expression. If the control system in biological rhythmicity is of greater than two dimensions and if it has significant non-linearities, then it is likely to express the properties of multifrequency, turbulent or periodic flows. The questions then are: what is the composition of such flows, how do they arise, and how do they function to generate a relatively precise time sense in biological systems?

It begins to appear that self-organization of complex chemical systems most often gives rise to chaotic chemical behaviour, but that the number of possible outcomes of this behaviour is much smaller than its initial state might suggest. Also, a few simple rules when applied to a chaotic system give rise to remarkably periodic behaviour of the collective.

Modelling Quantal Oscillators in the Cell Cycle

Using genetics to study the cell cycle is like using the electron
microscope to study quantum mechanics.
                              Anonymous 20th century dynamicist

The cell cycle is one of the more overt expressions of biological rhythmicity, yet the vast majority of investigations of cytokinetics have begun with either linear sequential or stochastic models and sought to find the necessary and sufficient constituents to step the cell forward on its track. In either of these approaches, time, as a continuous variable, has been factored out. What we see expressed in these two very disparate views of the cell cycle is the fact that in its temporal business the cell sometimes seems to be deterministically controlled, while at the other extreme its kinetic behaviour at times, is unpredictable. This could be expressed succinctly as "noisy periodicity", the characterization applied in the popular literature to describe deterministic chaos.

On the other hand, in my laboratory and those of others (Klevecz 1976; Edwards and Lloyd 1978; Chisholm and Costello 1980; Lloyd and Edwards 1984; Petrovic et al. 1984; Readey 1987), a considerable body of evidence has accumulated for an oscillator functioning as a cell cycle clock in eukaryotic cells. With a period that is a submultiple of the cell cycle (Klevecz et al. 1980a; Shymko and Klevecz 1981), this cellular oscillator has many of the properties classically associated with a biological clock, even though the period is considerably shorter than that of the circadian clock.

In an earlier effort at reconciling the deterministic and stochastic aspects of the cell cycle, simulations were performed using the Brusselator, a non-linear oscillator with just two variables, that displays limit cycle behaviour in certain regions of parameter space. To achieve the necessary dimensionality, a linear Z parameter was added. The time-keeping oscillator had no intrinsic indeterminism, but dispersion in generation times was accomplished by adding a random walk to each cell at each time step and additional quantal dispersion was seen owing to occasional random skipping of mitotic or chromosome replication cycle events due to subthreshold excursions, as well as delays and skipping of mitosis due to phase perturbations. Appropriate fit to experimental data was achieved by the choice of triggering threshold and random walk step size. The exact shape of the phase response curves could be varied by choosing appropriate values for oscillator parameters A and B.

This approach was taken because the passage of a cell through its division cycle contains a random component whose effects become apparent when cell generation times are analysed by time-lapse microscopy. The resulting individual cell intermitotic times have often been found to fall in a distribution skewed toward long generation times. Such distributions have been characterized as reciprocal normal (Kubitschek 1962) or more commonly, as negative exponential distributions (Burns and Tannock 1970; Smith and Martin 1973; Minor and Smith 1974; Brooks 1977). The exponential distribution can be simply explained in terms of the so-called "transition probability" model in which the cell cycle is divided into two parts: one which cells enter after mitosis and leave randomly at a constant rate per unit time; and a second state through which cells move uniformly, traversing this state in a constant time T_B. In this model, the plot of fraction of cells undivided versus cell age ("α-curve") has a shoulder of length T_B and an exponential tail due to the random exit from the "A-state" (Smith and Martin 1973). This predicted pattern does seem to be followed in some experimentally derived curves, and the analogous prediction that the distribution of differences in sister cell generation times (the "β-curve") should also be exponential has experimental support as well (Minor and Smith 1974; Shields 1978), although such data are somewhat limited. However, on some occasions intermitotic times have been seen to be skewed toward shorter generation times (Dawson et al. 1965).

In spite of the apparently random behaviour in cell populations, there is considerable evidence for more precisely controlled timing in the cell cycle. Our experiments on phase response in cultured cells perturbed by heat shock, ionizing radiation, or serum pulses have shown a characteristic tandemly repeated pattern of phase shift following the perturbation applied at different times in the cycle (Klevecz et al. 1980b; King et al. 1980). In rapidly growing cells with 8–9-h generation times, two such repeats are seen, suggesting that time keeping in the first half of the cycle is somehow similar to that in the second half, at least in terms of the factors responsible for the perturbation-induced phase shift.

Our observations on the distribution of the generation times in cultured cells using time-lapse videotape microscopy have indicated that intermitotic times tend to fall into clusters separated by slightly more than 4 h (Klevecz 1976). Such quantization of generation time, and the similarity of the patterns of phase response to different perturbing agents, has argued that the cell cycle may be timed by a macromolecular oscillator with a nominal 4-h period (Klevecz et al. 1980a). The simplest oscillator which can function independently of the events that it controls and yet spontaneously resume stable oscillation following a perturbation is termed a limit cycle. The detailed properties of these systems have been described elsewhere (Prigogine and Lefevre 1968; Pavlidis 1973). We have shown that if two loops of such a cycle

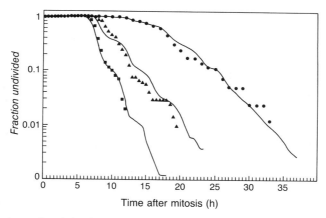

Fig. 3.4. Experimental and simulated generation time (T_g) distributions. To measure intermitotic times (IMT) of random cultures, cells were cultured at low densities and, after 18 h of growth, recording for analysis was begun using a time-lapse videotape recorder. Individual cells were tracked and a Markov chain or pedigree chart of generations was made; T_g values were constructed for each mother cell, its daughters, etc., for up to five generations. α-Plots and IMT distributions were calculated from IMT data. α-Curves of V79 and WI38 cells and their simulation by the oscillator model were obtained. α-Curves describing the undivided fraction of cells in T_g distribution curves are shown: 200 ($\pm$50) T_g values are represented in each curve. (▲) Distributions of T_g in V79 cells growing under suboptimal conditions; (■) the distribution of V79 cells growing under optimal conditions. Simulation of these curves (lines) shows that the distribution of T_g values is exponential but quantized within the exponential envelope. Under suboptimal conditions, threshold $\theta = 4.7$; under optimal conditions, $\theta = 4.5$. (●) T_g distributions of WI38 cells and the line gives the simulated distribution. Here parameters are the same except that cycle time is increased by requiring two rather than four oscillator loops in one cycle. Note that, with the long cell cycle generated by this model, the resulting distribution of T_g is smoother and approaches a straight line at long values of T_g.

are required for each 8–9-h cell cycle time, then an accurate representation of the repeated pattern in the observed phase response curves is generated (Klevecz et al. 1980a). Furthermore, if cells are assumed not to be rigidly confined to the cycle trajectory, but to deviate slightly from this trajectory due to random noise as they move around the cycle, then a distribution of generation times is predicted which is approximately a negative exponential, but quantized within this exponential envelope. The resulting α and β curves therefore have the same general shape as those predicted by the transition probability model, but for sufficiently short generation times have a regular undulating pattern in the exponential tail due to the quantization (see Fig. 3.4).

As further evidence of non-random behaviour in the cell cycle, there are data which are reported to show that sister cell generation times tend to be positively correlated (McQuilkin and Earle 1962; Minor and Smith 1974), and it has been suggested that there is also a negative correlation between mother and daughter cell generation times (Killander and Zetterberg 1965). A correlation between sister cell cycle times is inconsistent with the original simple transition probability model, which predicts a random negative exponential distribution of sister generation time differences (the β curve). Recognizing this problem, Brooks et al. (1980) have extended the transition probability model, proposing that a variable interval L begins in one cycle and is carried through division into the next, and is then followed in the latter part of the cycle by a purely random interval. The interval L, common to both sisters, correlates the sister cell generation times and the following random interval still generates the exponential distribution in generation time differences.

Neither our oscillator model as described above nor the transition probability model takes into account the relationship of cell growth to the timing of the cell cycle. Several models have been proposed in which it is assumed that some cell cycle event such as the initiation of DNA synthesis (Killander and Zetterberg 1965) is triggered when cell size reaches a critical level. In such models, cell cycle time and average cell size depend on the size threshold and the cell's growth rate. It is possible to relate the idea of a critical size threshold to the observation of sister–sister and mother–daughter generation time correlations. A cell whose mitosis is delayed without a change in its growth rate will be larger than normal at division, and consequently will produce two larger than normal daughters. Such large daughters will reach the size threshold in a shorter time than average; therefore they will both tend to have a shorter than average cycle time. Thus, mother and daughter generation times will be negatively correlated. Simultaneously, a positive correlation between sister cell generation times is predicted since the size of sisters at division will tend to be positively correlated.

In order to account for the observed generation time correlations in cell populations, we incorporated a size threshold requirement into the oscillator model. In that earlier work we showed that the statistical properties of cell growth, such as generation time quantization and pseudo-exponential generation time distributions given by the oscillator alone, are unaltered by the addition of a size threshold. We compared simulations of the patterns of cycle time correlations with the observed patterns in different cultured cell types, and discussed possible reasons for their similarities and differences (Klevecz and Shymko 1985).

The size-bounded oscillator system predicts that an individual cell within a given population which randomly becomes larger than average will have a shorter generation time in the immediate subsequent cycle and possibly for several cycles thereafter. Larger cell types with higher size thresholds are predicted to have longer generation times than do their smaller counterparts. A preliminary study of cell types immediately available suggested that this was generally true of established cells in culture, e.g. V79, L5178Y, CHO and HeLa. The exception to this was EMT-6 which is a large cell with a short generation time. However EMT-6 is tetraploid, suggesting that the critical size threshold may be scaled to the DNA content of the cell (Fantes and Nurse 1977). If, instead of varying the size threshold in the model system, growth rate is varied, keeping size threshold constant, the correlation between average cell size and average generation time in the population is reversed. That is, slowly growing cells will have longer generation times, but will now tend to be smaller than fast-growing cells. The reason for this is that a slowly growing cell which has just exceeded its size threshold will grow less than a fast cell during the nearly constant time required to traverse the last cycle loop, and will therefore be smaller at division than the fast cell.

For this process, the model described above uses a size-bounded oscillator in which the rate of increase in size was independent of time or cell cycle phase.

The cell was gated by the oscillator only at one point in the oscillator cycle and only if a critical value for size had been reached. In this model the developmental, more or less time-independent, properties of the cell cycle were identified with "size", recognizing that such a variable encompassed or obscured most of what is known about the growth of cells. Fortunately the model proved to be robust in this regard and, within the limits of available data, gave the same behaviour when size increased according to any number of monotonic functions, exponential, logistic or linear. The effect of combining a "noisy" oscillator with linear growth led to alternation of generation rates between mother cells and daughter, granddaughter or great granddaughter cells (Fig. 3.5). This interesting behaviour occurred because a cell having skipped a division, either due to a random excursion away from the triggering value of the oscillator or due to "perturbation" in the form of shifts in the value of the x and y variables, would result in a cell becoming unusually large and therefore being able to divide at the first possible gate. This size advantage tended to endure for several subsequent generations and so division pedigrees would be marked by sequences of long–short–short–short generation times. With relatively little manipulation of threshold, different cells showing very different pedigrees could be accurately modelled. The model accounts for some of the less familiar behaviours of cycling cells. In particular, delays induced by perturbation are handled in this model without resort to awkward or arcane descriptions such as "excess negative delay" or "delay for repair of sublethal damage" or "transient arrest".

Is the Cell a Chaotic Attractor?

Where the limit cycle model fails is in the somewhat ad hoc addition of a size threshold and the use of random walk to generate the polymodal distribution of generation times. In addition, some clocks behave or misbehave in ways that are not parsimoniously predicted by a limit cycle based model. Cells in situ and in culture commonly show quantal distributions in generation time while tumour cells in situ appear to undergo period halving in the course of progression (Klevecz and Braly 1987, 1991). Variability, period doubling and period addition phenomena seen in the generation times of cells emerge naturally from the behaviour of a chaotic attractor.

In the work reported here, all cells in the tissue ran the Rössler (1975) attractor using different initial values of the three variables in each cell. Explicitly, in the Rössler equations: $dx/dt = -y - z$, $dy/dt = x + ay$, $dz/dt = b + z (x - c)$. We generally begin with uniformly distributed random initial conditions of: $(-3 < x < 3; -3 < y < 3; 0 < z < 1)$. Numerical integration was performed using a fourth-order Adams predictor.

The Rössler attractor is an abstract chemical reaction scheme, well behaved and without "stiffness", showing deterministic non-periodic flow in a continuous system of equations. It consists of a set of three ordinary differential equations constructed originally from a two-variable limit cycle oscillator and a chemical hysteresis system. In Rössler's (1975) words it realizes "a prototypically simple chaos generating machine". He also pointed out prophetically that "the behavior of diffusion-coupled chaotic systems poses a challenging dynamical problem in its own right". In this connection others, including

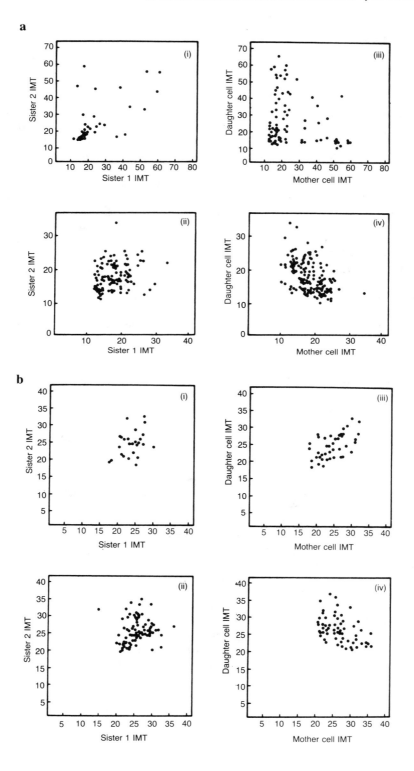

Winfree (1980), have anticipated the desirability of studying the "stochastic dynamics" of mutually coupled oscillators.

The choice of the Rössler attractor to represent the non-equilibrium chemistry of the system was intentional and demonstrates that quite complex patterns can be modelled to an unsettling degree without regard to specific chemistry. Most chaotic chemical oscillators are likely to be abstractions from the more complex chemistry of the living cell, and can be spoken of as thick closed surfaces though they have no volume. Thus the choice of one attractor over another may not affect the results seen thus far, so long as there is a single major basin of attraction.

Goldbeter has evolved a realistic model for periodic signalling in *Dictyostelium* but has not looked at its extension into discrete space (Goldbeter and Segel 1977; Goldbeter and Martiel 1987). The Goldbeter model is based on the activation of cAMP synthesis upon binding of cAMP to a cell surface receptor, with consequent desensitization of this receptor. He finds that instabilities arising from non-linearities associated with the refractory period of the receptor lead to autonomous oscillations of cyclic AMP. This system, whose behaviour is governed by a set of seven ordinary differential equations, seems particularly appropriate for the aims presented here and will be studied in the near future.

The x, y phase-plane portrait of the Rössler attractor for differing values of the parameter C, all in the chaotic and periodic domains, are shown in Fig 3.6. Note that by applying the rule developed in the previous section for the execution of specific cell cycle events, including initiation of DNA synthesis or mitosis, the distribution of these events in time will be very different depending on the choice of parameter values and the value and position of the triggering threshold. For a constant triggering area set in the surface of the attractor, both cell cycle time and its distribution will sometimes be very different for rather small changes in C. Phase response to perturbation is again similar to that seen using a limit cycle oscillator. Most interesting is the observation that the alternation of generation times, contrived in the original limit cycle model by the addition of a critical minimum size boundary for execution of events, falls naturally out of the behaviour of this three-variable system. When a cell takes the high z path, the subsequent times follow the long–short–short–short sequence discussed earlier. Alternation of generation times is an intrinsic feature of this system.

A bifurcation in the original limit cycle results in a doubling in the total length of the trajectory and therefore total time to traverse the loop. (A detailed description of bifurcation events in the Rössler attractor follows on pp. 63–65.) For further increases in C additional period doubling bifurcations occur for smaller and smaller increments in C. Eventually the graphical trace of the Rössler oscillator becomes so dense that it is impossible to resolve the individual paths. This is then an example of period doubling bifurcations

---

Fig. 3.5. a (i) Sister–sister (S–S) and (iii) mother–daughter (M–D) correlation plots for WI38 cells. Simulations (ii) and (iv) were done with $\theta = 4.7$, $\theta_s = 3$. Note that the time scales are different in experimental and simulation plots, indicating wider variability in experimental versus simulated generation times. IMT, intermitotic time. **b** S–S (i) and M–D (iii) correlations for HTC cells. Simulations (ii) and (iv) use $\theta = 4.7$, $\theta_s = 5$.

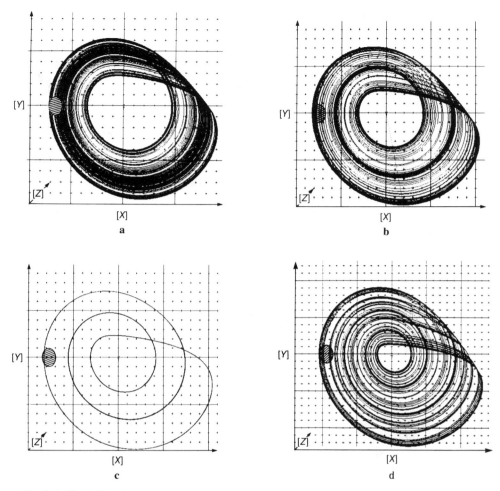

Fig. 3.6. Single Rössler attractors projected in x, y phase-space. Values of the parameter C vary from $C = 4.5$ in **a**, to 5.176 in **b**, to 5.184 in **c** and to 6.64 in **d**. Note particularly that the density of crossings through any portion of the grid varies greatly between attractors but will be statistically similar at different times in any one attractor. If we imagine the crossings as triggers for the execution of cell cycle events, a rather good approximation to the observed noisy periodicity results.

leading to chaos. The oscillator is very noisy, with spikes commonly at the fundamental or at low multiples of the fundamental. In the mathematical sense it is an aperiodic function. However, since the attractor is a closed loop there remains the semblance of periodicity in the x,y phase-plane. This can be seen in power series analysis using Fourier transform and in the x,y phase-plane trace as the more commonly visited region of state space. If we hypothesize that events are triggered to occur when limited regions of the three-dimensional state space are occupied, then a rich complexity of timekeeping is possible for a single value of the C parameter. This is in some way equivalent to the situation that results from hypothesizing that the range of cell cycle behaviours arises from tuning the C parameter and restricting the triggering state space to a single point on the attractor. It may be of interest in the future to include a size boundary for the execution of certain events into the chaotic attractor model, since there may yet be such a requirement.

Spatiotemporal Organization in a Tissue Composed of Chaotic Cellular Oscillators

After the construction of a model that reproduced many generic features of the dynamic behaviour of individual cells, what are the consequences of bringing a number of such cells together and causing them to communicate by sharing one of the chemical constituents of the oscillatory system?

A great deal of attention has been given to the role of intercellular communication during morphogenesis and to specifying the minimum constituents of cellular chemistry. Here it is shown that the spatial patterns and temporal behaviour seen in multicellular structures as part of the aggregation phenomenon in simple eukayotes or prokaryotes can be reproduced in a diffusively coupled array of simple chaotic attractors without reference to the specific chemistry. Spiral waves and concentric banding patterns appear as a general feature of spatial and temporal organization in prokaryotic and eukaryotic cells growing as aggregates. One of the most studied of these chemotaxis-mediated organizational phenomena is in *Dictyostelium*, where the spirals are formed by periodic inward motion of cells toward the rotor, which has been viewed as an organizing centre (Fig. 3.2b). Organization into spirals occurs autogenously so long as the initial distribution of cells is uniform through phase space so that the field of cells is locally heterogeneous with respect to phase.

A dynamic system comprising a large array of chaotic Rössler attractors coupled through one of the variables has been seen to organize spontaneously into concentric bands or archimedean spirals (Klevecz et al. 1991). Starting with random uniformly distributed initial conditions, a dynamic system comprising a large array of coupled chaotic Rössler attractors evolving on a torus was seen to organize spontaneously into paired inward spirals in which wave propagation was periodic. Individual attractors at the spiral centre appear to be periodic limit cycle oscillators, while those in the spiral arms were two-cycle periodic attractors.

Grey Scale Representations

Three graphic/grey scale representations of a tissue have been developed to facilitate the analysis and understanding of its system behaviour: (a) the attractor field (tissue) is simply the value, of each attractor (cell) in the array, for one of the variables (Fx, Fy, Fz) presented in normalized grey scale; (b) the

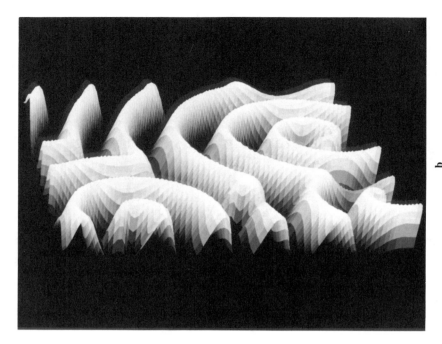

b

a

Fig. 3.7. Spiral formation from chaotic fields. **a** Starting with 90×90 fields of cells uniformly distributed in x, y and z, regions of organization continued to grow until spirals formed. White corresponds to the maximum x value in the field, black to the minimum. **b** Three-dimensional projection of the spiral in **a**.

system attractor (SA) is the phase portrait of all the cells in the field with normalized grey scale indicating the number of attractors in any region of paired variable (usually x,y) phase-space; (c) the system bifurcation (SB) diagram is a Poincaré section of the system attractor plotted against coupling strength (or other parameter), where grey scale is a representation of the frequency of visitation to any point on the section. SB analysis is thus useful for finding period changes in the system as a function of changing parameter values, in this case, coupling strengths, and can also be used to look at single cells or regions.

It is of interest that, for a restricted set of initial conditions with local concentration heterogeneity, the macroscopic behaviour of a field of cells can be closely represented by an abstract system of equations without reference to the actual underlying chemistry. It seems remarkable that these highly ordered states, so common to simple differentiating biological systems, can arise through diffusive coupling of a population of cellular oscillators that are individually not periodic.

Mechanism of Spiral Formation

There are striking morphological similarities between such disparate phenomena as the BZ reaction, colony growth of *Proteus* or *Escherichia coli*, zonation in *Nectria* and other ascomycetes, pulsatile amoeboid movement in *Dictyostelium*, and propagation of electrical activity in neural, retinal and cardiac tissues of higher eukaryotes. Does this similarity of field morphology alone give us sufficient cause to think that similar pattern-generating mechanisms are at work? Even within the biochemical and biological systems there are great differences in both time and space. Wave propagation in the BZ reaction takes place in a continuous, ostensibly homogeneous medium, while in the ascomycetes and *Proteus* the spiral is formed by the addition of new cells at the leading edge of growth. It seems likely that, in detail, the chemistry will be very different, but the importance of tight or loose coupling, or continuous or discontinuous fields, in a system with non-linear dynamics may generalize across the scale. What appear to be the common and perhaps necessary conditions for periodic spiral formation are the presence of unit oscillators that are chaotic, that these unit oscillators be continuous (not automata), but that the system be discrete and coupled by diffusion to permit "phase compromise". The need for local concentration heterogeneity (Klevecz et al. 1991) and the implied phase discontinuity does seem to hold, because the population of attractors must be somewhat symmetrically distributed around the steady state at the initiation of the simulation. The open question remains: how do cells behave inside the critical radius, $2\pi r = \lambda$, of the spiral. It seems irrefutable that a continuous system must have a phase singularity in the sense of Winfree (1980). However, according to Winfree phase singularities must exist for spirals to form and spirals form where singularities occur. In a discrete system coupled by diffusion, phase is not continuous. We believe that because of coupling, and the fact that chaotic attractors are of fractional dimension, local adjustment in phase space can be made to accommodate the discontinuities that occur.

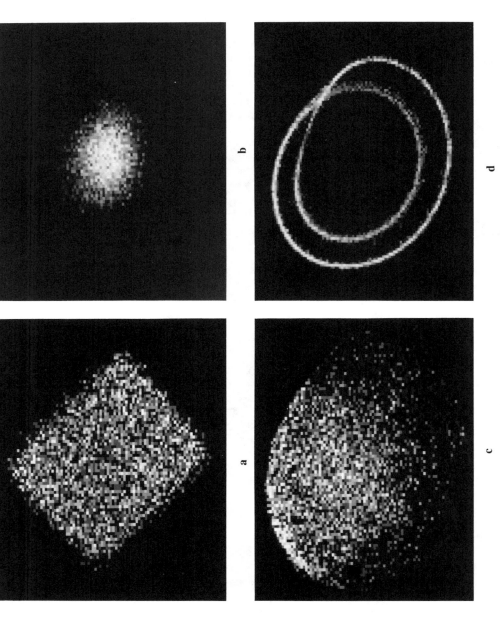

Fig. 3.8. The system attractor: a two-dimensional (x, y) phase plane histogram of the field of Rössler attractors. To generate the Rössler system each of the 8100 attractors in the field of Fig. 3.7 is placed into a histogram bin depending upon its x and y values. The brightness of the bin increases with increasing numbers of attractors in the bin. **a** The uniform distribution of the system of attractors at unit time 0.1. The attractors then collapse toward the singularity because of coupling. In the ensuing frames (**b–d**) the cells organize and spread away from the singularity, eventually forming a portrait whose surface appears to be hybrid between a two-cycle ($C = 3$) and the uncoupled chaotic attractor ($C = 5$)

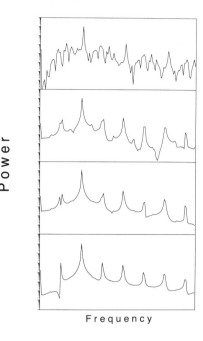

Power

Frequency

Fig. 3.9. Fast Fourier transform (FFT) analysis of spiral core and arms. Cells indicated in the spiral of Fig. 3.7 were analysed using FFT to generate a power spectrum. From top to bottom: the uncoupled Rössler attractor, a cell in the remote of the spiral, a cell in the near arm of the spiral, the central (pivot) cell of the spiral.

In a 90×90 cell tissue such as that shown in Fig. 3.7, two spirals can be seen. The self-organization of fields into spirals or concentric bands is perhaps the most interesting outcome from the physical perspective and, in the case of spirals, the most difficult to describe dynamically. Of all the possible configurations of the coupled array, only arrays forming spirals show an increased order in the form of permanent spatiotemporal periodicity compared to the uncoupled arrays. An analysis of various cells along a radius from the outer arms of the spiral to the core shows increasing periodicity as the centre of the spiral is approached (Fig. 3.9). The power spectrum of the spiral centre shows defined periodicity (note large spike indicating average period and its accompanying harmonics) and lacks the major broadband spectra associated with chaotic systems. Fast Fourier transforms (FFT) and phase plane portrait analysis of 100 or more oscillator cycles show cells lying inside a circumference, equal to the wavelength of the spiral, to be periodic with limit cycle trajectories, while cells in the spiral arm describe two- or four-cycle periodic or "noisy" periodic trajectories (Fig. 3.8).

The distinguishing characteristic of spiral propagating waves is the presence in the SA of low amplitude cells circling the singularity with near limit cycle trajectories. Interior cells were isolated by thresholding the spiral so that only the core remained. The SA for the spiral in its entirety (Fig. 3.9) can be compared with the behaviour of the six cells nearest the pivot (Fig. 3.10).

The amplitude of those cells within the one-wavelength circumference varies from nearly the full amplitude of the uncoupled attractor down to the very low amplitude trajectories taken by those cells closest to the pivot. This appears to be accomplished by occupying a minimal energy state compared to the system

a

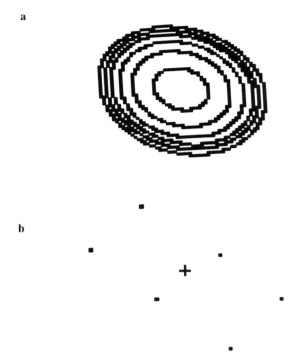

b

+

Fig. 3.10. Phase portraits of the spiral core. **a** Cumulative two-dimensional phase plane frequency histogram (x,y) of the six central cells generated as described in Fig. 3.8, but with the threshold set to exclude all cells whose x value ever exceeds 5.5 in the course of the simulation. Limit cycle behaviour is seen in these attractors over significant sampling intervals. **b** Strobed x,y phase position of the six central cells showing their relationship to the singularity, indicated by the plus sign.

as a whole, thus avoiding the high z return map. These interior cells can be examined more closely by reducing the threshold to $x = 5.5$ which eliminates all but six cells. By following the pivot cell in time it is possible to show that the spiral meanders, while the characteristic wavelength and form endure, so that phase information is passed coherently across the field. As one moves toward the pivot, and inside a region defined by a circumference equal to the wavelength of the spiral, the gradient of the phase space of any cell decreases, such that the difference between the gradients of neighbouring cells is minimized.

In the strict sense, the component oscillators in the spirals or bands may still be chaotic or they may be strange but non-chaotic, but at the level of a realistic biological system, the coupled field shows large-scale spatial and temporal periodicity. The cells that make up the centres of the spirals, meander very slowly through the field, but, on a shorter time scale they describe trajectories in phase space that are indistinguishable from limit cycles. In this discussion limit cycle is used to describe a three-space attractor in which the excitable component z has been severely damped in amplitude, with the understanding that this may not be a limit cycle in the strict sense.

Though spiral formation relies critically on initial conditions, once formed, spirals are resistant to destruction. Resetting a portion of the field variables to a single value either in the attractor surface or close to the steady state, and thus ablating the spiral core entirely, while leaving the early organizing source (EOS) region unaltered, allows regrowth of spirals. These newly formed spirals, not always stable, appeared in different locations with similar but non-identical morphology to the originals. Attempts to "transplant" the spiral cores by copying their variable and parametric values into an irregular field lacking the intrinsic capacity to form spirals or into a synchronous field, do not lead to formation of spirals. Wave propagation, though it appears to originate from the high amplitude EOS region of the field, is sustained by the elusive inter-action between the outward tug of the attractor surface and the amplitude-damping effect of coupling of neighbouring reaction domains. It is not clear whether there is a rotor driving the spirals.

Dependence on Initial Conditions

Dictyostelium discoideum is generally regarded as having a well-defined developmental programme with specific temporal phases and spatial patterns. However, in certain phases of this programme, most notably early aggregation, the selection among permitted patterns shows the strong dependence on initial conditions characteristic of processes controlled by strange or chaotic attractors. As Winfree (1973) has noted, specific local factors may be involved in main-taining the alternation of behaviours that leads to spiral formation, "but up to now their pertinence seems limited to determining such particulars as the numerical relationship between zone spacing and hyphal elongation (growth) velocity".

As mentioned above, in experiments with *Nectria*, Winfree (1973) found that the distribution of pattern types recurred similarly in spite of every attempt to alter the conditions of the experiment or to clone cells with a propensity to form one particular pattern.

In a series of simulations to test this property, seed values for initial con-ditions were chosen at random and a set of 20 to 30 simulations was done for all four boundary conditions, using identical parameter values. The occurrence of spirals, bands, etc. was then scored. In general the outcome in terms of frequency of single and multiple spirals and bands was similar to that seen in *Nectria*. For example, systems driven at a period of four times the fundamental $(4 \times 6\,UT) = 24\,UT$ (where UT is unit time), produced 55% circles, 32% single spirals, 10% double intercalated spirals, and 3% triple intercalated spirals. Though this must be numerology, the numbers are nearly identical with the percentages found in *Nectria* (Winfree 1973).

Period as a Function of Coupling Strength

Winfree (1973) has speculated that band interval must be a function of the fundamental period of the oscillator, while the handedness and the con-centricity, or lack thereof, is a property of the initial phase in a population of oscillators. In a chaotic array there appears to be a linear relationship between the value of the coupling constant and spatial wavelength of the resulting spiral (if one appears), and this wavelength is largely independent of the average oscillator period. A typical spiral at low coupling strength is capable of organizing for five or six turns, on average, thus higher couplings (which produce larger wavelengths) allow these spirals to cover more area and fewer turns.

Some Conclusions from Studies of Periodic Spiral Formation

Increased temporal precision arises from the collective behaviour of individually chaotic oscillators. An approach to the long-standing question of how the biological clock keeps precise time starting with sloppy components may have its realization in this model.

The number of outcomes is much smaller than the number of initial states, and perturbation of a significant fraction of the attractors does not alter the limiting pattern. For example, for a given set of boundary conditions and parameter values only spirals or bands are formed and these periodic solutions all have nearly identical wavelength.

The phase trajectory of the cells in a tightly coupled array is always greater than the fundamental period of the uncoupled attractor, except for those cells lying inside the core. Period doublings and period additions occur in the coupled array. In addition, wavelength in the field appears to change in a saltatory fashion with increasing coupling strength.

The spatial distribution of period increase in individual cells is interesting in that the core of the spiral is a limit cycle, the spiral arms close in are a two-cycle, and the spiral arms at their extremity, a four-cycle attractor.

Spiral waves might be an expected outcome from the coupling of limit cycle or multiply periodic attractors, and have been seen in cellular automata (Rössler 1975; Gerhardt et al. 1990) and simple reaction-diffusion models (Meinhardt 1986) but it does not seem that these results anticipated the finding of increased periodicity, period additions and period doubling bifurcations that emerge from the coupling of a system of continuous chaotic oscillators. More importantly, the mechanism of formation of spirals is more readily approached in a continuous surface with adequate spatial and temporal fine structure. In particular, the transitions from chaos to multiply periodic to limit cycle behaviour, as we move from non-spiral regions, to spiral arms and toward the pivot, and the observation that at its core the spiral is made up of cells showing low amplitude limit cycle behaviour would not seem to be possible in an automaton.

The emergent behaviour exhibited by this system presents an unavoidable metaphor for morphogenesis. Some spirals meander slowly through the field so that phase information is passed coherently from one group of cells to another.

Does the *per* Gene Product Tune a Strange Attractor?

Can a plausible transition in the application of this model be made from spatiotemporal organization of aggregates of unicellular eukaryotes and prokaryotes to the intact soma of a higher organism such as *Drosophila*? No detailed model of the *Drosophila* clock, built from the components, is yet possible. It is my intention, however, to suggest a few generic features that emerge from the two-dimensional tissue of cells described in the previous section.

In recent years a most productive approach to understanding biological clocks has taken the form of genetic and molecular genetic dissection of the circadian (Bargiello and Young 1984) and courtship song rhythms of *D. melanogaster* (Kyriacou and Hall 1980). Soon after their discovery in *Drosophila*, mutants affecting circadian rhythms were also found in *Neurospora* and *Chlamydomonas*. Very recently a mutant in the golden hamster (*Mesocricetus auratus*) has similarly been shown to have a period (20 h) significantly different from that of the wild-type. With regard to spectral analysis for precise characterization of period or periods, *Drosophila* is leagues ahead of the other systems (Dowse et al. 1987).

Bifurcation diagrams are a readily accessible graphical representation of attractor behaviour over a region state space. Multiple Poincaré sections are set side by side for incremental increases in the value of a parameter of the system, and the values of repeated crossings through the section, set say at $y = 0$, are shown for another variable, in this case x. By this means, regions of periodic and chaotic behaviour can be identified in relationship to the whole. The bifurcation diagram shown in Fig. 3.11 was constructed in this way by integration of the Rössler attractor for $a = b = 0.2$, while tuning the parameter C in steps of 0.01 ($1.9 < C < 15$). For small values of C the system shows limit cycle behaviour and passes repeatedly through the plane at only one point. For small increases in C in this region there is some expansion of the attractor that appears as a line sloping downward. At a value of C near 2.8, a bifurcation occurs such that a complete traverse of the Rössler cycle takes twice as long and the diagram now shows two sets of points diverging rapidly at first for increasing C. At C near 4.0, a second bifurcation occurs yielding a cycle that passes four times around the singularity before returning to the original point of crossing. The cycle is obviously four times as long as the fundamental period or limit cycle. These period doubling bifurcations continue with the onset of bifurcation occurring at smaller and smaller increments in C, approaching a ratio of values for succeeding bifurcations that is the Feigenbaum universality. For larger values of C the system never returns to the point of original crossing and hence has become a chaotic attractor. Note, however, that the passings through some point on the plane frequently occur at nearly equal time intervals and hence, for a sufficiently coarse criterion, it is periodic.

Further increases in C lead into regions embedded in chaos in which periodic behaviour of the attractor recurs, but with an interesting difference. At a $C = 5.2$, a periodic domain occurs in which the simplest cyclic behaviour is a three cycle. Again for increasing values of C, period doubling bifurcations occur in a sequence with periods equal to 3, 6, 12, etc. followed again by a chaotic domain.

Though the effect on period of coupling a large number of cells together should be considered, for simplicity the modelling starts here by considering a single Rössler element, without identifying its dynamics with any organizational level; it could be one fly, one organ or tissue, or one cell. Below, the effects of interactions within a population of attractors are examined and some specific requirements of the "real" system are suggested.

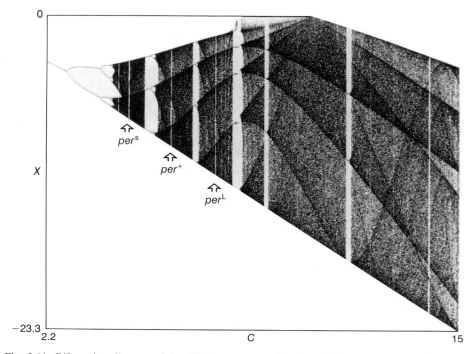

Fig. 3.11. Bifurcation diagram of the Rössler attractor with that of the *per* mutant series super-imposed. Bifurcation map is made as described in the text by continuously varying a parameter of the system, in this case C, and aligning the iterated values of x for fixed y. *per* mutants and wild-type are seen arrayed as a function of a change in the underlying attractor. The capacity to develop precise periodicity results from coupling among the cells of the pacemaker tissue. The value of the period is determined by the region of chaos in which the individual cells find themselves. So, for example, *per*$^+$, which is shown lying in a region of chaos in which four subbasins exist, would, when coupled to increased periodicity, show a circadian period that is four times the fundamental, here taken to be 6. *per*^{0} and *per*$-$, which lack this capacity, can have any of the *per* series backgrounds but remain chaotic due to the failure of coupling between cells.

The six unit times (UT; 100 time steps/UT), i.e. the time to complete a single loop of the attractor, is arbitrarily equated to hours. If it is supposed that the wild-type fly has a *per* product level such as to yield a C value near the embedded period 4, then the period of the circadian clock is 4×6 or 24 h (*per*$^+$). Increasing *per* slightly moves the system to the right into the period 5 domain, giving the fly a 30-h period (*per*^{L}), while lower levels move the system to the left into the period 3 or 18-h domain (*per*^{s}). For very low levels of *per* the system must be thought of as uncoupled and expressing the chaotic behaviour of the individual cell: 6 h or multiples of 6 h. Since the spectral analyses done thus far on the "arrhythmic" mutants give the same results as if the cell were a chaotic attractor, multiple 3-, 4- or 6-h nodes, these numbers should not be overspecified. Suffice to say that if the fundamental is 4 (and not 6) h then *per*^{s} is in the period 5 domain, *per*$^+$ in the period 6 and *per*^{L} in period 7 domain. The reason for first presenting this model, which

I might add is not correct to any detailed level, is to suggest by analogy how a bifurcation diagram of the coupled system might work. It happens that the bifurcation diagram of the Rössler model shows more clearly than the system bifurcation diagram the embedded multiply periodic nature of strange attractors. Since the detailed chemistry of the putative quantal clock is unknown, as is the exact nature of cell-to-cell communication, or the function of the *per* product, this was a necessary step. It is not possible to overlay the mutant series on to any region of the bifurcation diagram of the coupled system and show this simple progression of periods.

In the absence of specific knowledge of the transducing mechanism, we have set a threshold in x value in the bifurcation diagram for execution of events timed by the circadian clock, in a manner similar to that used in the cell cycle simulations shown in Fig. 3.6. In the above simulations, *per* has its primary effect directly on the dynamics of the ultradian oscillator, though it may also change coupling.

It should also be noted that lowering coupling strength below the value normally used to generate large periodic spirals results in spirals with maximum signal-to-noise ratios. That is to say, there is some optimal coupling that is associated with the shorter period (per^s). This suggests a simpler and more robust model in which the *per* gene product functions to increase the strength of diffusive coupling between cells.

It is possible to show that long and short circadian rhythms, or ultradian per^- rhythms arise from changes in coupling strength that lead to increases or decreases in period in a population of chaotic attractors whose fundamental rhythm is ultradian. A simple period doubling model alone, as suggested by Dowse and Ringo (1989) may be consistent with the predominance of ultradian components in per^0, but fails to explain what appears to be period addition (19.5, 24, 29-h τ) seen in other *period* mutants in *Drosophila*. Looking only at spiral wavelength, the simplest version of this model sets the *per* series from lowest to highest coupling as per^s, per^+, per^L! Since it is probably not correct to equate spiral wavelength with clock period we should not be too unsettled by this contrary outcome. Indeed, wavelength changes in the field do not necessarily result in period changes in the underlying system attractor. At high coupling strength, long wavelength spirals are not sustained, and the system collapses back to turbulent flow, while at very high coupling the system becomes synchronous and chaotic. Similarly, at very low coupling strengths, the system is chaotic. Thus at both extremes of coupling strength there is chaotic behaviour on the part of the tissue. How per^- and per^0 are viewed in this sequence depends on whether a chaotic attractor is regarded as aperiodic, which it is in the strict sense, or ultradian, as it would appear if one sampled some function of the putative oscillator over time. While this model has some attractive features it is not in accord with the observations of Coté and Brody (1986) that there is a logarithmic relationship between gene dosage and period, with increasing dosage giving shorter periods. Moreover, the measurements of Bargiello et al. (1987), which more directly assessed cell-to-cell communication, also would argue for highest coupling in the per^s flies. On the other hand, if a threshold for event execution is set in the surface of a cell or cluster of cells somewhere in the spiral, then the signal-to-noise ratio in *per* is higher and the period can be set appropriately by choice of the "correct" cell cluster. Deciding how to read the periodic output correctly will be a challenge.

Maximum signal-to-noise ratio appears to be a function of coupling strength, but it is also dependent upon which cell in the field is sampled. Moreover, whether an individual cell expresses 1-, 2- or higher cycles depends upon its location in the spiral field. When coupling is turned on, the appearance of the system attractor resembles a more periodic region in the bifurcation diagram of the single Rössler. One can envisage that period addition with the correct numerical outcome could be achieved in some region of parameter value, field element, thresholding and coupling constant. Detailed simulations have not been done. What is clear is that there can be an increase both in period and precision when an aggregate of cellularly based ultradian oscillations is allowed to share constituents.

A question naturally arises regarding the cell and tissue configuration of the "master oscillator". Only two periodic solutions are possible. In one, a tissue field of periodic attractors is coupled strongly enough to achieve synchronization. In the other, chaotic attractors are tuned to give optimal periodic output. This raises the question of the location within the tissue of the "master oscillator". In all the simulations described above except those with periodic individual attractors, the spiral is the only large stable periodic structure. In the spiral, periodicity is distributed throughout the tissue, and all phases are represented somewhere in the master oscillator/pacemaker. Overall, the system is almost perfectly asynchronous. Individual cells and cell clusters in the spiral arms are very periodic and show period addition or doubling phenomena. It seems then, a matter of transducing the periodic signal from various anatomical locations within the pacemaker to yield coordination of phases in various tissues. The time sense of the system will be very much a function of how the organism samples from its clock.

References

Allessie MA, Bonke FIM, Schopman FJG (1977) Circus movement in rabbit atrial muscle as a mechanism of tachycardia. Circul Res 41:9–18

Bargiello TA, Young MW (1984) Molecular genetics of a biological clock in *Drosophila*. Proc Natl Acad Sci USA 81:2142–2146

Bargiello TA, Saez L, Baylies MK, Gasic G, Young MW, Spray DC (1987) The *Drosophila* clock gene *per* affects intercellular junctional communication. Nature 328:686–691

Bourret JA, Lincoln RG, Carpenter BH (1969) Fungal endogenous rhythms expressed by spiral figures. Science 166:763–764

Brooks RF (1977) Continuous protein synthesis is required to maintain the probability of entry into S phase. Cell 12:311–317

Brooks RF, Bennett DC, Smith JA (1980) Mammalian cell cycles need two random transitions. Cell 19:493–504

Bruce VG (1972) Mutants of the biological clock in *Chlamydomonas reinhardii*. Genetics 70: 537–548

Burns FJ, Tannock IF (1970) On the existence of a GO-phase in the cell cycle. Cell Tissue Kinet 3:321–334

Chisholm SW, Costello JC (1980) Influence of environmental factors and population composition on the timing of cell division in *Thalassiosira fluviatilis* (Bacillariophyceae) grown on light/dark cycle. J Phycol 16:375–383

Collins JM (1978) Rates of DNA synthesis during the S-phase of HeLa cells. J Biol Chem 253:8570–8577

Coté GG, Brody S (1986) Circadian rhythms in *Drosophila melanogaster*: analysis of period as a function of gene dosage at the *per* (period) locus. J Theor Biol 121:487–503

Dawson KB, Madoc-Jones H, Field EO (1965) Variations in the generation times of a strain of rat sarcoma cells in culture. Exp Cell Res 38:75–84

Devreotes P (1982) Chemotaxis. In: Loomis WF (ed) The development of *Dictyostelium discoideum*. Academic Press, New York, pp 117–168

Dowse HB, Ringo JM (1987) Further evidence that the circadian clock in *Drosophila* is a population of coupled ultradian oscillators. J Biol Rhyth 2:65–67

Dowse H, Ringo JM (1989) Rearing *Drosophila* in constant darkness produces phenocopies of *period* circadian clock mutants. Physiol Zool 62:785–803

Dowse H, Hall J, Ringo J (1987) Circadian and ultradian rhythms in *period* mutants of *Drosophila melanogaster*. Behav Genet 17:19–35

Edmunds LN Jr (ed) (1984) Cell cycle clocks. Marcel Dekker, New York

Edmunds LN Jr (1988) Ultradian metabolic oscillators. In: Cellular and molecular bases of biological clocks. Springer-Verlag, New York, pp 305–321

Edwards SW, Lloyd D (1978) Oscillations of respiration and adenine nucleotides in synchronous cultures of *Acanthamoeba castellanii*: mitochondrial respiratory control in vivo. J Gen Microbiol 108:197–204

Fantes P, Nurse P (1977) Control of cell size at division in fission yeast by growth-modulated size control over nuclear division. Exp Cell Res 107:377–386

Feldman JF, Dunlap JC (1983) *Neurospora crassa*: a unique system for studying circadian rhythms. Photochem Photobiol Rev 7:319–368

Feldman JF, Hoyle MN (1976) Complementation analysis of linked circadian clock mutants of *Neurospora crassa*. Genetics 82:9–17

Feldman JF, Gardner G, Denison R (1979) Genetic analysis of the circadian clock of *Neurospora*. In: Suda M, Hayaishi O, Nakagwa H (eds) Biological rhythms and their central mechanism. Elsevier/North-Holland Biomedical Press, Amsterdam, pp 56–66

Forrest GL, Klevecz RR (1978) Tyrosyltubulin ligase and colchicine binding activity in synchronized chinese hamster cells. J Cell Biol 78:441–450

Gerhardt M, Schuster H, Tyson JJ (1990) A cellular automaton model of excitable media including curvature and dispersion. Science 247:1563–1566

Gerisch G (1971) Periodische Signale steuern die Musterbildung in Zeliverbanden. Naturwissenschaften 58:430–438

Goldbeter A, Martiel JL (1987) In: Degn H, Holden AV, Olsen LF (eds) Chaos in biological systems. Plenum Press, New York, pp 79–89

Goldbeter A, Segel LA (1977) Unified mechanism for relay and oscillation of cyclic AMP in *Dictyostelium discoideum*. Proc Natl Acad Sci USA 74:1543–1547

Goodwin BC (1963) Temporal organization in cells: a dynamic theory of cellular control processes. Academic Press, London

Higgins J (1967) The theory of oscillating reactions. Ind Eng Chem 59:18–62

Hoeniger JFM (1964) Cellular changes accompanying the swarming of *Proteus mirabilis*. Can J Microbiol 10:1–18

Holmquist GP, Gray M, Porter T, Jordan J (1982) Characterization of giemsa dark- and light-band DNA. Cell 31:121–129

Jauker F, Cleffmann G (1970) Oscillation of individual generation times in cell lines of *Tetrahymena pyriformis*. Exp Cell Res 62:477–480

Kapp LN, Millis AJT, Pious DA (1979) An altered S phase in synchronous Down's syndrome fibroblasts. In Vitro 15:669–672

Kauffman S, Wille JJ (1975) The mitotic oscillator in *Physarum polycephalum*. J Theor Biol 55:47–93

Keener JP (1986) A geometrical theory for spiral waves in excitable media. J Appl Math 46:1039–1056

Killander D, Zetterberg A (1965) Quantitative cytochemical studies on interphase growth. I. Determination of DNA, RNA and mass content of age determined mouse fibroblasts in vitro and of intercellular variation in generation time. Exp Cell Res 38:272–284

King GA, Archambeau JO, Klevecz RR (1980) Survival and phase response following ionizing radiation and hyperthermia in synchronous V79 and EMT6 cells. Radiat Res 84:290–300

Klevecz RR (1969a) Temporal order in mammalian cells. I. The periodic synthesis of lactate dehydrogenase in the cell cycle. J Cell Biol 43:207–219

Klevecz RR (1969b) Temporal coordination of DNA replication with enzyme synthesis in diploid and heteroploid cells. Science 166:1536–1538

Klevecz RR (1975) Molecular manifestations of the cellular clock. In: The cell cycle in malignancy and immunity. 13th Annual Hanford Biology Symposium, pp 1–19

Klevecz RR (1976) Quantized generation time in mammalian cells as an expression of the cellular clock. Proc Natl Acad Sci USA 73:4012–4016

Klevecz RR, Braly PS (1987) Circadian and ultradian rhythms of proliferation in human ovarian cancer. Chronobiol Int 4:513–523

Klevecz RR, Braly PS (1991) Circadian and ultradian cytokinetic rhythms in human cancer in situ. Ann NY Acad Sci 618:257–276

Klevecz RR, Kapp LN (1973) Intermittent DNA synthesis and periodic expression of enzyme activity in the cell cycle of WI-38. J Cell Biol 58:564–573

Klevecz RR, Ruddle FH (1968) Cyclic changes in synchronized mammalian cell cultures. Science 159:634–636

Klevecz RR, Shymko RM (1985) Quasi-exponential generation time distributions from a limit cycle oscillator. Cell Tissue Kinet 18:263–271

Klevecz RR, King GA, Shymko RM (1980a) Mapping the mitotic clock by phase perturbation. J Supramol Struct 14:329–342

Klevecz RR, Kros J, King GA (1980b) Phase response to heat shock as evidence for a timekeeping oscillator in synchronous animal cells. Cytogenet Cell Genet 26:236–243

Klevecz RR, Kauffman SA, Shymko RM (1984) Cellular clocks and oscillators. Int Rev Cytol 86:97–128

Klevecz RR, Pilliod J, Bolen J (1991) Autogenous formation of spiral waves by coupled chaotic attractors. Chronobiol Int 8:6–13

Konopka RS, Benzer S (1971) Clock mutants of *Drosophila melanogaster*. Proc Natl Acad Sci USA 68:2112–2116

Kubitschek HE (1962) Normal distributions of cell generation rate. Exp Cell Res 26:439–450

Kyriacou CP, Hall JC (1980) Circadian rhythm mutations in *Drosophila melanogaster* affect a short-term fluctuation in the male's courtship song. Proc Natl Acad Sci USA 77:6729–6733

Lloyd D, Edwards SW (1984) Epigenetic oscillations during the cell cycles of lower eukaryotes are coupled to a clock. In: Edmunds LN Jr (ed) Cell cycle clocks. Marcel Dekker, New York, pp 27–46

Martins-Ferreira H, DeOliveira-Castro G, Struchiner CJ, Rodrigues PS (1974) Circling spreading depression in isolated chick retina. J Neurophys 37:773–784

McQuilkin WT, Earle WR (1962) Cinemicrographic analysis of cell populations in vitro. J Natl Cancer Inst 28:763–799

Meinhardt H (1986) Hierarchical inductions of cell states: a model for segmentation in *Drosophila*. J Cell Sci 4 [Suppl]:357–381

Minor PD, Smith JA (1974) Explanation of degree of correlation of sibling generation times in animal cells. Nature 248:241–243

Muller SC, Plesser T, Hess B (1987) Two-dimensional spectrophotometry of spiral wave propagation in the Belousov–Zhabotinskii reaction. Physica 24D:71–86

Pavlidis T (1973) Biological oscillators: their mathematical analysis. Academic Press, New York

Pecora LM, Carroll TL (1990) Synchronization in chaotic systems. Phys Rev Lett 64:821–824

Petrovic AG, Oudet CL, Stutzmann JJ (1984) Temporal organization of rat and human skeletal cells. Circadian frequency and quantization of cell generation times. In: Edmunds LN Jr (ed) Cell cycle clocks. Marcel Dekker, New York, pp 325–349

Pittendrigh CS (1960) Circadian rhythms and the circadian organization of living systems. Cold Spring Harbor Symp Quan Biol 25:159–182

Pittendrigh CS (1965) On the mechanism of the entrainment of a circadian clock by light cycles. In: Aschoff J (ed) Circadian clocks. North-Holland Publ Co., Amsterdam, pp 277–297

Pittendrigh CS, Daan S (1976) A functional analysis of circadian pacemakers in nocturnal rodents. J Comp Physiol 106:223–252

Prigogine I, Lefevre R (1968) Symmetry breaking instabilities in dissipative systems. J Chem Phys 48:1695–1700

Ralph MR, Menaker M (1988) A mutation of the circadian system in golden hamsters. Science 241:1225–1227

Readey MA (1987) Ultradian photosynchronization in *Tetrahymena pyriformis* GLC is related to modal cell generation time: further evidence for a common timer model. Chronobiol Int 4: 195–208

Rössler OE (1975) Steps toward a temperature-compensated homogeneous chemical clock. In: Martin I (ed) Biomedical symposium. San Diego Biomedical Symposium Press, San Diego, pp 99–104

Shapiro JA (1988) Bacteria as multicellular organisms. Sci Am 258(6):82–89

Shields R (1978) Further evidence for a random transition in the cell cycle. Nature 273:755–758

Shymko RM, Klevecz RR (1981) Cell division gated by oscillatory timekeeping and critical size. In: Rotenberg M (ed) Biomathematics and cell kinetics. Elsevier/North Holland Biomedical Press, Amsterdam, pp 329–348

Smith JA, Martin L (1973) Do cells cycle? Proc Natl Acad Sci USA 70:1263–1267

Tyson JJ, Keener JP (1988) Singular perturbation theory of traveling waves in excitable media (A review). Physica D 32:327–361

Winfree AT (1967) Biological rhythms and the behavior of populations of coupled oscillators. J Theor Biol 16:15–42

Winfree AT (1970) Integrated view of resetting a circadian clock. J Theor Biol 28:327–374

Winfree AT (1973) Polymorphic pattern formation in the fungus Nectria. J Theor Biol 38:363–382

Winfree AT (1980) The geometry of biological time. Springer-Verlag, Berlin Heidelberg New York

Winfree AT (1987) When time breaks down. Princeton University Press, Princeton NJ

Chapter 4

Oscillations and Cancer

D. A. Gilbert and Heather MacKinnon

Science Fiction

An animal that never stops growing sounds like a creature out of science fiction. Yet the horror of unlimited cell growth is all too real for those suffering from cancer, particularly when the cells concerned invade local normal tissues and also spread to distant sites (metastasis) where they continue to create structural and functional havoc. Efforts to treat the disease, on the basis of the altered proliferative characteristics, have been developed largely by trial and error as the result of an inadequate understanding of both the transformation process and the nature of its effects on cell proliferation. The problem has been compounded by the adoption of invalid concepts of the processes underlying cell replication and even arguments which fall more within the realm of fiction than of science (see Gilbert 1981; MacKinnon and Gilbert 1992).

Proliferation is a most fundamental feature of cell biology. In an actively proliferating cell the reactions taking place in the interval between one division and the next is referred to as the cell cycle (see e.g. Lloyd et al. 1982; Edmunds 1988). The overt events are: the S phase, in which DNA synthesis (gene duplication) takes place; mitosis or M phase, wherein separation of the duplicated chromosomes can be seen; and the actual division process of cytokinesis. Generally, these all occur and do so in that order: it is therefore widely accepted that they form a sequence of reactions wherein each event depends on the completion of the previous reaction before it too can occur. Separating these overt events are the so-called G_1 and G_2 "gap" periods when nothing obvious takes place (though something must do so in order for the overt events to start). While rarely discussed explicitly, the impression is usually given that these are believed to consist of a sequence of reactions so that the whole interdivision period is considered to form one continuous series of dependent reactions and it appears to be generally assumed that these are genetic switch reactions. However, although there is suggestive evidence that the individual events do involve a succession of reactions, as indicated elsewhere (Gilbert 1981, 1988: MacKinnon and Gilbert 1992) there is no evidence that the overall cell cycle consists of such a sequence and, indeed, several theoretical arguments and also experimental observations are quite contrary to such a view, as we shall see.

If a sequence exists then it must be cyclic if the events are to recur, but to our knowledge no one has explained how such a series of reactions can be closed or how the progress through the sequence can be measured nor yet what constitutes the rate of movement through a series of dependent reactions. Moreover, neither has anyone accounted for all the observations referred to here. Arguments for a sequence are based largely on the different lengths of time delays following release from inhibition of proliferation by distinct reagents, but these data are totally invalid when not accompanied by a knowledge of the rates of traverse through the cycle, and perhaps not even then (Gilbert 1981; MacKinnon and Gilbert 1992). One can ask: (a) what stops a cell from existing at two stages of the cycle at the same time, say with two switches on simultaneously; and (b) whether such a concept implies that the reactions concerned are specifically associated with proliferation and, if so, how? If a particular reaction is totally blocked in a quiescent cell, can any of the other reactions in the proposed sequence still be operative? On the other hand, is it possible for them to be involved in cell function if they are not operative? Can they be exclusive to the cell cycle in such a highly integrated system? The view provides more questions than it answers and we now know that mitosis can be induced in intact cells before S phase has been completed. Moreover, it can occur repeatedly without division in a cell-free system (see e.g. Lhoka and Maller 1985; Hutchinson et al. 1987; Lewin 1990). These recent observations only confirm long-standing evidence that the events of the cycle can be dissociated. If we must reject the sequence view, can we continue to accept the notion that inhibition of replication results from the blockage (restriction) of such a sequence (Gilbert 1981)?

In short: cell proliferation is a dynamic process and to be completely understood, it requires an explanation in terms of reaction kinetics. There is, to our knowledge, no evidence to show that the dynamics can be adequately explained in terms of a sequence of dependent reactions (discontinuous or otherwise). On the other hand, as discussed in detail in the various articles to which reference is made in this text, the concept to be outlined here is consistent with, and provides an account of, all the major facets of replication.

Science Fundamentals

Before any problem can be resolved it is necessary to identify the issues and in this section we consider those presented by cancer. It is not sufficient to invoke a genetic (or any other) lesion for the disease without explaining how the defect is expressed as malignancy. Early views that cancer cells proliferate exceptionally fast is patently wrong when one compares them with host embryonic cells (or even regenerating liver). No evidence exists from culture studies to show that the processes determining proliferation have been lost or are replaced by a distinct mechanism, as is generally believed to be the case. Culture studies have failed to reveal any universal qualitative or simple quantitative differences between the proliferative reactions of normal and transformed cells.

Malignancy can be defined only with reference to the host: when isolated from tissues the cancer cells are liable to exhibit altered characteristics. However, it is unlikely that a common qualitative difference in behaviour could have been always obscured or missed. Nevertheless, we conclude that the same basic mechanism is operative in all mammalian cells and, as we describe below, it is one that can be produced in a number of ways.

Initially, it is important to note that there are many forms of the disease (in a very few of which the cells actually do stop growing) and there is marked variability among cells within most if not all tumours. The only factor common to the vast majority, and the prime malignant characteristic, is the apparent ability of the aberrant cells totally to ignore the normal regulatory growth restraints. As indicated, the cells do not replicate exceptionally fast but they do continue to proliferate when they should not do so. Since hyperplasia would be the only outcome if this were the sole characteristic we must invoke lesions which clearly also affect cell properties and function (Gilbert 1984b).

The above issue implies that reactions associated with proliferation are also involved in other aspects of cell metabolism. Since true replication (where all components in the cell are duplicated) apparently occurs, it would seem that the process is then essentially autocatalytic, again suggesting that each constituent is somehow able to influence, but not determine, proliferation.

There are also many diverse causes of the disease and we need to know how they are all able to produce the same kind of effect. (Can this be the case with the sequence model?) A rational approach to therapy through growth control thus requires an understanding of both the deviant proliferative behaviour and of the normal processes from which it deviates. However significant a genetic defect may be in a particular situation, it is essential that we correctly interpret its action at the metabolic level where the effects are expressed.

Clearly, these and other issues indicate the need to understand how proliferation is regulated but in order for us to achieve this aim we must look at the phenomenon in the proper way (Gilbert 1988; MacKinnon and Gilbert 1992) as its nature is not self-evident. In doing so we need to encompass relevant experimental observations on the multiplicative behaviour of cells in culture.

Science Facts

The ultimate question is concerned not with the difference between normal and malignant cells but with the distinction between cells which are actively proliferating and those which are quiescent. Cell proliferation is a highly dynamic process involving changes in the levels of components with time. Therefore, the phenomenon must be interpreted in terms of reaction kinetics, since these determine quantitative and temporal aspects of any problem. In what kinetic way are replicating cells dynamically distinct from their dormant counterparts?

Related questions are: what is the mechanism whereby a quiescent cell is induced to replicate (or, simply, to divide) and what governs when it does so? Of particular importance when considering all the issues is the response when cells are disturbed away from their steady-state behaviour, since this can enable us to distinguish between distinct mechanisms.

We can extend these thoughts and ask why it is that some cells are more easily stimulated than others. This last point raises a more general issue, namely that we are required to account not only for the distinct qualitative features of proliferation but also different quantitative aspects. Inter alia, these include: differences in behaviour between one cell species and the next, or, between one set of conditions and another; the dependence of replication rate on conditions; and the marked heterogeneity of proliferative properties within populations. These considerations apply irrespective of whether the cells concerned are normal or malignant.

Apart from the features to which reference has already been made, experimental studies of cell replication have yielded a wealth of information. As conflicting and puzzling as the facts are, no concept can be considered valid unless it accounts adequately for all such observations and shows how they are interrelated. No attempts have been made to explain such facets in terms of the sequence view, yet many workers still cling to that classical idea.

It is necessary to account not only for the ability of a wide range of agents to stimulate (and inhibit) replication but also how they interact. In some instances an agent can inhibit under one set of circumstances and stimulate under others. Then again, the effect of an agent can depend on the timing of its action (Gilbert 1974b, 1980).

All cells in culture cease to proliferate at some point, and this does not seem to be explicable in terms of complete exhaustion of any nutrient or growth factor (Gilbert 1977). What causes cells to stop dividing and what governs the number of cells at such times? There is evidence that transformation frequently results in the cells reaching higher densities in culture than their untransformed counterparts (cf. Gilbert 1977), a characteristic which, in at least one instance, has been associated with the rate of development of the corresponding tumour in the host (Aaronson and Todaro 1974).

Since there was no reason a priori to suppose that the underlying principles would be revealed through experimentation alone, several theoretical mechanisms have been proposed for the processes of proliferation, but most of these have been based on single features of replication whereas, as we have remarked, the phenomenon is multi-faceted. Several of the theoretical concepts are related to that to be discussed here, but their validity, already in doubt (Gilbert 1981), has been further questioned in view of recent observations on mitosis in cell-free systems (Murray and Kirschner 1989); these are concepts based on cell size, mass and membrane reactions for the origin of the cycle (cf. MacKinnon and Gilbert 1992).

Science Fallacy?

At this point we stress the fact that cell proliferation is a very dynamic process. It cannot therefore be explained solely in terms of the identity of components or merely by static considerations (Gilbert 1981, 1988; MacKinnon and Gilbert 1992). Can it be fallacious to believe that the way and, therefore, the means by which components vary with time is important in determining the features of proliferation? Surely, any acceptable concept must have a reaction kinetic basis (Gilbert 1988).

As the first step in this line of reasoning, the very old concept that the cell cycle has a rhythmic basis was revived (Gilbert 1968), and developed by Sel'kov (1970), who attempted to analyse in mathematical terms, the thiol/disulphide system of a cell and showed that the levels of the components could oscillate. Since then, one of us (D. A. G.) has shown that all major qualitative features of replication and several quantitative aspects can be understood in terms of the Sel'kov (1970) model (see Gilbert 1974b, 1977, 1978a,b, 1981, 1982a,b, 1988). As the model's behaviour is that expected for oscillating systems in general, we have used it to illustrate principles and arguments, irrespective of the validity of Sel'kov's analysis. (For consideration of other, related models, see Lloyd et al. 1982; Edmunds 1988.)

In outline, the control system concerned is believed to be one determining the formation, degradation and interconversion of coenzyme cellular components, which, by their nature, can influence their own levels by feedback regulation, thereby providing the necessary control action which enables the system to be autodynamic in the levels of those components. Then again, a metabolic reaction resulting in the conversion of the coenzyme components can affect both cell function and replication but, as we shall explain, without necessarily modifying proliferation characteristics. Such common processes thus allow proliferation to be linked to, for example, differentiation, contrary to what might be expected from any model based on cell cycle specific reactions.

The quiescent cell is thus considered to be one in which this control system is static (that is wherein the levels of the two forms of the coenzyme are effectively constant), and the proliferating cell is one in which the levels are oscillating, either continuously or as single cycles, as we shall soon explain. The basic argument (Sel'kov 1970) is that, as the levels of these coenzymes vary, as discussed below, they activate and then deactivate (and hence coordinate) the cell cycle events (see Fig. 4.1).

Not unreasonably, we believe, it is postulated that different cells with different patterns of metabolism have distinct sets of rates and kinetics for the control reactions and also that changes in conditions modulate these values (Gilbert 1978a). On this basis it is possible to consider the behaviour of different cells, or of cells existing in distinct conditions, by studying the behaviour of the system using different sets of rates and kinetics (parameter values; Gilbert 1980).

Accordingly we stress that the proliferative characteristics depend on the set of parameter values and not on any one alone (Gilbert 1978a). There are essentially five rates involved, two of which (the interconversion processes) being determined by numerous cellular reactions in which the coenzyme is concerned. There are thus many ways in which the behaviour can be modified and as many agents (including carcinogenic ones) which can modify the system kinetics. These reactions (in particular) also provide many ways in which proliferation can interact with cell function as indicated above. It is even possible, if not probable, for the cell cycle control system to become self-modulated through the feedback of "signals" resulting from the rhythmic changes in the levels of the coenzymes. It has been suggested (Gilbert 1980) that this process accounts for cellular (and possibly physiological) ageing (see e.g. Muella et al. 1979): Muella et al. also reported that cellular ageing was accompanied by a decreasing cell saturation density, as can be expected from

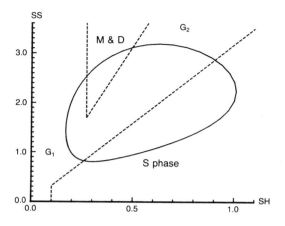

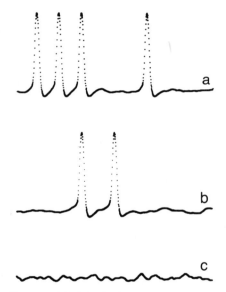

Fig. 4.1. The limit cycle oscillator model. *Top.* A phase plane plot of the limit cycle in the levels of the thiol (SH) and disulphide (SS) groups obtained from Sel'kov's (1970) equations for one set of parameter values. Also shown are hypothetical boundaries to indicate conditions where the cell cycle phases are permitted; in this diagram they are set by particular thiol values and different redox states (lines of positive slope corresponding to fixed ratios of the thiol and disulphide groups). Other components with coenzyme-like characteristics could be involved instead, and the thresholds for the cell cycle events could be different and variable in values though, from experimental data, Sel'kov (1970) has associated the various regions with the phases shown (see also Gilbert 1982a).

Bottom. Irregular triggering of individual cycles of the oscillation due to fluctuation in a particular set of parameter values (rates and kinetics) for the control system. The three plots show the decreasing frequency of triggering as the set of parameter values move away from the boundary separating static and autodynamic (sustained) oscillatory behaviour (**a** → **c**). As indicated in the text, it is believed that cancer is due primarily to an effective increase in the sensitivity of the cell cycle switch.

the present concept. The same mechanism must surely also contribute toward the interdependence of proliferation, differentiation, development and tumour progression. No such arguments are evident in respect of other concepts. It is not implied that all permanent disturbances in cell metabolism will alter proliferative characteristics in the way to be described and therefore cause cancer. Oncogene products are considered to be factors which are capable of modifying the kinetics of the control system directly so as to produce the altered properties but, again, it is not suggested that their expression will necessarily cause malignancy: as indicated, the behaviour depends on the whole set of parameter values and the effect of altered oncogene expression can be offset by other control reactions (Gilbert 1978a).

There are three main aspects to the regulation of proliferation: (a) the process of initiation; (b) the determination of the interdivision time for individual cells; (c) the determination of the population growth rate. Only the latter two facets have received significant attention. As indicated above, on the basis that the regulation of proliferation is achieved through discrete or dynamic changes in the rates and kinetics of the control system (its parameter values), it has been possible to provide detailed accounts of the various major features of proliferation. These include the mechanism of action and inter-action of regulators, i.e. the ability of such agents to have dual effects on the behaviour of the cell according to their concentration or time of action. As the concept requires these agents to exceed boundary (bifurcation) levels in order to achieve their effects, it has also been possible to explain their switch-like effect on replication and how its sensitivity is determined and thus how it can differ between cell species or according to the conditions, including the presence or absence of other mitogenic agents (Gilbert 1977, 1978a). It also explains the different facets of the density-dependent limitation of cell growth in culture, including the effects on the saturation level of the malignant trans-formation and, hence, tumour growth in vivo (see Gilbert 1984b and comments above). Evidence suggesting the existence of threshold conditions for mitogens was presented by Eagle et al. in 1961, and there are also results which indicate that it can be altered by changes in conditions (Basilico et al. 1974) as required by the theory, though little attention has been paid to the possible significance of the effect. Most studies are too crude to detect a threshold, and much needs to be done in this area.

The limit cycle oscillator concept can also explain, again in some detail, the different ways of achieving (partial) synchronization of a population of cells and why the degree of synchronization depends, inter alia, on the cell species (Gilbert 1980). Furthermore, it provides an explanation for phase response observations relating the moment of onset of the next division to the time of action of an agent during the cell cycle: this is usually a delay, up to a critical point, but beyond that the delaying effect is suddenly lost and may even change to an acceleration (Gilbert 1974b, 1980; see also Lloyd et al. 1982; Edmunds, 1988). Moreover, the concept reveals the interrelationship between these distinct aspects of proliferative behaviour.

Of particular concern here is the commitment phenomenon (Gilbert 1978b, 1982a), in which cells exhibit their potential ability to continue to pass through the cycle after a stimulus is removed (again, providing a critical time has been exceeded). It has been shown that this merely reflects the dynamics exhibited by such a control system for particular sets of parameter values. This phenomenon is related to the ability of the system to exhibit excitability, behaviour directly analogous to that shown by nerve cells, though involving a markedly different time scale. Excitability is evident when the state of the control system lies just outside the threshold set of conditions permitting continuous (self-generated) oscillations in the levels of the coenzyme forms. Under such circumstances, random fluctuations in the control system rates (as might be expected to arise primarily through interacting thermal gradients in the cell due to exothermic and endothermic reactions) can cause irregular triggering of single cycles of the oscillation (Fig. 4.1). When the cells are oscillating continuously the generation time is largely determined by the period of the rhythm, though there is some variation generated through modulation by random processes. When the set of parameter values causes the mean state of the system to lie outside the threshold conditions for continuous oscillation, triggering is entirely random and the interdivision time is totally irregular. Hence, in the latter situations, there will be a much greater variability in the apparent cell cycle times. The concept (Gilbert 1978b, 1982a,b) thus provides a detailed account for the distributions of generation (cell cycle) times within a population and how it varies with the mitogen concentration (predicting a subsequently reported deviation in behaviour at low stimulus levels; Brooks and Riddle 1988). The model also accounts for the fact that the minimum interdivision interval is nearly independent of the conditions (Gilbert 1982b). More important, instead of the switch being turned off suddenly as the parameter values alter, the effect of the random fluctuations is to cause a gradual change in triggering frequency.

These various ideas led to the suggestion (Gilbert 1974, 1977, 1978c) that the major cause of cancer is any change in metabolism which effectively increases the sensitivity of the cell cycle switch; that is, one which increases the frequency of random triggering or which, in the extreme, enables continuous oscillatory behaviour in the system. This has the effect of allowing the modified cells to continue to proliferate when they should not do so under the prevailing conditions; in essence, the switch is not turned off completely or remains on (a very malignant tumour in terms of growth rate; Gilbert 1978c). The concept thus invokes a quantitative defect rather than a qualitative one. Although the situation is a little more complex (Gilbert 1978b) a change in the pattern of temporal organization of parameter-modulating rhythms can be envisaged as causing a dynamic shift of the threshold.

We believe that this oscillator model accounts adequately for all the major proliferative facets of the disease while still allowing the cells to behave in qualitatively the same way in culture and without being inconsistent with other mechanisms, as in the autocrine view o˜ Sporn and Roberts (1985). However, present considerations (Gilbert 1977) suggest that an increased switch sensitivity may well be far more important with regard to cancer than an increased level of growth factor.

Despite being founded on observations, the limit cycle concept of cell proliferation has so far failed to gain widespread acceptance, but it is pertinent to point out that much attention has recently been concentrated on rhythmic changes in the levels of cyclins in relation to periodic mitoses in cell-free extracts (see e.g. Lewin 1990). These components undergo phosphorylation and dephosphorylation in relation to their function. Although the possible involvement of such reactions was considered much earlier (Gilbert 1974b),

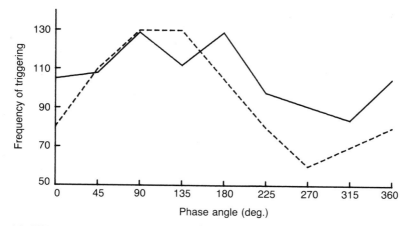

Fig. 4.2. Effect of oscillator phasing on random triggering. The coenzyme control system (see the text and Fig. 4.1) can exhibit excitability when the set of reaction rates (parameter values) cause the state of the cell to lie just outside the boundary conditions permitting sustained oscillatory behaviour. The excitability can give rise to irregular triggering of individual cycles of oscillation (cycles of cell division, cf. Fig. 4.1) if the rates fluctuate due to random disturbances (Gilbert 1978b, 1982a,b). If the rates are also modulated by short-period rhythmic processes, the frequency of random triggering can be dependent on the phasing between such oscillations as depicted here for two distinct sets of parameter values as indicated by broken and continuous lines (see also Gilbert 1980). Such data lend support for the concept that differentiation and cancer can result from changes in the pattern of temporal organization of cellular rhythms (Gilbert 1968).

the experimentalists have ignored theoretical arguments. However, it has yet to be shown that the system of cyclins forms the primary oscillator in intact cells and that the cyclins have properties which enable them to coordinate all cell cycle processes and to provide the necessary energy, as can the coenzymes (Sel'kov 1970; Gilbert 1988). Nevertheless, the observations should provide credence for the limit cycle model even if their ability to account for so many facets does not.

Science Focus

Having expressed our belief in the ultimate importance of what is generally a long period rhythm, we finally converge on the primary subject matter of this volume, namely the roles of high frequency oscillations. To begin, we note that Lloyd and Volkov (1990) have suggested that the periodic modulation of a continuous (relaxation) oscillator can account for the fact that (mean) cell cycle times appear to be multiples of some base period. By making the same assumptions, it was shown earlier (Gilbert 1980) that discrete changes in the phasing of rhythms modulating the parameter values of the model can alter the frequency of random triggering of the cell cycle oscillator (see Fig. 4.2).

These studies on the effect of high frequency rhythms on replication were undertaken in an attempt to provide support for an even earlier suggestion (Gilbert 1968, 1984a) that cells differentiate or become malignant as a result of changes in the pattern of temporal organization of intracellular rhythms and, in particular, alterations in their amplitudes, frequencies and phasings. It was emphasized (Gilbert 1973a,b, 1974b) that biochemical switch systems exhibit amplitude, frequency and phase discriminatory properties and that triggering of such processes, including the cell cycle, must be sensitive to changes in the pattern of temporal organization (see also Gilbert 1980). Cellular periodicities have been reported to occur in a wide range of cells (for a review see Edmunds 1988); we have described the existence of a number of oscillations in several mammalian cells, in particular, rhythmic variations in the activities and effective isozyme patterns of glycolytic enzymes (see e.g. Gilbert 1968, 1984a; Gilbert and Tsilimigras 1981; Tsilimigras 1982), in the glucose-6-phosphate dehydrogenase activity and in the rate of incorporation of amino acids into proteins (Tsilimigras 1982), in the amount of extractable protein (Tsilimigras

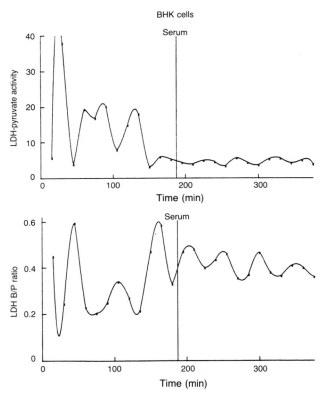

Fig. 4.3. Effect of serum on temporal organization in cells: I. Variations in the activity of lactate dehydrogenase (LDH) toward pyruvate (top) and in the effective LDH isozyme pattern (bottom) in BHK-C13 (baby hamster kidney) cells, before and after the addition of foetal calf serum (data from the same experiment). As in all the enzyme studies reported here, the cells were maintained overnight at low serum concentration before observations were started (for details of the methods see Gilbert 1984a). Effective isozyme patterns are determined from the ratios in the rates using either two different substrates or two concentrations of one substrate (see Gilbert 1968, 1969, 1984a). For LDH isozyme patterns the two substrates used are pyruvate (P) and α-ketobutyrate (B).

and Gilbert 1977; see also Brodsky, Chap. 2). Since then we have also observed oscillations in the activities of phospho amino acid phosphatases (Hammond et al. 1989a,b; H. Janura and D. A. Gilbert, unpublished results). It may be mentioned that the enzyme oscillations are not (entirely) due to the variations in the amount of extractable protein (Gilbert and Tsilimigras 1981).

More recently we have also presented new evidence for the occurrence of oscillations in cell morphology and redox state (Visser et al. 1990). It has been shown that different cells (including untransformed and polyoma virus-transformed BHK cells) can exhibit distinct patterns of temporal organization, in agreement with the concept just mentioned (Gilbert 1968, 1969, 1984a).

Extension of these views and observations gave birth to the belief that hormones and growth factors (including autocrines) may act by causing temporary alterations in the set of rhythms in a cell (Gilbert 1968, 1974a).

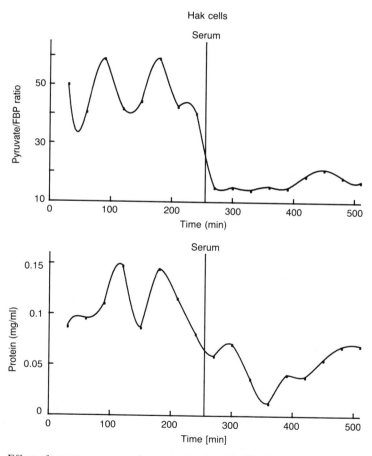

Fig. 4.4. Effect of serum on temporal organization in cells: II. The effect of serum on oscillations in the ratio of LDH-pyruvate activity to the aldolase-fructose bisphosphate (FBP) activity (top) and in the amount of extractable protein (bottom) using the hamster kidney (HaK) cell line. The two sets of data are again from the same experiment.

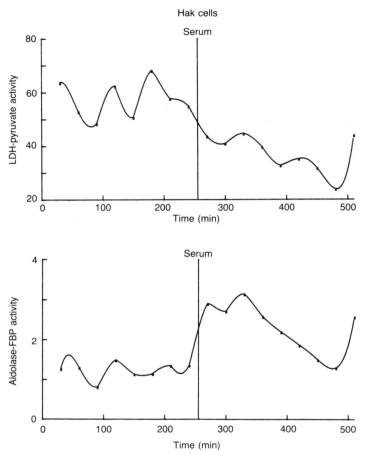

Fig. 4.5. Effect of serum on temporal organization in cells: III. In this diagram can be seen oscillations in the LDH-pyruvate activity (top) and in the aldolase-fructose bisphosphate (FBP) activity (bottom) and the effect thereon of serum (HaK cells, data from the same experiment). Apart from showing that this mitogenic stimulus disturbs the pattern of temporal organization, these diagrams confirm that the enzyme oscillations are not simply due to variations in the efficiency of extraction (see Gilbert and Tsilimigras 1981).

Some supporting evidence has already been presented (Gilbert 1984a). In Figs. 4.3–4.6, we give additional results indicating that the normal mitogenic stimulus (foetal calf serum) modifies enzyme oscillations in two cell lines. Elsewhere (Visser et al. 1990), we reported that insulin, another mitogenic agent, can disturb oscillations attributed to cell surface movements (ruffling) in a frequency-dependent manner. In the latter article it was shown that the hormone decreased the amplitudes of morphological oscillations of certain frequencies but it was stated that rhythms having other periods are affected differently. Fig. 4.7 confirms this point because certain oscillations in the data are stimulated while another is apparently unaffected by the hormone.

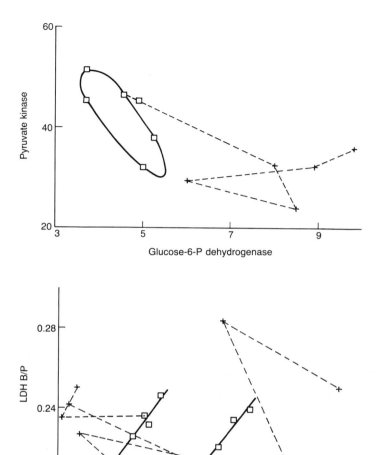

Fig. 4.6. The effect of serum on temporal organization: IV. Here we use the phase plane method of examining oscillatory behaviour (see Gilbert 1974a, 1984a) before (□) and after (+) the addition of serum. In the top diagram it can be seen from the initial loop structure that the oscillations in the LDH and glucose-6-phosphate dehydrogenase activities were nearly in antiphase (see Gilbert, 1974a, 1984a) but this temporal pattern was rapidly affected by the serum. In the lower diagram it is evident that the two isoenzyme pattern rhythms were totally in phase with each other at the start of the observations, but that switching (see Gilbert 1984a) was apparently occurring in the hexokinase system. Immediately after the addition of serum there is a shift in the mean level of the hexokinase pattern but no change in the phase relationship. Eventually, however, the behaviour pattern changed rapidly. The sampling interval was 5 min in the top figure. For the other experiment, a "smoothing" method was used: in this case, each set of three consecutive 5-min samples were combined to give an effective sampling time of 15 min. This removes higher frequency components in the data. B, α-ketobutyrate; P, pyruvate; LoG, low glucose conc.; HiG, high glucose conc.

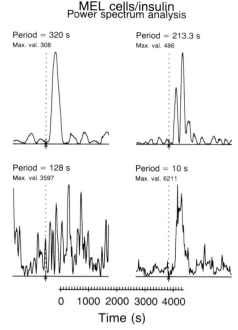

Fig. 4.7. The effect of insulin on morphological dynamics. Shown here are the temporal variations in the power values for four oscillations (of the given periods) obtained by power spectrum analysis of fluctuations in the intensity of light scattered by murine erythroleukaemic cells in suspension (see Visser et al. 1990). Insulin was added at the point shown by the dotted lines. As can be seen, there is a frequency-dependent effect; three of the rhythm patterns are transiently increased in amplitude while the other appears to be essentially unaffected. As reported by Visser et al. (1990), the predominant effect is a stimulation of the rhythms but some are damped (or, perhaps, changed in frequency). Also obvious are the periodic modulations of the power values discussed previously (Visser et al. 1990), showing that the primary oscillations tend to occur in bursts. The lower frequency scatter changes are believed to reflect rhythmic changes in gross morphology and the high frequency components reflect rapid surface movements. In each diagram the values plotted are given as a ratio to the maximum value observed for the given frequency: the individual maximum power values are as indicated.

Here, we present our evidence showing that the predominant effect is one of stimulation of the surface fluctuations. Also evident in Fig. 4.7 is the periodic modulation of the oscillations reported previously (Visser et al. 1990) which gives rise to burst-like characteristics.

We therefore confirm our belief that modification of oscillatory cellular dynamics plays at least a part in the proliferative response to some, if not all, stimuli, that is in the triggering of the cell cycle switch. The morphological fluctuations (cf. Satoh et al. 1985) are also considered to be important for cell locomotion (and hence tumour metastasis) (cf. Visser et al. 1990). They may also have a more general effect on metabolism (D. A. Gilbert and G. Visser, unpublished results).

Under normal culture conditions cells divide several times despite receiving only a single stimulus; one explanation is that the primary response is a damped periodic process which repeatedly activates replication, though we favour the shift in the threshold view referred to above. Nevertheless, we are convinced that, in one way or another, high frequency rhythms have a marked influence on cell proliferation and therefore on the growth of cancers. As the general properties and behaviour of cells must reflect the existence of any metabolic fluctuation (Gilbert 1984a), we further believe that the short period and cell cycle rhythms both affect other malignant characteristics, namely tumour progression and metastasis.

A paradoxical possibility arising from these studies is that tumour growth might be regulated by the use of agents capable of stimulating replication (Gilbert 1978a). Unfortunately, while such treatment could produce the minimum of side effects, the considerations do not yet allow one to predict which agents should be used in particular circumstances. Could the local application of chemotactic agents (which include insulin and other mitogens) dissuade tumour cells from metastasizing to distant sites and so make them more amenable to effective treatment by the above or other methods?

Summary

Cancer primarily involves a change in the ability of cells to proliferate and form other cells with similar, altered properties. In order to understand the aberrant behaviour it is necessary to know what makes any cell divide. Classical views are dismissed and the major features of cancer and the various behavioural properties of replicating cells are outlined. It is then pointed out that a mathematical model of a cellular control system can mimic the various features of proliferation. It is concluded (a) that a cell proliferates when a control system determining the levels of coenzyme components, is induced to undergo a cycle of oscillation in their concentrations, and (b) that malignancy results from any discrete or temporal metabolic or genetic change which affects the rates of the control system such that the state of the cell is effectively shifted closer to, or within, the threshold separating static and oscillatory modes of operation. Theoretical studies indicate that permanent disturbances in the pattern of temporal organization of shorter period cellular rhythms can alter the replication frequency and thereby cause cancer. Experimental evidence is presented showing that serum and insulin can disturb cellular periodicities and the suggestion is made that the altered dynamics may be involved in their mitogenic action.

Acknowledgements

Thanks are due to Dr Carolyn Tsilimigras for the enzyme data and Mr Geoffrey Visser for the morphological results. D. A. G. is in receipt of a grant from the Medical Research Council of South Africa in respect of the latter studies.

References

Aaronson SA, Todaro GJ (1974) Basis for the acquisition of malignant potential by mouse cells cultured in vitro. Science 162:1024–1026

Basilico C, Renger HG, Burstin SJ, Toniolo, D (1974) Host cell control of viral transformation. In: Clarkson B, Baserga R (eds) Control of proliferation in animal cells. Cold Spring Harbor Laboratory Press, Cold Spring Harbor, NY, pp 167–176

Brooks RF, Riddle PN (1988) The 3T3 cell cycle at low proliferation rates. J Cell Sci 90:601–612

Eagle H, Piez KA, Levy M (1961) The intracellular amino acid concentration required for protein synthesis in cultured human cells. J Biol Chem 236:2039–2042

Edmunds LN Jr (1988) Cellular and molecular bases of biological clocks. Springer, Berlin Heidelberg New York

Gilbert DA (1968) Differentiation, oncogenesis and cellular periodicities. J Theor Biol 21:113–122

Gilbert DA (1969) Phase plane analysis of periodic isozyme pattern changes in cultured cells. Biochem Biophys Res Commun 37:860–866

Gilbert DA (1973a) The malignant transformation as a metabolic steady state transition: the possible significance of the phasing of enzyme synthesis and related aspects. BioSystems 5:128–139

Gilbert DA (1973b) Biochemical phase discrimination in relation to differentiation and development. S Afr J Sci 69:348–349

Gilbert DA (1974a) The temporal response of the dynamic cell to disturbances and its possible relationship to differentiation and cancer. S Afr J Sci 70:234–244

Gilbert DA (1974b) The nature of the cell cycle and the control of cell proliferation. BioSystems 5:197–206

Gilbert DA (1977) Density dependent limitation of growth and the regulation of cell replication by changes in the triggering level of the cell cycle switch. BioSystems 9:215–228

Gilbert DA (1978a) The mechanism of action and interaction of regulators of cell replication. BioSystems 10:227–233

Gilbert DA (1978b) The relationship between the transition probability and oscillator concepts of the cell cycle and the nature of the commitment to replication. BioSystems 10:234–240

Gilbert DA (1978c) The malignant transformation: the nature of its effects on cell replication characteristics. S Afr J Sci 74:48–49

Gilbert DA (1980) Mathematics and cancer. In: Getz W (ed) Mathematical modelling in biology and ecology. Springer, Berlin Heidelberg New York, pp 97–115 (Lecture notes in biomathematics, vol 33)

Gilbert DA (1981) The cell cycle 1981: one or more limit cycle oscillations? S Afr J Sci 77:541–546

Gilbert DA (1982a) Cell cycle variability: the oscillator model of the cell cycle yields transition probability alpha and beta type curves. BioSystems 15:331–339

Gilbert DA (1982b) The oscillator cell cycle model needs no first or second chance event. BioSystems 15:331–339

Gilbert, DA (1984a) Temporal organisation, re-organisation and disorganisation in cells. In: Edmunds LN Jr (ed) Cell cycle clocks. Marcel Dekker Inc, New York, pp 5–25

Gilbert DA (1984b) The nature of tumour cell proliferation. Nature 311:610

Gilbert DA (1988) On G_0 and cell cycle controls. BioEssays 9:135–136

Gilbert DA, Tsilimigras CWA (1981) Cellular oscillations: the relative independence of enzyme activity rhythms and periodic variations in the amount of extractable protein. S Afr J Sci 77:66–72

Hammond KD, Cloutman L, Mindel B, Gilbert DA (1989a) Temporal changes in phosphoamino acid phosphatase activities in murine erythroleukemic cells. Int J Biochem 21:197–201

Hammond KD, Cloutman L, Gilbert DA (1989b) Cancer reversal: regulation of phosphotyrosine phosphatase. Biochem Soc Trans 17:1048–1049

Hutchinson CJ, Cox R, Drepaul RS, Gomperts M, Ford CC (1987) Periodic DNA synthesis in cell-free extracts of *Xenopus* eggs. EMBO J 6:2003–2010

Lewin B (1990) Driving the cell cycle: M-phase kinase, its protein and substrate. Cell 61:743–752

Lhoka MJ, Maller JL (1985) Induction of nuclear envelope breakdown, chromosome condensation and spindle formation in cell free extracts. J Cell Biol 101:518–523

Lloyd D, Volkov EI (1990) Quantized cell cycle times: interaction between a relaxation oscillator and ultradian clock pulses. BioSystems 23:305–310

Lloyd D, Poole RK, Edwards SW (1982) The cell division cycle. Academic Press, London

MacKinnon H, Gilbert DA (1992) To divide or not to divide? That is the question. In: Bittar EE (ed) Medical cell biology, vol 7. JAI Press (in press)

Muella SN, Rosen EM, Levine EM (1979) Cellular senescence in a cloned strain of bovine fetal aortic endothelial cells. Science 207:889–891

Murray AW, Kirschner MW (1989) The role of cyclin synthesis and degradation in the control of maturation promoting factor. Nature 339:275–280

Satoh H, Ueda T, Kobatake Y (1985) Oscillations in cell shape and size during locomotion and in contractile activities of *Physarum polycephalum, Dictyostelium discoideum, Amoeba proteus* and macrophages. Exp Cell Res 156:79–90

Sel'kov EE (1970) Two alternative, self-oscillating stationary states in thiol metabolism – two alternative types of cell division, normal and malignant ones. Biophysika 15:1065–1073

Sporn MB, Roberts AB (1985) Autocrine growth factors and cancer. Nature 313:745–747

Tsilimigras CWA (1982) The effect of viruses and other agents on the temporal organisation of cells. PhD thesis, University of the Witwatersrand

Tsilimigras CWA, Gilbert, DA (1977) High frequency, high amplitude oscillations in the amount of protein extractable from cultured cells. S Afr J Sci 77:123–125

Visser G, Reinten C, Coplan P, Gilbert DA, Hammond KD (1990) Oscillations in cell morphology and redox state. Biophys Chem 37:383–394

Chapter 5

Genetic and Molecular Analysis of Ultradian Rhythms in *Drosophila*

C. P. Kyriacou, Mary L. Greenacre, M. G. Ritchie,
A. A. Peixoto, G. Shiels and J. C. Hall

The genetic analysis of ultradian rhythms in *Drosophila* is a relatively new area of study and was stimulated by the observation of a short-term rhythmic fluctuation in the courtship song of the male fly (Kyriacou and Hall 1980). This cycle has a 60 s periodicity in the species *D. melanogaster* and 40 s in the sibling species *D. simulans* (Kyriacou and Hall 1980, 1986). Remarkably it was observed that the *period* rhythm mutations, per^{L1}, per^{s}, and per^{01}, which lengthen, shorten or obliterate free-running circadian behaviour (Konopka and Benzer 1971) also had parallel effects on the period of the male's courtship song cycle (Kyriacou and Hall 1980). Thus began a series of genetic studies which have sprung several surprises, some controversy, and have in some respects led the way towards the molecular dissection of ultradian and circadian systems.

Historical Background to the Male Courtship Song

A male "serenades" a female by extending one wing and vibrating it to produce a courtship song. This song consists of two components in *D. melanogaster*: a train of pulses, interrupted by interpulse intervals (IPI) which vary from 15 to 100 ms within an individual courtship; and a sinusoidal hum (see Fig. 5.1). The average length of IPIs generated by a *D. melanogaster* male will usually fall in the 30–40 ms range if the courtship proceeds at 25 °C. In *D. simulans* the mean IPI will fall between 40 and 80 ms depending on the strain of fly used (Kyriacou and Hall 1980, 1986; Kawanishi and Watanabe 1980; Cowling and Burnet 1981). Thus it could be that females respond best to the mean IPI produced by their conspecific male, thereby helping to preserve the species boundary. Some early "playback" experiments suggested that *D. melanogaster* females did indeed respond best to a mean IPI of 34 ms, as opposed to a mean IPI of 48 ms, characteristic of *D. simulans* males (Bennet-Clark and Ewing 1969; Schilcher 1976). However, these results were difficult to reproduce (Cowling 1979; Kyriacou and Hall 1980; Schilcher 1989; for comments on these experiments,

Hum   Pulse

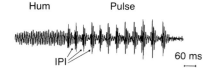

Fig. 5.1. Burst of courtship song showing *hum*, *pulse* song, and interpulse intervals (*IPI*).

see Hall and Kyriacou 1990). Therefore up to 1982, there was little direct experimental evidence that the mean IPI plays any role at all in species "signatures". The discovery of cycles in the IPI of *D. melanogaster* and *D. simulans* (Kyriacou and Hall 1980, 1986) and the subsequent demonstration that the IPI and the cycle period were important factors in the females' discrimination between different songs (Kyriacou and Hall 1982, 1984, 1986; Greenacre 1990) suggested a role for both IPI and cycle periods in species identification.

Cycles in the Male's Song

The initial observations of Kyriacou and Hall (1980), which led them to conclude that cycles were present in *Drosophila* songs, were made by dividing courtship time into 10-s time bins and calculating the mean IPI in each bin. Kyriacou and Hall (1980) showed that the mean IPIs appeared to lengthen and to shorten with a regular pattern (see Fig. 5.2). By using curvilinear regression techniques they were able to fit sine waves accounting for between 30% and 80% of the variance to the mean IPI points. The periodicity of these sine waves was between 50 and 65 s in most strains of *D. melanogaster* (Kyriacou and Hall 1980), and individual males appeared to initiate their own particular song rhythm at almost any phase of the cycle. When males interrupted their courtship song, as they frequently do, they resumed their IPI production in phase with the song cycle they had initially generated (Kyriacou and Hall 1980, 1985). Consequently these results led to the conclusion that an oscillator was controlling the generation of IPI lengths in *D. melanogaster* songs. This was extended to *D. simulans* males, who in spite of having generally longer IPIs (Cowling and Burnet 1981) showed a shorter, 35–40 s IPI cycle (Kyriacou and Hall 1980, 1986; Wheeler et al. 1991), and *D. yakuba* males, who have a more complex pulse song which nevertheless shows IPI periodicities in the 75–80 s range (Thackeray 1989).

Hybrid males between *D. melanogaster* and *D. simulans* sing with IPI cycles that have rhythm periods determined by the X chromosome they carry. Thus a hybrid male with an X chromosome from *D. simulans* will sing with the typical *D. simulans* 40 s cycle. The reciprocal hybrid male carrying a *D. melanogaster* X chromosome will sing with the *D. melanogaster* 60 s cycle (Kyriacou and Hall 1986). These two types of hybrid are genetically identical with respect to their autosomes (chromosomes II, III and IV), only their X chromosomes are different. Thus the species-specific difference in courtship song cycle maps to a gene or genes on the X chromosome. Cytoplasmic differences between the reciprocal hybrids which could also give apparent sex-linkage in this experi-

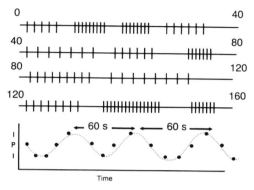

Fig. 5.2. Diagrammatic representation of a courtship song cycle over a 160-s time period. The mean IPIs from the burst of pulses are plotted in 10-s time bins, giving an approximately 60-s cycle.

ment can be ruled out because of the types of cross used in these experiments (see Kyriacou and Hall 1986).

A candidate gene for determining this species difference in behaviour is the *period* (*per*) gene. This gene was identified by Konopka and Benzer (1971) in a screen for mutations of the circadian system. Using the pupal–adult eclosion phenotype, they isolated three mutants: per^s, which showed a 19 h cycle; per^{L1}, which shows a 29 h cycle; and per^{01}, which appeared to be arrhythmic. The effects of these mutant genes on the period of circadian locomotor behaviour of individual flies was similar, per^s shortened, per^{L1} lengthened and per^{01} obliterated the cycle (Konopka and Benzer 1971). The *per* locus maps to a genetically well-defined region of the X chromosome (Smith and Konopka 1981), and this proved to be helpful in the subsequent molecular cloning of the gene. Thus, the three mutants isolated by Konopka and Benzer turned out to be different alleles of the same locus.

Kyriacou and Hall (1980) subsequently examined whether the 60 s courtship song cycle was also affected by these mutations. Surprisingly perhaps, they observed that the per^s song cycle was shortened to 40 s, that of the per^{L1} male was lengthened to 80 s, and per^{01} songs showed little evidence of cycling (Fig. 5.3). Thus a correlation appeared to exist between the effects of the *per* mutations in the circadian and the ultradian domain. Other Clock genes such as *Andante* (*And*), which lengthens the circadian period (Konopka et al. 1991), also lengthens the song period (Greenacre 1990). However, the *Clock* mutation, which has short circadian periods, does not have a corresponding effect on the courtship song cycle (Dushay et al. 1990). This is of particular interest because the *Clock* mutation maps very close to *per*, and perhaps may

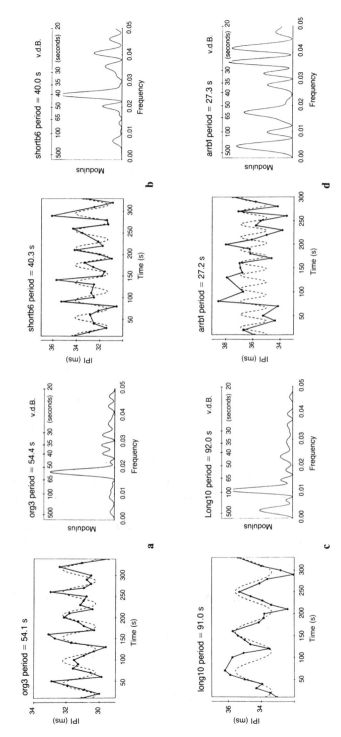

Fig. 5.3. Mean IPI/time plots and spectral analyses of songs from: **a** wild-type males (*org3*), **b** *per^s* male (*shortb6*), **c** *per^{L1}* male (*long10*) and **d** *per^{01}* male (*arrb1*). The dotted line represents the best curvilinear regression through the mean IPIs. The period of the cycle is given above the panel. The spectral analysis of the data is given to the right of each IPI/time plot. Note that the peaks in the spectrogram correspond to periods similar to those obtained in the regression analysis. In the *per^{01}* spectrogram, the highest peak corresponds to very fast frequencies (after Kyriacou et al. 1990b).

even be a mutation within the *per* regulatory elements. If this is the case, it is not too difficult to see that a mutation in an "enhancer" region for circadian behaviour may leave an adjacent courtship song "enhancer" sequence intact. Finally, the blind *disco* mutant, in which neural connections from the eyes are disconnected from the brain (Steller et al. 1987), is also arrhythmic in circadian behaviour (Dowse et al. 1987; Dushay et al. 1989). However, this mutant has an approximately normal song cycle, although the periodicity may be slightly longer than normal (Greenacre 1990).

Other phenotypes affected by the *per* mutants include the 10-day developmental cycle from egg to adult of *D. melanogaster* (at 25 °C), which is significantly lengthened in per^{L1} and shortened with both per^s and per^{01} under a variety of experimental conditions (Kyriacou et al. 1990a).

The conclusion from this brief description of the phenotypes affected by *Clock* mutations is that the *per* gene encodes a protein which appears to perform a general timing function in the organism, irrespective of the temporal domain in which the individual "oscillator" acts. The Per^s protein shortens the periods of some oscillators; Per^{L1} lengthens and Per^{01} can obliterate them. Consequently a relationship between the circadian and ultradian domain has been demonstrated, something which has been proposed in the past (Pavlidis 1969), but has proved difficult to demonstrate.

Do Ultradian Song Cycles Really Exist?

Crossley (1988) and Ewing (1988) questioned the existence of song cycles and suggested that they were an artefact of the statistical method used to detect them. They presented song data which when analysed with a particular spectral method apparently revealed that 60-s cycles did not exist in songs from per^+ insects (Crossley 1988). Furthermore the genotype-specific song periods observed in *per* mutants were also questioned (Ewing 1988). In response, Kyriacou and Hall (1989, 1990) reanalysed their original 1980 data using two sophisticated spectral methods, and were able to confirm their original conclusions, which were based on curvilinear regression. Furthermore they were able to demonstrate that the "off-the-shelf" spectral analysis used by their critics was inappropriate for use with any data which had "missing" values. That is, a song record, which is a series of mean IPIs in adjacent 10-s time bins, usually has several bins where there are no data, in that the male did not sing during this time. Crossley's (1988) and Ewing's (1988) analyses required every bin to contain data, and so they made a crude estimation of the IPI value for the empty time bins, and then performed their analyses. As most song records have a number of "missing" bins this introduced a serious artefact into their methodology which is not present in either of Kyriacou and Hall's spectral methods, or their regression technique (Kyriacou and Hall 1980, 1989, 1990; Hall and Kyriacou 1990; Kyriacou et al. 1990b). This was further emphasized by a following observation made by Kyriacou and Hall (1989) on Crossley's song results, which are presented in some detail in her paper (Crossley 1988). Crossley had included eight vigorous songs where she did not have to estimate any missing data. In other words, her spectral analysis was appropriate for this subset of songs. Remarkably, and completely unnoticed by

Crossley, six of these eight songs gave spectral frequencies in the 50–65-s range, which is characteristic of wild-type songs (Hamblen et al. 1986; Kyriacou and Hall 1989, 1990). The probability of six or more of eight songs having these spectral periods by chance is about 40 in a million (Kyriacou and Hall 1989, 1990). In other words, when Crossley's analysis was free from the artefact of introducing missing data, she had obtained results almost identical with those of Kyriacou and Hall (1980, 1989). A final encouraging note to add to this story is that Crossley sent Kyriacou and Hall a set of mean IPIs for a song she considered to be arrhythmic (Crossley 1988). Her spectral analysis had indicated that it had a period in the 150-s range. Using Crossley's mean IPIs for this song, both of Kyriacou and Hall's more sophisticated spectral analyses revealed that in fact the period for this song was 60 s, not 150 s. Crossley, by inserting a number of "missing" data points in this song had obtained a spurious period of 150 s (Kyriacou and Hall 1989, 1990). The correct period was 60 s, and again typical of the *D. melanogaster* species-specific period.

Ewing (1988) was apparently aware of the problem of missing data points, and in order to minimize these, built a courtship chamber that was extremely small. He reasoned that if the female could not escape from the male, the male would sing more or less continuously and therefore this might solve the problem of missing data. Unfortunately, it is well known that female movement appears to stimulate the male to court vigorously (Tompkins et al. 1982). Consequently the result of Ewing's experiment was that the extremely small and unrealistic (in the ethological sense) chamber appeared to inhibit the male song. This was formally demonstrated by Kyriacou and Hall (1989), who recorded songs in chambers of the same size as that of Ewing. Kyriacou and Hall (1989) also noticed that there was a severe amount of interference between the male and female, giving extremely noisy and difficult-to-interpret song records (see also Schilcher 1989). The IPI means obtained from such small chambers by Kyriacou and Hall (1989) did not oscillate, which was not surprising given this poor experimental design.

Thus, the conclusion to this "controversy" appears to be that Crossley's (1988) and Ewing's (1988) criticisms of the Kyriacou and Hall (1980) results were ill founded. However the debate did have the positive effect of clarifying some of the statistical and methodological details that are necessary to demonstrate this behavioural rhythm. It was also extremely interesting to see how Crossley's (1988) paper, which methodologically was a good and serious attempt to follow Kyriacou and Hall's behavioural techniques, and which superficially appears to contradict their results, does, on closer examination, support their conclusions. Readers interested in the statistical aspects of this debate might wish to read papers by Kyriacou and Hall (1980, 1989, 1990), Crossley (1988, 1989), Ewing (1988, 1989), Logan and Rosenberg (1989), Schilcher (1989), Bennet-Clark (1990), Hall and Kyriacou (1990) and Kyriacou et al. (1990b).

Is There a Song Reception Mechanism in the Female Which Responds to IPI Cycles?

Because *Drosophila* females respond best to simulated songs bearing the correct species-specific song cycles, 55 s or 35 s for *D. melanogaster* and

D. simulans, respectively (Kyriacou and Hall 1982, 1986), the question arises as to whether the females match up the song cycle of the male with an internal "rhythm" template. There are several ways in which the female may be processing the incoming IPIs which may have nothing to do with cycles. For example, imagine that each *Drosophila* female has a preferred IPI value, say a 35-ms IPI for female A, 37-ms IPI for female B, 38-ms IPI for female C etc., and each of them requires to hear a certain number of these "preferred" IPIs in order for mating to be "triggered" (Kyriacou and Hall 1982). Let us imagine also that *D. simulans* females need to hear fewer IPIs than do *D. melanogaster* females before they will mate. In the natural situation, we might expect that a male sings briefly to a female and if he is unsuccessful he moves on quickly to the next available female. Under such circumstances it would be advantageous for the male to sing in cycles, because he will scan the range of IPIs preferred by a number of conspecific females. The different periodicities of *D. melanogaster* and *D. simulans* males may simply reflect the receptivity of the different female species. The *D. simulans* males can scan the range of acceptable IPIs more quickly if the *simulans* females require fewer "critical" IPIs. The slower IPI cycle of *D. melanogaster* males implies that the *melanogaster* female requires to hear more individual IPIs at her critical IPI value.

The above is pure conjecture and other hypotheses can account for why females might prefer cycling IPIs (Kyriacou and Hall 1982). Perhaps females do have an oscillator, and they use this to match up the incoming cycle. However, this is unlikely, as we imagine (although we do not know) that, in the wild, courtship interaction might be quite brief. Under this scenario, there would not be enough time in any one courtship interaction for the female to get a measure of the periodicity of the incoming song cycle.

Irrespective of what kind of song reception mechanism the female uses, perhaps the *per* gene also plays a role in building this female structure. Alexander (1962) has pointed out that a gene which affects the male output (song) and also determines the female input (song reception mechanism) would be expected to be an important evolutionary factor in speciation. A new mutation which affected both the male song output and female song input simultaneously might be expected to produce eventual sexual isolation between the carriers of the new gene and the non-carriers. This is termed "genetic coupling", and there is some rather sketchy evidence in favour of it from studies of hybrids between different species of cricket and tree-frog. With these organisms, female hybrids between the two species prefer hybrid songs (Hoy et al. 1977; Doherty and Gerhardt 1984). In fact females that are hybrid between *D. melanogaster* and *D. simulans* also prefer songs with intermediate periods (45 s) and intermediate IPIs (Kyriacou and Hall 1986). However, at best this is very indirect evidence for genetic coupling (Butlin and Ritchie 1989), as single genes affecting both characters cannot be identified unless segregation analysis is performed. The general inviability of such hybrids makes this kind of genetic analysis very difficult.

However, with the *per* gene, we do have single gene mutations which change the periodic feature of the male song. Do these mutations produce a complementary change in the female's reception? For example, do *per*[s] females prefer the 40-s cycling *per*[s] song, and do *per*[L1] females prefer 80-s cycling *per*[L1] songs? The answer appears to be "no". Both mutant *per*[s] and *per*[L1] females prefer 55-s cycles over 80- and 40-s cycles (Greenacre 1990). Thus, there appears to be no genetic coupling of male song output and female song input with respect to *per* in the manner proposed by Alexander (1962).

Further Genetic Analysis of *per*

Using a genetic trick to generate diplo-X males, Kyriacou and Hall (1980) showed that the per^s mutant was semi-dominant to per^+ for the song cycle. A similar semi-dominance of per^s was observed by Konopka and Benzer (1971) for the period of circadian locomotor behaviour in female per^s/per^+ heterozygotes. The per^{L1} mutation was recessive to the per^+ allele in the song cycle, but shows a semi-dominant effect in locomotor activity cycles (Konopka and Benzer 1971; Kyriacou and Hall 1980). Similarly per^{01}/per^+ heterozygotes have slightly longer than normal circadian cycles in the song phenotype, and per^{01}/per^+ heterozygous males gave short, per^s-like song cycles. This suggested that perhaps, with the song phenotype, per^{01} was acting like a dominant or semi-dominant mutation in the "fast" direction, implying that the per^{01} phenotype might be ultra-short. We will return to this point in a later section. Furthermore flies carrying per^+ in trans with a *per* deficiency sang with a period similar to that of per^+/per^{01} heterozygotes, indicating that the per^{01} mutation might be an amorph, i.e. lacking a gene product (Kyriacou and Hall 1980).

Dosage studies with *per* and the song phenotype revealed that male flies carrying two copies of per^+, one on the X chromosome and one on the Y, sing with slightly shorter than wild-type song cycles (Hamblen et al. 1986). Similar speeding up of the circadian clock was observed when multiple doses of per^+ were crossed into single individuals (Smith and Konopka 1982). However, by simply stacking up several doses of the per^+ gene one cannot produce a fly with an extremely short per^s-like 19-h cycle. Consequently this suggests that the per^s mutation is not simply an overproducer of the per^+ protein. Rather it is (and indeed turns out to be) a mutation which qualitatively changes the *per* protein. The per^s mutation, however, can be considered to be, at least formally, a hypermorph. The per^{L1} mutation is a hypomorph, and acts as if less Per protein is being produced. The circadian period can be lengthened by placing a per^+ gene over a *per* deletion – slightly longer periods are obtained (Smith and Konopka 1981). Thus, the absence of the *per* gene in one of the X chromosomes in these heterozygotes produces a longer, weaker circadian cycle.

Localization of the Ultradian Song Clock

Studies of genetic mosaics, which are flies that are part per^s and part per^+, suggest that the song oscillator lies in the fly's thorax and the circadian oscillator in the fly's brain (R. Konopka, C. P. Kyriacou and J. C. Hall, preliminary results cited by Hall (1984); Konopka et al. 1983). Antibody staining of the Per protein also reveals that thoracic regions stain positively for *per* (Saez and Young 1988; Siwicki et al. 1988). Perhaps it is in these regions that the song "clock" resides. More elaborate mosaic and antibody studies might conceivably identify cells which generate this thoracic rhythm, and attempts are currently being made in this direction. It is more than likely that the song oscillator will involve a neuronal network partly because the cycle period is so short, 60 s. Also, Kyriacou and Hall (1985) showed that temperature-sensitive paralytic mutations which block action potentials at restrictive temperatures appeared to phase shift the song cycle when neuronal conductance was blocked (Suzuki et al. 1971; Wu and Ganetsky 1980). The phase shift appeared to be rather a simple one in that a 15-s temperature-induced paralysis would give an approximately 15-s phase delay in the male's song, and longer periods of paralysis would give correspondingly longer delays. With *Aplysia*, the marine mollusc, blocking neuronal conductance with pharmacological agents tends not to phase shift the circadian cycle in the eye (Eskin 1977). Similar results are obtained with the mammalian circadian pacemaker, the suprachiasmatic nucleii, whose phase appears to be resistant to sodium channel blockers (Schwartz et al. 1987; Schwartz 1991). If a similar situation is found with the *Drosophila* circadian system, i.e. that circadian cycles cannot be phase shifted by blocking action potentials, then this could indicate fundamental differences in how the Per protein is used in the ultradian and circadian mechanisms. However, as yet the circadian system of the fly has not come under such "neuronal scrutiny".

Molecular Analysis of the *per* Locus

The *per* gene has been cloned and sequenced in a number of *Drosophila* species (for reviews, see Hall 1990; Kyriacou 1990). The three *per* mutations have been mapped to single nucleotide changes. The per^{L1} and per^s mutant phenotypes are caused by single amino acid substitution, a valine-to-aspartate in the case of the former and a serine-to-asparagine substitution for the latter (Baylies et al. 1987; Yu et al. 1987a). The protein's primary sequence is composed of about 1200 amino acid residues, and in the case of per^{01} it is perhaps not surprising that this variant encodes a stop codon after encoding approximately 460 residues (Baylies et al. 1987; Yu et al. 1987a). Consequently it is unlikely that the translated per^{01} protein has biological activity.

The Threonine-Glycine Region

In the middle of the gene is encoded a run of alternating threonine-glycine (Thr-Gly) residues (Jackson et al. 1986; Citri et al. 1987). This region is polymorphic in length, both in natural and in laboratory populations (Yu et al. 1987b; Costa et al. 1991). The minimum number of Thr-Gly pairs encoded is about 14 and the maximum about 30 (R. Costa, personal communication). *D. simulans* strains are also polymorphic for the number of Thr-Gly pairs encoded (Wheeler et al. 1991; R. Costa, personal communication), and the one strain of *D. yakuba* in which the entire *per* coding sequence has been determined, reveals 15 Thr-Gly pairs (Thackeray and Kyriacou 1990). Other species in the *melanogaster* subgroups, such as *D. erecta* and *D. teissieri*, which are more

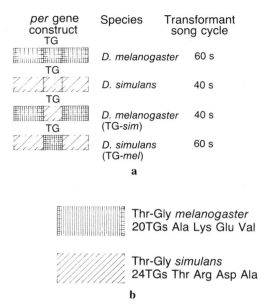

per gene construct	Species	Transformant song cycle
TG	*D. melanogaster*	60 s
TG	*D. simulans*	40 s
TG	*D. melanogaster* (TG-*sim*)	40 s
TG	*D. simulans* (TG-*mel*)	60 s

a

Thr-Gly *melanogaster*
20TGs Ala Lys Glu Val

Thr-Gly *simulans*
24TGs Thr Arg Asp Ala

b

Fig. 5.4. a Diagrammatical representation of the results obtained by Wheeler et al. (1991) when the *Thr-Gly* (*TG*) region was "swapped" between *D. melanogaster* (*D. mel.*) and *D. simulans* (*D. sim.*). **b** The species-specific amino acid residue differences encoded in the Thr-Gly fragments of the two species. Species-specific song cycle differences map to one or more of these amino acid changes.

closely related to *D. yakuba* than are *D. melanogaster* and *D. simulans*, appear to encode about 15 Thr-Gly pairs (R. Costa, personal communication).

The Thr-Gly region is extremely interesting with respect to the ultradian song cycle, because removing the perfect Thr-Gly repeat and transforming per^{01} flies with this Thr-Gly-deleted construct produces male flies which have short courtship song cycles of approximately 40 s (Yu et al. 1987b). Furthermore, transformants carrying the Thr-Gly-deleted gene have temperature-sensitive circadian periods (Ewer et al. 1990). The dramatic change in the courtship song cycle suggests that the Thr-Gly region is critical in its determination. This has been confirmed in recent experiments where the Thr-Gly and some surrounding regions of the *per* genes of *D. simulans* and *D. melanogaster* have been substituted into each other's *per* genes (Wheeler et al. 1991). The result is that transformed per^{01} (*melanogaster*) flies carrying a largely *D. melanogaster* gene, but a *D. simulans* Thr-Gly region, sing with a 40-s *simulans* cycle. per^{01} hosts carrying the reciprocal construct, i.e. *D. simulans* *per* material flanking the *D. melanogaster* Thr-Gly region, sing with 60-s cycles (Wheeler et al. 1991). Thus, the species-specific difference in behaviour "maps" to the small 700 base-pair region of *per* which was "swapped" between the two species (Fig. 5.4).

This 700 base-pair DNA fragment, carries 24 Thr-Gly pairs in *D. simulans* and 20 Thr-Gly pairs in *D. melanogaster*, and in addition encodes four further species-specific amino acid differences (Wheeler et al. 1991). It was quickly determined that the absolute number of Thr-Gly pairs in the range 17 to 23 pairs made no difference to the species specificity of the song cycle. *D. simulans* males, with 23 Thr-Gly pairs, sing with the typical *simulans* 40-s cycle and *D. melanogaster* males with 23 Thr-Gly pairs sing with their characteristic 60-s cycle (Wheeler et al. 1991). Thus, it appears that any one (or all) of the four species-specific amino acids, substitution of alanine for threonine, lysine for arginine, glutamine for aspartate and valine for alanine (*melanogaster* to *simulans* substitution), is the critical factor for determining the different species rhythms (see Fig. 5.4).

Thus we have an initial clue as to the types of amino acid change that will speed up or slow down an ultradian oscillation. Unfortunately, this knowledge is not particularly helpful because of our current state of ignorance about the biochemical nature of this oscillation. It should be added that these species-specific amino acid substitutions do not appear to change circadian behaviour in any species-specific manner (Wheeler et al. 1991; Hall et al. 1992).

Relationship of the Circadian and Ultradian Cycles

We have seen above that there is both a molecular independence and a dependence between ultradian and circadian cycles with respect to *per* gene structure. The *per^s*, *per^{L1}* and *per^{01}* mutations affect both types of cycle, whereas changes in amino acids encoded near the Thr-Gly region appear to affect the ultradian cycle only (Wheeler et al. 1991). How might the *per* protein control such very different oscillations?

One attractive hypothesis which may explain the similar effects of the three *per* mutations on the two cycles was put forward by Bargiello et al. (1987). They proposed, and provided some supporting experimental evidence, that the Per protein might act as a "coupler" between cells and thereby be involved in cell-to-cell communication. They suggested that the 24-h circadian period is the product of the oscillating characteristics of networks of many cells. Each of these cells has an ultradian periodicity, perhaps 5–15 h, and when these cells are coupled together by the Per protein, the final 24-h output is produced. The probability that the circadian oscillation can be demultiplied into its constituent ultradian cycles has been proposed before (Pavlidis 1969) and thus the protein which links these "oscillating" cells together takes on a special significance. Bargiello et al. (1987) demonstrated that *per^s* salivary cells were more tightly coupled together than *per^+* cells, and *per^{01}* cells appeared to communicate poorly. Thus the *per^s* mutation appears to accelerate the flow of information between cells. Extrapolating this finding to the hypothetical neural network which determines circadian periodicity, we can see how faster intercellular communication might speed up the final clock "output", giving 19-h periods. In *per^{01}* individuals, the lack of Per protein might be expected to produce arrhythmicity under this model, because there would be little coupling between cells. Equally likely, however, is that the ultradian cycles may express themselves weakly in *per^{01}* flies. In fact, sophisticated mathematical analysis of the "arrhythmic" *per^{01}* locomotor activity patterns, reveals significant short period cycles of 5–15 h (Dowse et al. 1987). The song of *per*[01] males also reveals a higher than expected proportion of fast, but weak, "cycles" in the 20–30-s range (Kyriacou and Hall 1989; Hall and Kyriacou 1990; Kyriacou et al. 1990b). Remember too that genetic analysis of the *per*[01] mutation with respect to the song phenotype suggested that *per*[01] might be acting as a very short period variant for this character (Kyriacou and Hall 1980). Thus, we could be having a glimpse here of the ultradian oscillation which underlies both song and circadian cycles in *per*[01] flies.

Another observation related to the above is that the primary amino acid sequence of Per suggests that the protein may have the properties of a vertebrate proteoglycan called serglycin (Ruoslahti 1988). This vertebrate protein has a series of Ser-Gly repeats which are at least superficially similar to the Thr-Gly repeats of *per*. Attachment of sugar molecules to the serine residues in serglycin produces a highly glycosylated product, and it is conceivable that a similar post-translational modification occurs at the threonine residues of *per*. Furthermore there are also some Ser-Gly repeats in *per* which are dotted around the protein and which might also serve as glycosylation sites. Proteoglycans have been observed to modulate gap junctional communication between cells (Spray et al. 1987). Therefore these proteins can be involved in cell-to-cell exchanges, and this observation indirectly supports the possible "coupling" role for *per*. Antibodies to the Per protein in flies stain cell boundaries, which is where one might expect to find an intercellular coupling protein (Bargiello et al. 1987). However, nuclear and cytoplasmic regions also stain (Saez and Young 1988; Siwicki et al. 1988; Zerr et al. 1990). The *per*/ proteoglycan/gap junction story provides an elegant and "global" explanation of how a mutation such as *per*[s] will affect both the song and circadian circuits in similar ways simply because there will be a tighter coupling of the hypothetical "oscillating" cells which make up the circadian and song neuronal networks. However, there are, as with all theories, some problems, and they have been discussed at length by Hall and Kyriacou (1990). A couple of relevant objections are that:

1. Almost the entire Thr-Gly encoding repeat can be removed from the *per* gene but robust, if not completely normal, circadian and song rhythmicity is maintained (Yu et al. 1987b). A more abnormal phenotype might have been expected if glycosylation at the Thr-Gly region is critical for cell–cell interactions and in turn for circadian periodicity.

2. There is little evidence that threonine residues are ever used for sugar molecule attachment in proteoglycans. However, it may be that a glycosylating enzyme exists in *Drosophila* which does recognize Thr-Gly repeats. This enzyme has yet to be identified.

Further discussion of the proteoglycan-*per* hypothesis can be found in the review by Hall and Kyriacou (1990). One of the implications of this hypothesis is, of course, that *per* does not affect the actual "ticking" mechanism. Other genes presumably give these hypothetical "oscillating cells" their own periodic characteristics. Consequently it should be possible, for example, to speed up the *per*[L1] clock by inducing a mutation in another gene which "speeds up" the ticking of the "oscillating" cells of the neural network. This would compensate for the poorer cell-to-cell communication in *per*[L1]. Such second-site suppressors of *per*[L1] might yield to a mutagenesis programme.

Cycling of the Per Protein and mRNA

Antibodies to the per protein reveal a wide distribution of per in different tissues of adult *Drosophila* (Saez and Young 1988; Siwicki et al. 1988; Zerr et al. 1990). Perhaps this is to be expected, given the number of temporal phenotypes affected by the *per* mutants. The brain and visual systems have been found to stain much more intensively during the night phase (Siwicki et al. 1988; Zerr et al. 1990). These *per* immunoreactivity cycles are seen both in light–dark cycles and in constant darkness, with a period of about 24 h (Siwicki et al. 1988). In *per*^{s} adults, the rhythm cycles with an approximately 20-h period, and the peak of the cycle occurs much earlier in *per*^{s} in light–dark conditions than in the wild-type (Zerr et al. 1990). The implications of this observation are that the protein product must influence its own cycling rate. This was confirmed with the observation that the *per* mRNA in the head also showed a cycling in abundance, with a period of 24 h in the wild type and a rather shorter period in *per*^{s}. There was no *per* mRNA cycling observed in *per*^{01} (Hardin et al. 1990).

These results are fascinating because they can be used to propose a rather different model for the way *per* determines circadian behaviour. The simplest hypothesis is that the Per protein itself acts on its own promotor, either directly or indirectly via intermediates to give cycling *per* mRNA levels (Zweibel et al. 1991). These fluctuating protein levels are then somehow "geared" into the behavioural phenotypes. How they could be scaled down to account for 60-s song cycles is difficult to imagine. Therefore this kind of model begins with the circadian cycle and then attempts to scale down to the ultradian domain – quite the opposite of the "*per* as coupler" story, whereby ultradian oscillations are geared upwards to give circadian cycles.

Summary

The *per* gene is thus an important component of the ultradian and circadian machinery. As yet we know very little about the Per protein's biochemical properties. However, antibodies to the fly's Per protein label the circadian pacemaker cells in the molluscan eye (Siwicki et al. 1989), suggesting that parts of the Per protein have been widely conserved in the circadian oscillators of different species. Consequently the analysis of the *per* gene in *Drosophila* at the behavioural, molecular and biochemical level should contribute significantly to our understanding of how biological rhythms are generated. As yet we have had only a tantalizing glimpse into the molecular machinery of this mysterious biological phenomenon.

Acknowledgements

We thank the SERC for grants to C.P.K. and USNIH for a grant (GM.21473) to J.C.H.

References

Alexander BD (1962) Evolutionary change in cricket acoustic communication. Evolution 16:443–467

Bargiello TA, Saez TA, Baylies MK, Gasic G, Young MW, Spray DC (1987) The *Drosophila* gene *period* affects intercellular junctional communication. Nature 328:686–691

Baylies MK, Bargiello TA, Jackson FR, Young MW (1987) Changes in abundance or structure of the *period* gene product can alter periodicity of the *Drosophila* clock. Nature 326:390–392

Bennet-Clark HC (1990) Do the song pulses of *Drosophila* show cyclic fluctuations? Trends Ecol Evol 5:93–96

Bennet-Clark HC, Ewing AW (1969) Pulse interval as a critical parameter in the courtship song of *Drosophila melanogaster*. Anim Behav 17:755–759

Butlin RK, Ritchie MG (1989) Genetic coupling in mate recognition systems: what is the evidence? Biol J Linn Soc 37:237–246

Citri Y, Colot HV, Jacquier AC, Yu Q, Hall JC, Baltimore D, Rosbash M (1987) A family of unusually spliced and biologically active transcripts is encoded by a *Drosophila* clock gene. Nature 326:42–47

Costa R, Peixoto AA, Thackeray JT, Dalgleish R, Kyriacou CP (1991) Length polymorphism in the threonine-glycine-encoding repeat region of the *period* gene in *Drosophila*. J Mol Evol 32:238–246

Cowling DE (1979) Genetic and behavioural studies of courtship song in the *melanogaster* species subgroup of *Drosophila*. PhD thesis, University of Sheffield

Cowling DE, Burnet B (1981) Courtship songs and genetic control of their acoustic characteristics in sibling species of the *D. melanogaster* subgroup. Anim Behav 29:924–935

Crossley SA (1988) Failure to confirm rhythms in *Drosophila* courtship song. Anim Behav 36:1098–1109

Crossley SA (1989) On Kyriacou and Hall's defence of courtship song rhythms in *Drosophila*. Anim Behav 37:861–863

Doherty JA, Gerhardt HC (1984) Acoustic communication of hybrid tree frogs: sound production by males and selective phonotaxis by females. J Comp Physiol [A] 154:319–330

Dowse HB, Hall JC, Ringo JM (1987) Circadian and ultradian rhythms in *period* mutants of *Drosophila melanogaster*. Behav Genet 17:19–35

Dushay MS, Rosbash M, Hall JC (1989) The *disconnected* visual system mutations in *Drosophila melanogaster* drastically disrupt circadian Rhythms. J Biol rhythms 4:1–27

Dushay MS, Konopka RJ, Orr E, Greenacre ML, Kyriacou CP, Rosbash M, Hall JC (1990) Phenotypic and genetic analysis of *Clock*, a new circadian rhythm mutant in *Drosophila melanogaster*. Genetics 127:557–578

Eskin A (1977) Neurophysiological mechanism involved in the photo entrainment of the circadian rhythm from *Aplysia* eye. J Neurobiol 8:273–299

Ewer J, Hamblen-Coyle M, Rosbash M, Hall JC (1990) Requirement for *period* gene expression in the adult and not during development for locomotor activity rhythms of imaginal *Drosophila melanogaster*. J Neurogenet 7:31–73

Ewing AW (1988) Cycles in the courtship song of male *Drosophila melanogaster* have not been detected. Anim Behav 36:1091–1097

Ewing AW (1989) Reply to Kyriacou and Hall 1989. Anim Behav 37:860–861

Greenacre ML (1990) Genetic analysis of courtship behaviour and biological rhythms in *Drosophila*. PhD thesis, University of Leicester

Hall JC (1984) Complex brain and behavioral functions disrupted by mutations in *Drosophila*. Dev Genet 4:355–378

Hall JC (1990) Genetics of circadian rhythms. Ann Rev Genet 24:659–697

Hall JC, Kyriacou CP (1990) Genetics of biological rhythms in *Drosophila*. Adv Insect Physiol 22:221–298

Hall JC, Hamblen-Coyle MJ, Moroz L, Wheeler DA, Rutila JE, Yu Q, Rosbash M (1992) Circadian rhythms of *D. melanogaster* transformed with DNA from the *period* gene of *D. simulans*. Dros Inf Serv (in Press)

Hamblen M, Zehring WA, Kyriacou CP, Reddy P, Yu Q, Wheeler DA, Zwiebel LJ, Konopka RJ, Rosbash M, Hall JC (1986) Germ-line transformation involving DNA from the *period* locus in *Drosophila melanogaster*. Overlapping genomic fragments that restore circadian and ultradian rhythmicity to *per*[0] and *per* mutants. J Neurogenet 3:249–291

Hardin PE, Hall JC, Rosbash M (1990) Feedback of the *Drosophila period* gene product on circadian cycling of its messenger RNA levels. Nature 343:536–540

Hoy RR, Hahn J, Paul RC (1977) Hybrid cricket auditory behaviour: evidence for genetic coupling in animal communication. Science 195:82–84

Jackson FR, Bargiello TA, Yun SH, Young MW (1986) Product of *period* of *Drosophila* shares homology with proteoglycans. Nature 320:185–188

Kawanishi M, Watanabe TK (1980) Genetic variations of courtship song of *Drosophila melanogaster* and *Drosophila simulans*. Jap J Genet 55:235–240

Konopka RJ, Benzer S (1971) Clock mutants of *Drosophila melanogaster*. Proc Natl Acad Sci USA 68:2112–2116

Konopka RJ, Wells S, Lee T (1983) Mosaic analysis of a *Drosophila* clock mutant. Mol Gen Genet 190:284–288

Konopka RJ, Smith RF, Orr D (1991) Characterization of *Andante*, a new *Drosophila* clock mutant and its interaction with other clock mutants. J Neurogenet 7:103–114

Kyriacou CP (1990) The molecular ethology of the *period* gene in *Drosophila*. Behav Genet 20:191–211

Kyriacou CP, Hall JC (1980) Circadian rhythm mutations in *Drosophila* affect short-term fluctuations in the male's courtship song. Proc Natl Acad Sci USA 77:6929–6933

Kyriacou CP, Hall JC (1982) The function of courtship song rhythms in *Drosophila*. Anim Behav 30:794–801

Kyriacou CP, Hall JC (1984) Learning and memory mutations impair acoustic priming of mating behaviour in *Drosophila*. Nature 308:62–65

Kyriacou CP, Hall JC (1985) Action potential mutations stop a biological clock in *Drosophila*. Nature 314:171–173

Kyriacou CP, Hall JC (1986) Inter-specific genetic control of courtship song production and reception in *Drosophila*. Science 232:494–497

Kyriacou CP, Hall JC (1989) Spectral analysis of *Drosophila* courtship song rhythms. Anim Behav 37:850–859

Kyriacou CP, Hall JC (1990) Song rhythms in *Drosophila*. Trends Ecol Evol 5:125–126

Kyriacou CP, Oldroyd M, Wood J, Sharp M, Hill M (1990a) Clock mutants alter developmental timing in *Drosophila*. Heredity 64:395–401

Kyriacou CP, Berg MJ van den, Hall JC (1990b) *Drosophila* courtship song cycles in normal and *period* mutant males revisited. Behav Genet 20:617–644

Logan IG, Rosenberg J (1989) A referee's comment on the identification of cycles in the courtship song of *Drosophila melanogaster*. Anim Behav 37:860

Pavlidis T (1969) Populations of interacting oscillators and circadian rhythms. J Theor Biol 22:418–436

Ruoslahti E (1988) Structure and biology of proteoglycans. Ann Rev Cell Biol 4:229–255

Saez L, Young MW (1988) In situ localisation of the *period* clock protein during development of *Drosophila melanogaster*. Mol Cell Biol 8:5378–5385

Schilcher F von (1976) The function of sine song and pulse song in the courtship of *Drosophila melanogaster*. Anim Behav 24:622–625

Schilcher F von (1989) Have cycles in the courtship song of *Drosophila* been detected? Trends Neurosci 12:311–313

Schwartz WJ (1991) Further evaluation of the tetrodotoxin-resistant circadian pacemaker in the suprachiasmatic nucleii. J Biol Rhythms 6:149–158

Schwartz WJ, Gross RA, Morton MT (1987) The suprachiasmatic nucleii contain a tetrodotoxin-resistant circadian pacemaker. Proc Natl Acad Sci USA 84:1694–1698

Siwicki KK, Eastman C, Petersen G, Rosbash M, Hall JC (1988) Antibodies to the *period* gene product of *Drosophila* reveal diverse tissue distribution and rhythmic changes in the visual system. Neuron 1:141–150

Siwicki KK, Strack S, Rosbash M, Hall JC, Jacklet JW (1989) An antibody to the *Drosophila period* protein recognizes circadian pace-maker neurons in *Aplysia* and *Bulla*. Neuron 3:51–58

Smith RF, Konopka RJ (1981) Circadian clock phenotypes of chromosome aberrations with a breakpoint at the *period* locus. Mol Gen Genet 183:243–251

Smith RF, Konopka RJ (1982) Effects of dosage alterations at the *period* locus on the circadian clock of *Drosophila*. Mol Gen Genet 185:30–36

Spray DC, Fujita M, Saez JC, Choi H, Watanabe T, Heitzberg L, Rosenberg C, Reid LM (1987) Proteoglycans and glycosaminoglycans induce gap junction synthesis and function in primary liver cell cultures. J Cell Biol 105:541–551

Steller H, Fischbach KF, Rubin G (1987) *disconnected*: a locus required for neuronal pathway formation in the visual system of *Drosophila*. Cell 50:1139–1153

Suzuki DT, Grigliatti R, Williamson R (1971) Temperature-sensitive mutations in *Drosophila melanogaster* – mutation (*para*) causing reversible adult paralysis. Proc Natl Acad Sci USA 68:890–893

Thackeray JR (1989) Molecular analysis of behavioural rhythms in *Drosophila*. PhD thesis, University of Leicester

Thackeray JR, Kyriacou CP (1990) Molecular evolution in the *Drosophila yakuba period* locus. J Mol Evol 31:389–401

Tompkins L, Grass AC, Hall JC, Gailey DA, Siegal RW (1982) The role of female movement in the sexual behaviour of *Drosophila melanogaster*. Behav Genet 12:295–307

Wheeler DA, Kyriacou CP, Greenacre ML, Yu Q, Rutila JE, Rosbash M, Hall JC (1991) Molecular transfer of a species-specific courtship behaviour from *Drosophila simulans* to *Drosophila melanogaster*. Science 251:1082–1085

Wu C-F, Ganetsky B (1980) Genetic alteration of nerve membrane excitability in temperature-sensitive paralytic mutations of *Drosophila melanogaster*. Nature 286:814–816

Yu Q, Jacquier AC, Citri Y, Colot HM (1987a) Molecular mapping of point mutations in the *period* gene that stop or speed up biological clocks in *Drosophila melanogaster*. Proc Natl Acad Sci USA 84:784–788

Yu Q, Colot HV, Kyriacou CP, Hall JC, Rosbash M (1987b) Behaviour modification by in vitro mutagenesis of a variable region within the *period* gene of *Drosophila*. Nature 326:765–769

Zerr DN, Rosbash M, Hall JC, Siwicki KK (1990) Circadian rhythms of period protein immunoreactivity in the CNS and the visual system of *Drosophila*. J Neurosci 10:2749–2762

Zweibel LJ, Hardin PE, Liu Xin, Hall JC, Rosbash M (1991) A post-transcriptional mechanism contributes to circadian cycling of a per-β-galactosidase fusion protein. Proc Natl Acad Sci USA 88:3882–3886

Chapter 6

Do Ultradian Oscillators Underlie the Circadian Clock in *Drosophila*?

H. B. Dowse and J. M. Ringo

Ultradian rhythms are being discovered to be nearly as ubiquitous as circadian rhythms. Here, we define ultradian periods as the range between about 1 h and 18 h. Oscillations faster than that reflect metabolic processes (Lloyd and Edwards 1984; Edmunds 1988), and 18 h is an hour less than the mean period of the shortest clearly circadian clock mutant in *Drosophila melanogaster* (hereafter simply *Drosophila*: Konopka and Benzer 1971; Hall and Kyriacou 1990). Ultradian oscillations have been found in organisms as disparate as unicells (Lloyd et al. 1982; Michel and Hardeland 1985) and mammals (Daan and Aschoff 1981). They occur in both the presence and the absence of normal circadian rhythms, and now are found also in animals that lack circadian rhythms but do exhibit tidal or lunar periodicity (Dowse and Palmer 1990, 1992).

In this chapter, we focus on three fundamental questions about ultradian rhythms. First, are the rhythms evidence of underlying oscillations or are they simply epiphenomena owing their existence to interactions among circadian clocks or "random noise"? Second, what is the function of these high frequency clocks? Third, is there any functional relationship between circadian and ultradian rhythms? To address these questions, we review experiments, primarily those using *Drosophila* as a model system.

Ultradian Rhythms in *Drosophila*

Ultradian rhythms in *Drosophila* locomotor activity were first reported by Dowse et al. (1987) in mutant animals lacking a functional allele of the *period* (*per*) gene. Mutations of this sex-linked gene affect rhythmicity in *Drosophila* in striking ways (Konopka and Benzer 1971; Hall and Kyriacou 1990). per^{Long} (per^L) slows the clock to a period of about 29 h, while per^{short} (per^s) accelerates the clock to a period of about 19 h. The non-functional (*null*) allele mentioned above, per^0, along with heterozygous overlapping deletions that eliminate the locus entirely (here per^-) (Smith and Konopka 1981; Bargiello and Young 1984; Reddy et al. 1984) also eliminate normal circadian rhythms, while revealing a wealth of higher-frequency periodicities (Dowse et al. 1987).

When per^0 or per^- flies are raised in a light–dark (LD) cycle and their movement is recorded in constant darkness (DD), there is no consistent organization of the activity into clear bouts (Konopka and Benzer 1971; Smith and Konopka 1981). This contrasts with wild-type, per^+, flies, which exhibit modulation of the amplitude of activity into obvious cycles of about 24 h (Konopka and Benzer 1971). Simple inspection of event recorder data, and the use of insensitive digital signal analysis techniques such as the "periodogram" (Enright 1965, 1990) are inadequate to reveal the underlying patterns. We (Dowse et al. 1987) used a sensitive spectral analysis technique, MESA (maximum entropy spectral analysis; Dowse and Ringo 1989a, 1991), coupled with a robust test for significance, the correlogram (Chatfield 1989; Dowse and Ringo 1989a), and discovered that more than half of the per^0 and per^- animals tested had clear-cut ultradian rhythms in the range 4 to 18 h. Usually, the spectral analysis revealed more than one periodicity. The periods of these rhythms were seldom simple multiples (e.g. 10.0 h and 17.5 h), and thus could not have arisen from harmonics of some longer period. The presence of ultradian rhythms is not limited to per^0 and per^- mutants. We found that per^L flies also displayed strong ultradian rhythms in spectra of their activity; wild-type did as well, although to a lesser extent (Dowse and Ringo 1987). per^s animals display much less power in the ultradian range (Dowse and Ringo 1987); alternating current signal power is equivalent to the variance in the data with the mean (direct current) component removed and comes from an electronic signal analogy (Beauchamp and Yuen 1979). When wild-type *Drosophila* are reared in constant darkness and are then tested as adults in DD, having never been exposed to light let alone any cyclic illumination, they frequently resemble per^0 in being either arrhythmic or in displaying multiple ultradian rhythms (Dowse and Ringo 1989b). Some of these DD-reared animals have residual wild-type circadian rhythmicity, some have per^L-like rhythms, and some are unusual in displaying short, weak circadian rhythms ($\tau \approx 20$–22 h; Dowse and Ringo 1989b). Rearing animals in constant light (LL) followed by testing in LL also produces per^0-like behaviour, with multiple ultradian periodicities (Fig. 6.1). However, we have observed LL-animals with only arrhythmia or ultradian rhythmicity (J. Power, unpublished results), unlike the situation with DD-reared insects.

disconnected (*disc*) mutants possess eyes that fail to connect to the brain developmentally, and are missing other portions of the nervous system, including portions of the optic lobes (Steller et al. 1987). The precise extent of the neural lesion is probably not yet known. These animals are blind, and are per^0-like in rhythmic behaviour (Dushay et al. 1989). They also display strong ultradian rhythms (Dowse et al. 1989). Blindness alone is insufficient to block normal rhythmicity, as the mutant *no receptor potential A* (*norpA*) blocks retinal light transmission everywhere, not just in the compound eyes (Pak 1975, 1979), yet permits normal rhythms entrainable to light cycles (Dushay et al. 1989). Similarly, the complete lack of eyes or optic lobes does not yield the *disc* result, since *small optic lobes* (*sol*) and *sine oculis* (*so*) mutations are very close to wild-type in their rhythmic behaviour (Helfrich and Engelmann 1983; Helfrich 1986; Dushay et al. 1989). It has been hypothesized that this effect is analogous to the destruction of circadian organization in mammals after

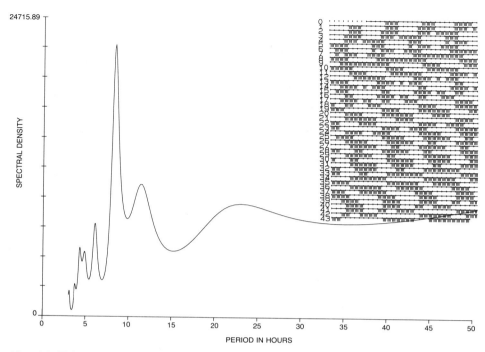

Fig. 6.1. This plot is a combination of maximum entropy spectral analysis (MESA) and an actogram (inset) for a wild-type *Drosophila melanogaster* adult free-running in constant bright illumination. The spectral analysis reveals strong periodicity in the ultradian range, at 8.4 h. This is evident in the actogram, which is plotted modulo 8 h rather than 24 h, the usual convention. For clarity, the data are plotted four times in the usual manner.

lesioning of the suprachiasmatic nucleus (SCN) of the hypothalamus (Rusak 1977; Rosenwasser and Adler 1986). Presumably there is a "master clock" in some region of the nervous system that is eliminated by the *disc* lesion (Liu et al. 1988).

Measurement of *per* expression in the photoreceptor cells of the eye is a convenient method of looking for rhythmicity at the cellular level. This can be accomplished by fusing a reporter gene, whose product is not found in the host (in this case, the *lacZ* gene from *Escherichia coli*, which encodes β-galactosidase) to the *per* promoter and transforming the germline DNA of the appropriate strain of *Drosophila* with this construct (Liu et al. 1988). In these *per-lacZ* transformants, whenever the cellular signal is present for *per* to turn on, the reporter gene will become active as well; staining the tissue for the reporter gene product reveals the location and extent of this gene activity. *per* activity assessed by this means remains strong and rhythmic in *disc* individuals, despite the lack of behavioural activity rhythms, favouring the possibility that the *disc* lesion removes a neural region critical for organizing locomotor activity, but not necessary for all rhythmicity per se. Further, through use of this cellular labelling technique, it has been found that certain "lateral neurons" of uncertain function near the neuropil of the optic lobe are missing in *disc* mutants (Siwicki et al. 1988). These are now being actively investigated to determine whether they constitute the putative SCN-like master pacemaker in the fly (J. Hall, personal communication).

Other environmental manipulation may also enrich the ultradian portion of the spectrum. The activity rhythms of *Drosophila* that had been administered deuterium oxide (^2H_2O) showed that treating flies with deuterium oxide via drinking water revealed increased power in the ultradian range; this was true for all the *per* alleles, including wild-type (White et al. 1991). Curiously, the effects on ultradian power and signal strength were not strictly dose dependent. Even the lowest concentration of ^2H_2O used in the experiment (10%) caused severe interference with normal circadian clock function. There was also a general loss of strength in the circadian signal as evidenced by a lowering of the signal-to-noise ratio (SNR). And, yet, the flies were not sick. Far from it: survival was higher at 10% and 20% ^2H_2O than at 0%. The ability to induce ultradian rhythmicity pharmacologically is a new and important tool for investigation of high frequency periodicity in this system.

Do Fundamental Oscillations Underlie Ultradian Rhythms in *Drosophila*?

The first general question to be addressed is whether or not ultradian behavioural rhythms are a result of fundamental oscillations of the same period or are produced by disturbed circadian clocks or perhaps even shorter cellular clocks. Ultradian periodicities are well documented in mammalian systems. For example, they occur naturally in the form of bouts of rapid eye movement (REM) sleep, replaced with periodicity in the state of awareness and psychomotor proficiency during waking periods (Webb and Dube 1981). In rats, the extent of ultradian rhythmicity in the activity record is under genetic control, with some strains showing very pronounced high frequency bouts of activity superimposed on the circadian rhythm (Büttner and Wollnik 1984; Wollnik et al. 1987; Schwartz and Zimmermann 1990). Ultradian rhythmicity can also be induced surgically. In test animals, the circadian system typically breaks down after SCN lesions, leaving only ultradian rhythms (Rusak 1977; Rosenwasser and Adler 1986). Surgery on the SCN may also affect ultradian rhythmicity in strains in which it occurs naturally, since SCN lesions abolish ultradian as well as circadian rhythms in rats (Wollnik and Turek 1989).

For this system, it has been proposed that these ultradian periods are a result of uncoupling multiple circadian oscillators (Rosenwasser and Adler 1986). Spectral analysis of the activity of mammals may indicate ultradian rhythms, but inspection of the raw data plotted in actograms indicates multiple circadian rhythms still at work. How could multiple oscillations interact to produce other, distinct rhythms? In a simple example, if two circadian clocks act in phase to produce a single peak, and subsequently uncouple, two things may happen. The two rhythms may remain uncoupled and drift apart. Assuming no destructive interference occurs, a sufficiently long record with a high resolution analysis would show both periodicities in the spectrum, neither ultradian. A spurious long period would also appear at the "beat" frequency, the point at which mutual reinforcement occurs. Beat frequencies must always be longer than both contributing cycles (Tyson et al. 1976). Alternatively, the two oscillators may recouple at 180° out of phase. Spectral analysis would show an "ultradian" 12-h peak, but the fundamental oscillations would be circadian. If three oscillators are at work, they might produce a trimodal pattern with an 8-h peak in the spectrum.

It is more difficult to explain ultradian rhythms in *Drosophila* as the result of interacting circadian rhythms. Actograms from individuals displaying ultradian rhythms, e.g. per^0, do not show multiple circadian components and peaks in spectra are not submultiples of 24 h. Continuing the example above to explain these results, assuming the coupling is not 180°, but perhaps 105°, then two intervals will occur during each 24-h period: 7 h and 17 h. But it does not follow that spectral analysis will result in two spectral lines at 7 and 17 h. Only half the number of peaks are present needed for 7-h and 17-h rhythms, as every other one is "missing". Practically speaking, when analysing *Drosophila* records with two activity peaks per *ca* 24-h day aligned as described above, routinely we see only a single spectral line at the circadian period (H. B. Dowse and J. M. Ringo unpublished results). To explain multiple periods resulting from interacting circadian periods thus requires "Ptolemaic reasoning", multiplying the number of components to ridiculous complexity. A 7-h period would need four 28-h clocks coupled 7-h out of phase, and there is no way to explain the 17-h period other than two 34-h clocks 180° antiphase. A far simpler explanation is that the periodicities in the record reflect the operation of oscillators of the same period. To demonstrate this conclusively will require elucidation of the state variables of the oscillators themselves, which remain as mysterious as the circadian clock itself. We are conducting research into the formal properties of these rhythms, and hope to determine whether they are temperature compensated, sensitive to deuteration, phase-entrainable by light, and phase-shiftable by light and other known zeitgebers.

We can exclude another explanation offered by Enright (1990), namely that these rhythms are short-lived phenomena that appear in the data record for just one or two cycles. These could either be short-lived transient ripples in a steady state, or truly spurious random variations devoid of any biological significance. However, when actograms from records rich in ultradian rhythms are plotted modulo 24 h – the normal procedure for circadian rhythms – ultradians are seldom obvious. If, however, ultradian periodicities are detected by spectral analysis, and the data are plotted modulo τ for a strong ultradian rhythm period, that ultradian period becomes obvious in the actogram (Dowse et al. 1987). Ultradian rhythms plotted in this way appear as strong and regular as circadian rhythms, and, critically, they continue unchanged in period or phase for the entire duration of some records (Fig. 6.2). This is scarcely what would be expected if only one or two cycles were present, as Enright (1990) has suggested. Observation of actograms also argues strongly against the interpretation that interacting circadian clocks are at the root of ultradian rhythms, since at no point do bona fide circadian rhythms appear in the record when the data are plotted modulo 24 h or modulo some appropriate ultradian period.

It should be noted that, if, as we argue below, the circadian clock itself is composed of coupled ultradian oscillators, it may also be that the observed ultradian rhythms in *Drosophila* behaviour records reflect interactions between

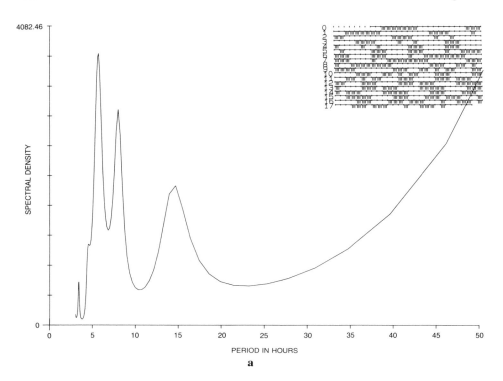

a

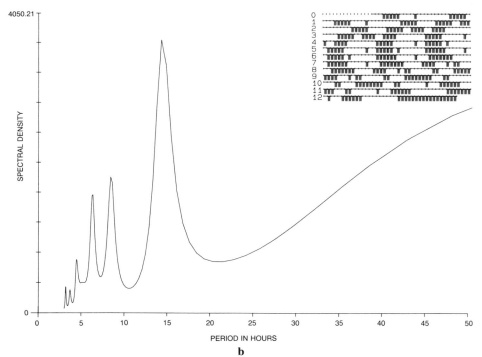

b

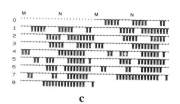

c

Fig. 6.2. Analysis of data from locomotor activity of per^0 flies reared in LD cycles and tested in DD. **a** This plot displays an actogram (inset) and MESA for a per^0 fly in DD that shows multiple peaks in the spectrum at 5.8 h and 8.0 h, as well as a slightly weaker one at 14.7 h. There is no obvious simple relationship among these periodicities. The actogram is plotted modulo 7 h, and reveals strong rhythmicity, but the complexity of the record is clear, as the various cycles all appear in the record. The actogram is quadruply plotted. **b** This per^0 fly, free-running in DD, displays a single peak at 14.3 h in its spectrum. The triply-plotted actogram, modulo 14 h, shows a clean, long-lived series of activity peaks throughout the entire record, rather than just a few transient activity peaks. Such records are typical of this genotype. **c** When the same data analysed for **b** are plotted modulo 24 h, the most common technique in searching for periodicity in activity data, no obvious ultradian periodicity is evident. This explains, in part, the failure to discover ultradian periodicity in these flies for more than a decade.

subpopulations of even higher-frequency oscillators. This would account for the high variability that we have seen in the periods of these rhythms. There is no favoured ultradian period in *Drosophila* – values are spread fairly uniformly through the 4-h to 18-h region of the spectrum. This leaves open the question of whether some very high frequency quantal oscillators may yet be found that can account for both the longer observed ultradian periods and the circadian clock itself.

A further question needs to be answered: what is the cellular location for these ultradian oscillations (or the quantal oscillator mentioned above)? It must be stressed again here that we are talking about behavioural rhythms. It is therefore reasonable to suppose that these rhythms either originate in the nervous system or are modulated neurally. As has been noted, *disc* putatively eliminates the portion of the central nervous system (CNS) required by the circadian clock for its normal function, leaving principally ultradian oscillations (Siwicki et al. 1988; Dowse et al. 1989). We ask whether only the expression of the circadian rhythm is affected, or whether the master pacemaker for behavioural rhythms is deleted. As has been noted, there is still residual pacemaker activity in other cells (Siwicki et al. 1988). We believe that every cell in a metazoan has a competent pacemaker, and that the role of structures such as the SCN and the cells deleted by *disc* is to coordinate activity globally (Moore-Ede and Sulzman 1981). Where then do the ultradian rhythms come from? How might cellular ultradian rhythms be coupled to activity at the level of complex locomotor behaviour? We can only speculate, but we propose at this point, that the quantal oscillators alluded to above may couple randomly to form overt ultradian composites at some point either upstream or downstream from the missing master circadian control centre on the path to initiation of locomotor activity. This would also explain ultradian rhythmicity in animals that have had their circadian clocks disrupted by constant light, DD-rearing, and lesions of the *per* locus. Thus it may be that only certain metazoan cells are constituted properly to assemble quantal oscillators into a true circadian clock, while the ultradian clocks are sufficient for the intracellular housekeeping needs. This might account for the fact that the product of *per*, Per, is found in only some cells. Cells of *Drosophila* that contain Per apparently process 24-h information; Per may be coupling ultradian rhythms intercellularly to produce a 24-h rhythm. A unicell would, as a matter of course, have a complete system for maintaining a 24-h clock using an ensemble of intracellular oscillators.

Do Ultradian Rhythms Have a Function?

It is next appropriate to ask what purpose these ultradian rhythms (oscillators?) serve in *Drosophila*. We are unaware of any relevant geophysical correlate; further, the overt rhythms are characterized by their wide variability (Dowse et al. 1987), making a search for a single phase-locked environmental variable fruitless. Behaviourally, a 24-h timer is justifiable, constraining eclosion and adult activity to the cool moist times near dawn and dusk, since *Drosophila* are highly susceptible to desiccation (Pittendrigh 1958). It seems unlikely that organization of *Drosophila* activity into discrete short bouts could have the same adaptive value as has been postulated for vertebrates (Daan 1981) when there is such a high interindividual variation both in period and in level of expression.

If ultradian rhythms lack an adaptive function in *Drosophila*, and yet if they are real (not "random noise"), then they must be by-products of the organizing timekeeping system, which afford a window into the physiological processes of this system. Bünning (1964) hypothesized long ago that maintenance of internal temporal order was a primary duty of any cellular clock. Mutually incompatible cellular processes that cannot be separated in space must be separated in time. It is necessary for products of the cell's metabolism to be at the right place at the right time. Oatley and Goodwin (1971) also noted that oscillations around a steady state are inevitable, and that these oscillations may be acted upon by natural selection to produce any period of use. We propose, then, that the multiple ultradian oscillations we see in *Drosophila* are manifestations of the internal cellular-housekeeping clocks. As argued above, their appearance in the activity record is an artefact of disrupted intracellular organization in the region of the nervous system responsible for delivering circadian timing information to the upper behavioural centres in the CNS. If the ultimate quantal clock could be discovered, it would presumably have a much shorter period, and would be found in every cell of the body. A cell lacking this sort of timekeeper would die.

For unicells, one specific cellular process that has been linked to ultradian oscillations is the cell cycle (see Lloyd, Chap. 1). Extensive evidence has been accumulated on long-lived, temperature-compensated oscillations in a number of unicells, (Lloyd and Edwards 1984). The cell cycle in these organisms is temperature dependent, and thus seems not to be a candidate for putative control by these higher-frequency oscillators. But if they are quantal in nature, like building blocks, they might be added or subtracted to lengthen or shorten the period of the controlled cycle as the temperature changes, and the discrepancy would be resolved (Lloyd and Edwards 1987; Lloyd and Kippert 1987). Using the same logic, Lloyd and Edwards (1987) have extended this argument to the circadian clock. Klevecz et al. (1984) have pursued a similar line of reasoning for the mammalian cell cycle.

Is There a Relationship Between Ultradian and Circadian Rhythms?

The final question to be addressed concerns the relationship between the hypothetical quantal oscillators, the observed behavioural ultradian rhythms, and the circadian system. We have argued at length (Dowse and Ringo 1992) that the circadian clock is actually a "meta-oscillator", i.e. there is no time-keeping mechanism with a genuine 24-h escapement. We have hypothesized (Dowse et al. 1987; Dowse and Ringo 1992), in company with many other investigators (e.g. Pavlidis 1971; Winfree 1980; Klevecz et al. 1984; Lloyd and Edwards 1987), that the "clock" is composed of many higher-frequency escapements. The time constants of these high-frequency components are closer to those of metabolic processes than to 24 h, which is two to three orders of magnitude slower (Lloyd and Edwards 1987; Edmunds 1988). Similarly, the phenomena of temperature compensation, precision and resistance to pharmacological meddling may be easier to explain in populations of coupled oscillators. We further hypothesized that *per* in *Drosophila* and *per*-like analogues in other organisms are required for assembly of these components into the circadian system. We briefly summarize this argument, but the reader is referred to the review by Dowse and Ringo (1992) for full coverage, particularly of the molecular evidence.

We postulate a population of oscillators that we have dubbed "U-oscillators" (Ur, Ultradian, or Uhr). These are coupled by weak, non-linear, inhibitory phasic links. The strength of this coupling is of primary importance to the countenance of the resulting clock. Normally, the coupled system has a 24-h output. This will appear in virtually every aspect of the organism's metabolism at all levels. If coupling is overly robust, the period will shorten, and vice versa. We maintain that this is evident in the *per* allelic series, and have circumstantial evidence to support the claim. We looked at the proportion of ultradian rhythmicity in flies homozygous for each of four alleles of *per*: wild-type, *Long*, *short*, and *null*. We found that the *null* allele has rhythms almost entirely in the ultradian range, per^L has many ultradians, per^+ has a few ultradians, and per^s has virtually no ultradian periods in the spectrum. We thus find that period relates to the amount of ultradian power in a simple way. We reasoned that "tightness of coupling" would make itself manifest in the signal strength of the circadian rhythm itself. A poorly coupled meta-oscillator should have multiple "freelance" peaks in the ultradian range of the spectrum. We quantified this as the SNR, calculated using a waveform-independent technique of our own devising (Dowse and Ringo 1989a). We were thus able to look at correlation between putative coupling strength and period, and found a log-linear dependence of frequency on SNR. We predicted from the regression equation that, if per^0 were to have a circadian rhythm at all, it would appear in the 35- to 40-h range. This was confirmed dramatically by the analysis of a new per^0 allele, per^{04}. This is a hypomorphic mutation, in contrast to the original amorph, per^{01} (there are no per^{02} or per^{03} alleles). Periods of approximately 40 h were uncovered with a spectral analysis technique specifically set up to look for such long cycles (Hamblen-Coyle et al. 1989).

Evidence of a purely theoretical nature supporting the meta-oscillator hypothesis has also recently appeared. Klevecz et al. (1991) have devised a computer model of the metazoan clock as a population of chaotic (Rössler) attractors (see Chap. 3). When a state variable of the system is coupled throughout a large population of individual attractors, the system enters a periodic modality in which spirals or concentric rings appear. The strength of the coupling can be varied, and period depends critically on the coupling constant, which may be seen as analogous to coupling by *per*. The hypothesis is, then, that there are individual ultradian oscillators of uncertain period in each cell of an appropriate tissue, and that intercellular coupling, facilitated by *per*, organizes the population into an accurate clock.

We also hypothesized (Dowse and Ringo 1989b) that in the *Drosophila* system light or some other phasically active stimulus is required for proper initialization of the circadian clock. DD-reared animals display a wide range of periodicities in their activity, as has been noted above, and random and partial assembly of subunits into various levels of organization is the only currently tenable hypothesis for the wide range of results observed. Furthermore, we find parallels between a lack of circadian beahvioural rhythms in DD-reared flies and activity of the *per* gene at the cellular level (L. Wood, unpublished results). We have used a β-galactosidase reporter gene fused to the *per* promoter to monitor the activity of the *per* gene in wild-type animals raised in both DD and LD, and tested in DD. *per* is expressed rhythmically in the CNS (and a few other regions) in wild-type animals raised in LD and tested in DD. In the DD-reared stocks that are never exposed to light, we find that the cellular rhythms disappear. Crucially, *per* activity is low at all times of the objective day, compared with a robust periodicity in the LD-reared animals.

The meta-oscillator hypothesis is consistent with all the evidence. It suffers from the same problem as all other clock hypotheses, however, in that no underlying mechanisms are known. Forty years or more of serious work on the circadian clock per se has failed to yield any state variable of an escapement. This may be a symptom of the fact that there is no 24-h escapement. Every cellular process that varies rhythmically behaves like a "hand" as opposed to a "gear" of the clock. This would be expected if each oscillator were only a part of a large whole, and, when that process is interfered with in some way, the balance of the oscillators is sufficient to maintain near-normal circadian timing. The net result would be the appearance of a hand.

This argument must be pushed down to the level of the ultradian rhythms that we observe in *Drosophila*. The wide variability of the periods is a strong argument for thinking them to be meta-oscillators in their own right. The ultimate source of the timing could be very high frequency oscillations, perhaps in the range 2–4 h. Looking for evidence of such oscillators and elucidating their molecular mechanisms seems to us to offer the best chance of unravelling the 24-h clock.

Summary

Ultradian rhythms (having periods in the range 4–18 h) have been observed in the locomotor activity of *Drosophila melanogaster*. These ultradian activity rhythms are especially prominent when strong circadian rhythms are absent, which is characteristic of two unrelated clock mutants *per*[0] and *disc*, as well as of flies that have never been exposed to light (DD-reared flies). Ultradian activity rhythms occur also in flies with normal circadian rhythms, e.g. LD-reared, wild-type flies tested in DD; in these normal flies, though, ultradian rhythms are usually weak and are present only in some individuals. In general, the strength of the ultradian rhythms of individuals varies inversely with the strength of their circadian rhythms. In many of the genotypes on which we have reported thus far, the periods of these rhythms are distributed evenly throughout the range 4–18 h. Often, they are strong and long lived, but, since no geophysical correlate is evident, their function in regulating behaviour is a puzzle. We hypothesize that these rhythms reflect interactions among yet higher-frequency cellular oscillators, and that these oscillators underlie the circadian timing system as well. We present circumstantial evidence favouring this hypothesis.

Acknowledgements

This work was made possible, in part, by grant GEN 5 R01 NS26412 from the National Institutes of Health (to H. B. D.).

References

Bargiello TA, Young MW (1984) Molecular genetics of a biological clock in *Drosophila*. Proc Natl Acad Sci USA 81:2142–2146

Beauchamp K, Yuen C (1979) Digital methods for signal analysis. George, Allen, Unwin, Boston

Bünning E (1964) The physiological clock. Academic Press, New York

Büttner D, Wollnik F (1984) Strain differentiated circadian and ultradian rhythms in locomotor activity of the laboratory rat. Behav Genet 14:137–151

Chatfield C (1989) The analysis of time series. Chapman and Hall, London

Daan S (1981) Adaptive daily strategies in behavior. In: Aschoff J (ed) Biological rhythms. Plenum Press, New York, pp 275–298 (Handbook of behavioral neurobiology, vol 4)

Daan S, Aschoff J (1981) Short-term rhythms in activity. In: Aschoff J (ed) Biological rhythms. Plenum Press, New York, pp 491–498 (Handbook of behavioral neurobiology, vol 4)

Dowse H, Palmer J (1990) Evidence for ultradian rhythmicity in an intertidal crab. In: Hayes DK, Pauly JP, Reiter RJ (eds) Chronobiology: its role in clinical medicine, general biology, and agriculture, pt B. Wiley–Liss, New York, pp 691–697

Dowse H, Palmer J (1992) Comparative studies of tidal rhythms. XI. Ultradian and circalunidian rhythmicity in four species of semiterrestrial, intertidal crabs. Mar Behav Physiol (in press)

Dowse H, Ringo J (1987) Further evidence that the circadian clock in *Drosophila* is a population of coupled ultradian oscillators. J Biol Rhythms 2:65–76

Dowse H, Ringo J (1989a) The search for hidden periodicities in biological time series revisited. J Theor Biol 139:65–76

Dowse H, Ringo J (1989b) Rearing *Drosophila* in constant darkness produces phenocopies of *period* circadian clock mutants. Physiol Zool 62:785–803

Dowse H, Ringo J (1991) Comparisons between "periodograms" and spectral analysis: apples are apples after all. J Theor Biol 148:139–144

Dowse H, Ringo J (1992) Is the circadian clock a "metaoscillator?" Evidence from studies of ultradian rhythms in *Drosophila*. In: Young M (ed) Molecular approaches to circadian clocks. Marcel Dekker, New York, pp 195–220

Dowse H, Hall JC, Ringo J (1987) Circadian and ultradian rhythms in *period* mutants of *Drosophila melanogaster*. Behav Genet 17:19–35

Dowse H, Dushay M, Hall JC, Ringo J (1989) High-resolution analysis of locomotor activity rhythms in *disconnected*, a visual system mutant of *Drosophila melanogaster*. Behav Genet 19:529–542

Dushay MS, Rosbash M, Hall JC (1989) The *disconnected* visual system mutations of *Drosophila melanogaster* drastically disrupt circadian rhythms. J Biol Rhythms 4:1–27

Edmunds LN Jr (1988) Cellular and molecular bases of biological clocks. Springer, Berlin Heidelberg New York

Enright JT (1965) The search for rhythmicity in biological time-series. J Theor Biol 8:662–666

Enright JT (1990) A comparison of periodograms and spectral analysis: don't expect apples to taste like oranges. J Theor Biol 143:425–430

Hall JC, Kyriacou CP (1990) Genetics of biological rhythms in *Drosophila*. Adv Insect Physiol 22:221–298

Hamblen-Coyle M, Konopka RJ, Zwiebel LJ, Colot HV, Dowse HB, Rosbash M, Hall JC (1989) A new mutation at the *period* locus of *Drosophila melanogaster* with some novel effects on circadian rhythms. J Neurogenet 5:229–256

Helfrich C (1986) Role of the optic lobes in the regulation of the locomotor activity rhythm of *Drosophila melanogaster*. Behavioral analysis of neural mutants. J Neurogenet 3:321–343

Helfrich C, Engelmann W (1983) Circadian rhythm of the locomotor activity in *Drosophila melanogaster* mutants sine oculis and small optic lobes. Physiol Entomol 8:257–272

Klevecz RR, Kaufmann SA, Shymko RM (1984) Cellular clocks and oscillators. Int Rev Cytol 5:97–126

Klevecz RR, Pilliod J, Bolen J (1991) Autogenous formation of spiral waves by coupled chaotic attractors. Chronobiol Int 8:6–13

Konopka RJ, Benzer S (1971) Clock mutants of *Drosophila melanogaster*. Proc Natl Acad Sci USA 68:2112–2116

Liu X, Lorenz L, Yu Q, Hall JC, Rosbash M (1988) Spatial and temporal expression of the *period* gene in *Drosophila melanogaster*. Genes Dev 2:228–238

Lloyd D, Edwards S (1984) Epigenetic oscillations during the cell cycles of lower eukaryotes are coupled to a clock. In: Edmunds LN Jr (ed) Cell cycle clocks. Marcel Dekker, New York, pp 27–46

Lloyd D, Edwards S (1987) Temperature-compensated ultradian rhythms in lower eukaryotes: timers for cell cycles and circadian events? In: Pauly JE, Scheving LE (eds) Advances in chronobiology, pt A, Alan R Liss, New York, pp 131–151

Lloyd D, Kippert F (1987) A temperature-compensated ultradian clock explains temperature-dependent quantal cell cycle times. In: Bowler K, Fuller BJ (eds) Temperature and animal cells. Cambridge University Press, Cambridge, pp 135–155 (Symp Soc Exp Biol 41)

Lloyd D, Edwards SW, Fry JC (1982) Temperature compensated oscillations in respiration and cellular protein content in synchronous cultures of *Acanthamoeba castellanii*. Proc Natl Acad Sci USA 79:3785–3788

Michel U, Hardeland R (1985) On the chronobiology of *Tetrahymena*. III. Temperature compensation and temperature dependence in the ultradian oscillation of tyrosine aminotransferase. J Interdiscipl Cycle Res 16:17–23

Moore-Ede M, Sulzman F (1981) Internal temporal order. In: Aschoff J (ed) Biological rhythms. Plenum Press, New York, pp 215–241 (Handbook of behavioral neurobiology, vol 4)

Oatley K, Goodwin BC (1971) The explanation and investigation of biological rhythms. In: Colquhoun WP (ed) Biological rhythms and human performance. Academic Press, New York, pp 1–38

Pak WL (1975) Mutants affecting the vision of *Drosophila melanogaster*. In: King RC (ed) Handbook of genetics, vol 3. Plenum Press, New York, pp 703–733

Pak WL (1979) Study of photoreceptor function using *Drosophila* mutants. In: Breakfield XO (ed) Neurogenetics: genetic approaches to the nervous system. Elsevier/North Holland, New York, pp 67–79

Pavlidis T (1971) Populations of biochemical oscillators as circadian clocks. J Theor Biol 33:319–338

Pittendrigh CS (1958) Adaptation, natural selection, and behavior. In: Roe A, Simpson GG (eds) Behavior and evolution. Yale University Press, New Haven, CN, pp 390–416

Reddy P, Zehring WA, Wheeler DA, Pirotta V, Hadfield C, Hall JC, Rosbash M (1984) Molecular analysis of the *period* locus in *D. melanogaster* and identification of a transcript involved in biological rhythms. Cell 38:701–710

Rosenwasser A, Adler N (1986) Structure and function in circadian timing systems: evidence for multiple coupled circadian oscillators. Neurosci Biobehav Rev 10:413–448

Rusak B (1977) The role of the suprachiasmatic nucleus in the generation of circadian rhythms in the golden hamster, *Mesocricetus auratus*. J Comp Physiol 118:145–164

Schwartz W, Zimmermann P (1990) Circadian timekeeping in BALB/c and C57BL/6 inbred mouse strains. J Neurosci 10:3685–3694

Siwicki K, Eastman C, Petersen G, Rosbash M, Hall J (1988) Antibodies to the *period* gene product of *Drosophila* reveal diverse tissue distributions and rhythmic changes in the visual system. Neuron 1:141–150

Smith RF, Konopka RJ (1981) Circadian clock phenotypes of chromosome aberration with a breakpoint at the *per* locus. Mol Gen Genet 185:243–251

Steller H, Fischbach KF, Rubin G (1987) *disconnected*: a locus required for neuronal pathway formation in the visual system of *Drosophila*. Cell 50:1139–1153

Tyson JJ, Alivisatos SGA, Richter O, Grün F, Schneider FW, Pavlidis T (1976) Mathematical background report. In: Hastings JW, Schweiger H-G (eds) The molecular basis of circadian rhythms. Dahlem Konferenzen, Berlin, pp 85–108

Webb W, Dube M (1981) Temporal characteristics of sleep. In: Aschoff J (ed) Biological rhythms. Plenum Press, New York, pp 499–522 (Handbook of behavioral neurobiology, vol 4)

White L, Ringo J, Dowse H (1991) A circadian clock of *Drosophila*: effects of deuterium oxide and mutations at the *period* locus. Chronobiol Int (in press)

Winfree AT (1980) The geometry of biological time. Springer, Berlin Heidelberg New York

Wollnik F, Turek F (1989) SCN lesions abolish ultradian and circadian components of activity rhythms in LEW/Ztm rats. Am J Physiol 25:R1027–R1039

Wollnik F, Gârtner K, Büttner D (1987) Genetic analysis of circadian and ultradian locomotor activity rhythms in laboratory rats. Behav Genet 17:167–178

Part II

The Neuroendocrinal and Developmental Level

It is the pattern maintained by this homeostasis, which is the touchstone of our personal identity. Our tissues change as we live: the food we eat and the air we breath become flesh of our flesh and bone of our bone, and the momentary elements of our flesh and bone pass out of our body every day with our excreta. We are but whirlpools in a river of ever-flowing water. We are not the stuff that abides, but patterns that perpetuate themselves.

A pattern is a message, and may be transmitted as a message . . .

Norbert Wiener, The human use of human beings

The rapid progress in new assessment methods for measuring ultradian rhythms in endocrinology has caught writers of medical textbooks short, as well as those in the popular press. The medical profession and layman have been made well aware of the significance of circadian rhythms in daily life but to most the word "ultradian" still has a strange ring. Most people are incredulous, for example, when they are told that the various periods throughout the day when they feel tired implies something more than personal inadequacy. The task of the middle section of this volume, therefore, is to build a bridge between the molecular, genetic and cellular levels covered in the previous section to the behavioural and psychosocial levels of the next section.

We have long known that the word "neuroendocrinal" implies a close relationship between the neural and hormonal levels of control in the organism. Only recently, however, have we come to understand how molecular mechanisms within single cells of the hypothalamus of the brain (and a few other places throughout the body) can transduce information between neural and the endocrinological levels. From an informational perspective it is now clear that the neural and endocrine systems are one. The current unification of the neural, endocrine and immune systems into one cybernetic network on the molecular level is an astonishing validation of the scientific intuition of the unity of mind and body by theorists of previous generations.

The original research reports by Brandenberger (Chap. 7) and Veldhuis (Chap. 8) from two independent laboratories on opposite sides of the Atlantic are therefore of crucial significance in documenting the central role of current neuroendocrinology in this new awareness. The data they present integrate endocrinology and chronobiology with new levels of mathematical precision that will be of central importance for any future theory of psychobiology. The pulsate release of most hormones in ultradian rhythms is the connecting link between the molecular–genetic–cellular levels that can be manipulated in the laboratory and the psychosocial levels that are best studied in naturalistic field studies of human behaviour.

Chapter 9 by Toke Hoppenbrouwers, on the ontogenesis of ultradian rhythms in humans, is an excellent example of how a balanced approach between laboratory control and field observations can provide a well-rounded understanding of chronobiological variables in human development. Her appreciation and utilization of non-linear mathematical methods for conceptualizing unique developmental events such as sudden infant death syndrome are a practical clinical realization of the new dynamics of chronobiology. These new dynamics are lucidly illustrated by Garfinkel and Abraham (Chap. 10) writing on phase plots of temporal oscillations. The generations of chronobiologists who have grown myopic in their efforts to comprehend the intricate complexities of their graphs will find joy in the brilliant clarity of Garfinkel and Abraham's work. The way they replot the traditional raster graph of chronobiology with a more simple model that immediately sharpens our perception of the dynamics of desynchronization and entrainment leaves no doubt that this is the way of the future in understanding complex behaviour.

It is therefore entirely appropriate that we conclude this section with Maurice Stupfel's search (Chap. 11) for a unifying paradigm in the complex behaviour of ultradian rhythms in metabolism, endocrinology and behaviour. Stupfel's wide-ranging perspective gained from his lifetime of devoted chronobiological research surveys the entire panorama of scientific principles, from the thermodynamic to the psychosocial. His contributions, together with those of other pioneers such as Nathaniel Kleitman and Rütger Wever, represented in Part III, will enable us to take another step in search of a unifying paradigm in chronobiology.

Chapter 7

Endocrine Ultradian Rhythms During Sleep and Wakefulness

G. Brandenberger

Episodic hormone secretion appears to be a common characteristic of several endocrine systems, including the hypothalamic–pituitary axis, adrenal glands, gonads and the pancreas. Alterations in episodic secretion are associated with the pathophysiology of certain diseases. Correction of these pulsatile disorders with exogenous hormonal replacement therapy frequently cures the defect, the most notable examples of this phenomenon being the hypogonadotrophic disorders. While there is substantial evidence indicating that the central nervous system is at the origin of hormone pulsatility, the organization of the corresponding pacemaker structures remains to be elucidated. In particular, the question of unity versus multiplicity of pacemakers responsible for the ultradian rhythms in the 80–120-min range is debated (Lavie and Kripke 1981). In humans, included in this range is the time of recurrence of the episodic pulses of many hormones as well as a number of rhythms of apparently unrelated physiological and behavioural processes, such as the rate of urine flow, gastric motility and cognitive variables. One of the most prominent ultradian rhythms is the alternation of rapid eye movement (REM) and non(N)REM stages of sleep, which is accompanied by similar cycles of dreaming, penile erection and cardiac variability. Kleitman proposed the concept of a basic rest–activity cycle (BRAC), suggesting that the periodic recurrence of REM sleep reflects a fundamental physiological periodicity which also modulates brain functions during wakefulness (Kleitman 1963, 1982).

Hormone release results from neuronal messages reflecting circadian mechanisms which interact with the episodic secretion of a series of releasing factors. The hormone pulses can be endogenously generated by a variety of processes, or alternatively they may reflect external, environmental and sleep-related influences. The purpose of this chapter is to review current knowledge of the regulation of episodic hormone release in relation to sleep and to the intrinsic circadian and ultradian rhythmicities which interact in producing the 24-h secretory profiles. More precisely I focus on the patterns of hormone release during two distinct intervals of the 24-h period: (a) the sleep–wake cycle; (b) the REM–NREM sleep cycle.

Several experimental protocols have been used in an attempt to differentiate between the endogenous and the sleep-related influences on the hormone profiles. These include acute or repeated shifts in the timing of the sleep–wake cycle; partial, selective and total sleep deprivation; and pharmacological treatments which alter either sleep or the hormone secretion. Alternatively, certain pathological cases have provided models to see whether the relation between sleep and hormones persists when either sleep or hormones are altered.

Objective comparison between the sleep stages and the frequency or the magnitude of the hormone pulses gives rise to methodological problems, since the circulating hormone levels have to relate to rapid changes in central nervous system activity. First, it would be optimal to obtain peripheral blood samples using a time-interval that permits the detection of all major secretory episodes. Merriam and Wachter (1984) have proposed a guideline consisting of sampling five to six times during the expected interval between peaks. For most of the hormones, blood sampling at 10-min intervals enables the major secretory episodes to be detected and gives information concerning the circadian pattern of hormone secretion. The pulses can then be characterized using one of the variety of pulse detection programmes available which takes into account the detection limit of the analytical procedure and the variable precision of the assay for the different concentration ranges. For each significant pulse identified, the ascending and the declining phases, the total duration and the time of their maximum are identified. A recent study has compared the various programmes for pulse identification and has evaluated the performances of each of them (Urban et al. 1988). Following pulse characterization, the relation between the different phases of the hormone pulses and the dominant sleep stage can be analysed.

Pulsatile Hormone Release in Relation to the Sleep–Wake Cycle

Hormones Strongly Influenced by Sleep: Prolactin and Thyrotrophin

Plasma prolactin and thyrotrophin (TSH) levels vary widely across the day, and pulsatile release has been identified for both hormones. Under basal conditions, the 24-h prolactin profiles show low concentrations during the daytime, with meal-related peaks (Quigley et al. 1981), and high levels after sleep onset. The episodic pulses of prolactin probably result from intermittent hypothalamic inhibition for which dopamine has been identified as one of the major inhibitory factors (MacLeod 1976), as the nocturnal rise is abolished by dopaminergic agents such as L-dopa and bromocriptine (Frantz 1979). TSH secretion has been reported to have a circadian rhythm, with a steep increase in the evening, before the onset of sleep (Patel et al. 1972; Vanhaelst et al. 1972; Weeke and Gundersen 1978). Besides the circadian variation, small short-term fluctuations have also been observed, with no regularity in their occurrence (Brabant et al. 1986; Greenspan et al. 1986). Both pulsatile and circadian TSH release is predominantly controlled by thyrotrophin-releasing hormone (Brabant et al. 1991).

Sleep exerts an important modulatory role on the organization of prolactin and TSH pulses. Studies of the 24-h prolactin profiles following a shift in the

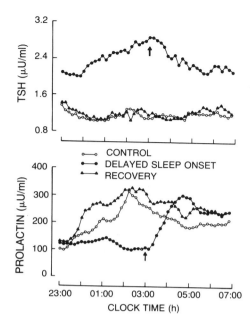

Fig. 7.1. Effect of delayed sleep onset on the mean nocturnal levels of TSH and prolactin.

normal sleep period have consistently demonstrated increased prolactin secretion associated with sleep (Sassin et al. 1973). The strength of this association has led many investigators to describe the 24-h rhythm of plasma prolactin as being entirely dependent on sleep. However, more recent studies have attenuated this concept and have indicated the existence of an additional circadian influence on prolactin release. For example, studies on the effects of jet lag have shown that changes in prolactin secretion occur in temporal association with anticipated, rather than actual, time of sleep (Désir et al. 1982).

Similarly, sleep strongly influences the circadian TSH secretory pattern. Unlike prolactin, sleep withdrawal augments TSH secretion and an inhibitory effect of sleep on TSH release has repeatedly been suggested (Parker et al. 1987; Brabant et al. 1990). Figure 7.1 shows the opposing effect of delayed sleep onset on the nocturnal prolactin and TSH profiles in the same subjects. On the control nights and the recovery nights, the lights were turned off at 23:00 h, enabling the subjects to fall asleep but, on the experimental night, the subjects were kept awake until 03:00 h. Delaying sleep onset had opposite effects on prolactin and TSH: prolactin levels became depressed and TSH levels rose. This opposite response contrasts with the close association described for these hormones in other situations (Delitala et al. 1987).

Hormones Poorly Influenced by Sleep

The Pituitary–Adrenal Axis and Melatonin

The profiles of cortisol and adrenocorticotrophic hormone (ACTH), considered to represent the best examples of endogenous circadian rhythms, differ

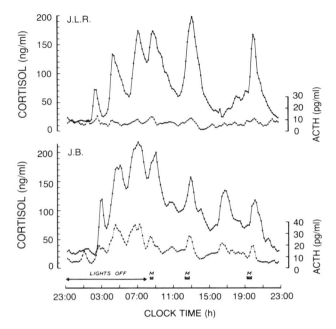

Fig. 7.2. Individual 24-h plasma ACTH (*dotted*) and cortisol profiles from samples collected at 10-min intervals. M indicates meal intake.

from the sleep-related rhythms of prolactin and TSH. They arise from a succession of secretory pulses rather than from continuous release with super-imposed secretory episodes (Van Cauter and Honinckx 1985; Veldhuis et al. 1991). The spontaneous cortisol peaks are preceded by increases in ACTH, although they are not quantitatively linked throughout the 24-h period. Power spectra analysis of individual ACTH and cortisol profiles gives evidence of a predominant periodicity between the oscillations for both hormones. These periodicities vary between individuals, ranging from 50–140 min for ACTH and 90–180 min for cortisol, indicating that on occasion a single cortisol peak may be initiated by two ACTH peaks (Follenius et al. 1987). In normal conditions, the acrophase of the pituitary–adrenal rhythm occurs between 07:00 and 10:00 h and the quiescent periods last from 21:00 to 02:00 h. During the daytime, in addition to apparent random fluctuations, there are two peaks related to the main meals (Fig. 7.2): a large peak that coincides with the midday meal and a smaller peak following the evening meal with a large inter- and intra-individual variability (Quigley and Yen 1979; Follenius et al. 1982).

The temporal organization of the secretory episodes within the 24-h period is relatively independent of sleep, since it is unaltered by short-term manipulations such as sleep reversal, sleep deprivation, and abrupt shifts of the sleep period. However, some studies have provided evidence that sleep slightly modifies the cortisol profile (Weitzman et al. 1983; Born et al. 1988). Results from studies in depressed patients (Linkowski et al. 1985) and after transmeridian flights (Désir et al. 1981) indicate that cortisol secretion is under multifactorial control and that the evening nadir and the morning acrophase are synchronized by different mechanisms (Van Cauter and Refetoff 1985).

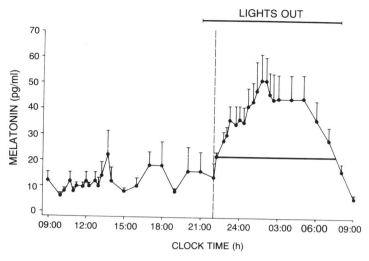

Fig. 7.3. Mean melatonin levels of samples collected at 20- or 60-min intervals from eight male volunteers over 24 h. (Reproduced, with permission, from Waldhauser and Steger 1987.)

Melatonin is a pineal gland hormone that may be involved in such different areas as human sexual maturation, circadian time keeping and sleep. An endogenous circadian rhythm, with low values during the daytime and high values during the night-time has been clearly established (Fig. 7.3; Claustrat et al. 1986; Waldhauser and Steger 1987; Rivest et al. 1989), and is highly reproducible within the same individual. Attempts to assess its pulsatile secretion have provided inconsistent results, the suggested pulse frequency varies from 5 to 6 pulses/h, 1 pulse every 60–80 min, and 4–5 pulses during the night. In a recent study no clear pulses were revealed by an objective pulse detection programme and it was suggested that the pineal is neither under a hypothalamic pulse generator which provokes the pulsatile release of many pituitary hormones nor under an intrinsic pulse generator located in the human pineal (Trinchard-Lugan and Waldhauser 1989).

Very little is known of the effect of sleep on melatonin secretion. Birkeland (1982) found low melatonin values associated with REM sleep but this relationship was not confirmed. In psychiatric research, a variety of abnormalities in the circadian rhythm of melatonin have been described. There is some evidence for a decline in the amplitude of the circadian rhythm in depression, but not all studies are consistent. Phase-delayed and phase-advanced rhythms have been described, and some authors have proposed appropriate rhythm-shifting treatments, particularly using bright light. This approach is of considerable interest but remains controversial at the present time (Arendt 1989).

The Gonadotrophin Axis: Testosterone and Luteinizing Hormone

Considerable data regarding gonadotrophins and testosterone rhythms have been collected and the importance of their ultradian pulsatile release has been clearly demonstrated (Knobil and Hotchkiss 1985). Pulse disturbances leading to a down-regulation of the pituitary–gonadal system can affect fertility, so that assessment of the pulse characteristics in normal and infertile men provides the basis for a therapeutic approach to infertility. Marked variations between individuals have been described. The presence of circadian variations in gonadotrophins and testosterone seems to be generally accepted; however, this finding has not been uniformly reported (Spratt et al. 1988) and any effect of the sleep–wake cycle has yet to be clearly demonstrated. Most authors have reported that testosterone levels are not directly influenced by sleep and are not related to any of the sleep stages (Judd et al. 1974; Miyatake et al. 1980). For luteinizing hormone (LH), studies have provided different results, depending on whether they were conducted in pubertal or prepubertal children, or in adult men or women (Boyar et al. 1972; Kapen et al. 1974, 1976). The influence of sleep on LH levels as well as the 24-h profiles seems to depend on the various influences of ultradian, circadian and infradian (menstrual, seasonal) rhythms which interact continually to assure the maturation of the reproductive system.

Pulsatile Hormone Release in Relation to Internal Sleep Structure

REM–NREM Sleep Cycles: The Renin–Angiotensin–Aldosterone System

The renin–angiotensin system is a complicated hormonal system regulating blood pressure, blood volume and sodium retention. One of the key substances in this system is the enzyme renin, a protease secreted by the juxtaglomerular cells of the kidney. Circulating renin release cleaves its substrate, angiotensinogen, to generate the decapeptide angiotensin I, which in turn is converted in angiotensin II by a converting enzyme. Angiotensin II is a potent vasoconstrictor that is required for the regulation of blood pressure. Plasma renin activity (PRA) is currently measured as an index of renin release.

Earlier studies in normal men have consistently demonstrated a circadian rhythm of PRA, culminating in the early morning (Modlinger et al. 1976; Bartter et al. 1979). However, simultaneous monitoring of PRA levels and the changes in the sleep stages gave evidence of a strong relation between PRA levels and the sleep cycles (Mullen et al. 1980; Brandenberger et al. 1985). NREM sleep is invariably linked to increasing PRA levels, and declining levels are observed when sleep becomes lighter; spontaneous and provoked awakenings blunt the rise in PRA normally associated with NREM sleep. So, PRA curves exactly reflect the pattern of sleep stage distribution. When the sleep cycles are regular, PRA levels oscillate at a regular ultradian periodicity and for incomplete sleep cycles, PRA curves reflect all irregularities in the sleep structure (Brandenberger et al. 1988). It does not appear that this association can be broken. In normal men, modifying the renal renin content only modulates the amplitude of the nocturnal oscillations without disturbing their relation to sleep stages (Brandenberger et al. 1990), and in the case of sleep disorders, such as narcolepsy or sleep apnea, the renin profiles reflect all disturbances of the internal sleep structure (Follenius et al. 1991).

During the daytime, PRA fluctuates in a damped and irregular manner, and power spectra applied to the diurnal data are generally split into two or three

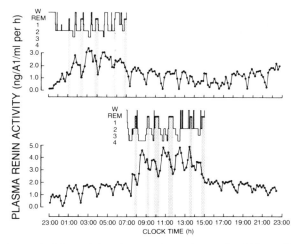

Fig. 7.4. Effect of an 8-h delay of the sleep–wake cycle on the 24-h profile of plasma renin activity.

peaks, indicating no predominant periodicity (Brandenberger et al. 1987). However, for some of the subjects, PRA oscillates at a regular periodicity throughout the 24 h (see Fig. 7.4). Stimulating renin release by a diuretic or an angiotensin-converting enzyme inhibitor reveals the episodic nature of renin release even in awake subjects (Lédée et al. 1990). It is possible to speculate that the different modes of stimulation induce pulsatile renin release via processes that are linked to the REM–NREM sleep cycles in asleep subjects and to the BRAC during wakefulness.

In the light of these findings, the existence of an endogenous circadian rhythm for PRA was re-evaluated. A study with a shift in the normal sleep time clearly demonstrated that sleep stage alternation is responsible for the increased nocturnal amplitude of PRA oscillations. An example of the 24-h profiles obtained after sleep reversal is illustrated in Fig. 7.4. Under basal conditions, an increase of PRA oscillations occurred during sleep; after a shift in the sleep period, a sleep-associated increase was clearly apparent during daytime hours, indicating the absence of any inherent circadian component.

The secretion of aldosterone also has been referred to as episodic, with a general increase in mean levels during the last hours of sleep. The two major hormonal systems involved in the regulation of aldosterone are the renin–angiotensin system and the adrenocorticotrophic system. In basal conditions, the relative importance of these two hormonal systems in controlling the ultradian aldosterone secretory patterns is unclear. Modulating renin levels by either consuming a low sodium diet or administration of a β-blocker enabled this relationship to be clarified (Krauth et al. 1990). The renin–angiotensin system can be seen to dominate aldosterone secretion when it is stimulated by a low sodium diet. Aldosterone oscillations reflect PRA oscillations with a delay of about 20 min and the relationship of aldosterone to sleep stages is dependent on the relationship of PRA to sleep stages. Atenolol depresses PRA and in this case aldosterone reflects the fluctuations of ACTH with a

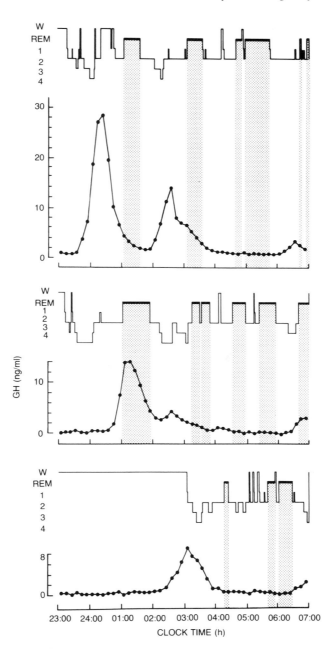

Fig. 7.5. Nocturnal GH profiles in three individuals for two reference nights (top, middle) and for a night of partial sleep deprivation (bottom). The main secretory episode can occur either during (top), after (middle), or before (bottom) the first SWS episode.

delay of about 10 min. From these results, it appears that the relationship of aldosterone to sleep stages is dependent on the relationship of both renin and ACTH to sleep stages.

Slow Wave Sleep

Growth Hormone

In basal conditions, the 24-h growth hormone (GH) profile consists of low levels abruptly interrupted by secretory pulses and the first slow wave sleep (SWS) episode at the beginning of the night has been frequently associated with the major GH secretory pulse (Sassin et al. 1969; Weitzman et al. 1975). A sudden shift in the timing of sleep immediately shifts the major secretory pulse, and the absence of sleep prevents its occurrence. However, the possibility of a weak circadian component modulating the ultradian distribution of GH pulses cannot be excluded (Van Cauter and Refetoff 1985).

Despite a large number of studies examining the relation between GH and SWS, the underlying mechanisms have not been clearly identified. Various studies, when either sleep or GH secretion was modified, have reported a clear dissociation between the first phase of SWS and the major secretory peak. Unexplained temporal dissociation has also been found in normal subjects (Steiger et al. 1987). Generally, the GH surges follow sleep onset and overlap with SWS (Fig. 7.5, top), but GH secretory bursts can occur after SWS episodes, or even before sleep onset, as shown with an acute partial sleep deprivation procedure (Fig. 7.5). A recent study has also been unable to demonstrate any consistent relationship between delta wave activity and sleep-related growth hormone secretion (Jarrett et al. 1990).

Cortisol

While it is accepted that the 24-h rhythm of plasma cortisol represents a "paradigm of a circadian rhythm", studies correlating specific sleep stages with the nocturnal cortisol secretory episodes have provided contradictory results. Decreasing cortisol levels have frequently been associated with REM sleep and increasing levels with prolonged intra-sleep awakenings. Considering the low values during the first hours of sleep, when SWS is present to its greatest degreee, an inhibitory role of SWS on cortisol release has been proposed, even though the quiescent period of adrenocortical activity begins a few hours before sleep (Weitzman et al. 1983). Contrary to these results, Born et al. (1988) found that the latency of the first secretory episode was delayed when delaying sleep onset.

When the cortisol profiles, obtained following 4 h of sleep deprivation, were analysed (sleep began at 03:00 h), the SWS episodes which occurred mainly in the second part of the night, coincided with a period of high cortisol secretory activity. It could then be seen that SWS was generally associated with the declining phases of cortisol secretory episodes (Fig. 7.6), which tends to imply that cortisol-releasing mechanisms or even cortisol itself may be involved in the regulation of sleep.

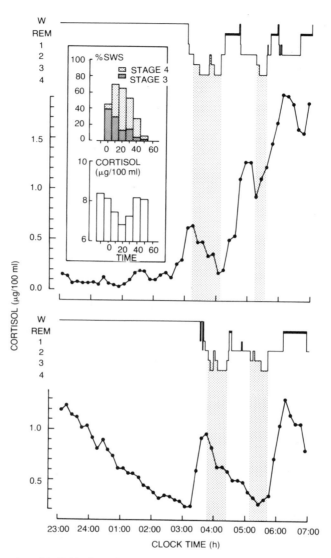

Fig. 7.6. Examples of individual cortisol profiles after a sleep delay of 4 h. SWS episodes represented by shaded areas lie in descending phases of the secretory episodes. The inset gives the mean cortisol levels ($n = 12$) when SWS was taken at point 0.

Thyrotrophin

Similarly, detailed analysis of the pulsatile pattern of TSH secretion revealed that, in addition to the well-established inverse correlation between TSH secretion and the quantity of sleep, there is a consistent relationship between SWS and the descending phases of TSH pulses (Fig. 7.7). Such a relation-

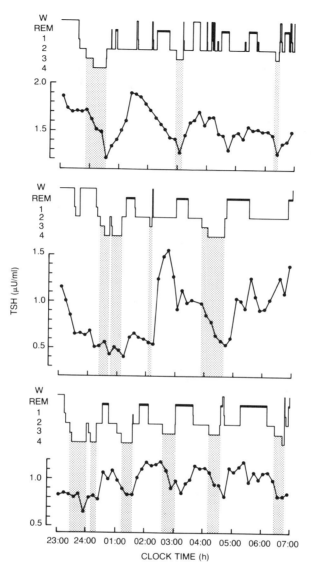

Fig. 7.7. Three individual TSH profiles with the concomitant sleep-stage patterns. SWS lies in the descending phase of the secretory episodes.

ship was not reported in earlier studies. Could these data suggest that some inhibitory effect on TSH release occurs during SWS? Or, conversely, that increased TSH secretion prevents the occurrence of SWS? And is the well-known inhibitory effect of sleep dependent on the presence of SWS sleep? These questions remain unanswered.

REM Sleep

REM Sleep Onset

The changes associated with the periodic recurrence of REM sleep represent a marked shift in cerebral activity. Despite some methodological problems linked to the fact that peripheral hormone levels have to be related to rapid changes in the activity of the central nervous system, it appears that REM sleep generally begins during the declining phase, at the peak levels, or at nadirs of the secretory episodes of certain anterior pituitary hormones, which all reflect a decrease in pituitary secretory activity. Plasma prolactin, ACTH, LH, TSH and GH levels, measured in 10-min blood samples were very seldom in an increasing phase at the beginning of REM sleep (Follenius et al. 1988). This observation is consistent for a number of hormones with widely differing regulatory mechanisms and with different relationships to the sleep–wake cycle. For example, those such as ACTH and LH are poorly influenced by sleep, whereas prolactin and TSH are strongly influenced by sleep, although in a contrastive manner, and GH secretion is linked to the first SWS episode. A reduction of sympathetic activity and disruption of vegetative function at the beginning of REM sleep may be implicated in this finding of reduced anterior pituitary activity.

REM Sleep

On the whole, REM sleep episodes do not share common features such as are observed for the main pituitary hormones with the beginning of REM sleep. Prolactin often increases very quickly after REM sleep onset. Figure 7.8

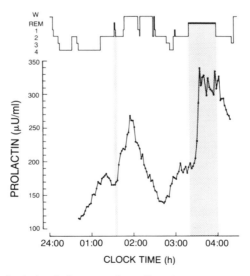

Fig. 7.8. Plasma prolactin levels from samples collected at 2-min intervals. REM sleep begins in the non-ascending phase of the secretory episodes.

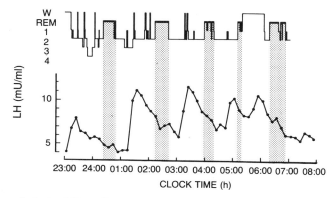

Fig. 7.9. Nocturnal plasma LH profile of one subject with the corresponding sleep-stage pattern. REM sleep occurs in the shaded areas.

illustrates one example of the abrupt changes in prolactin levels immediately after REM sleep onset.

ACTH and cortisol can also increase within a few minutes after REM sleep onset; contrastingly, LH generally shows a regular descending trend during the whole duration of REM sleep (Fig. 7.9). Finally, as renin and GH increases are associated with, respectively, NREM sleep or SWS, their levels consequently decrease during REM sleep.

The available results indicate that either increases or decreases of hormone levels may occur during REM sleep episodes, which gives little support for an active role of REM sleep or REM sleep-generating processes in influencing hormone secretion. It is likely that, apart from the short period of REM sleep onset, the pulsatility of hormone release is entrained by a complex ultradian clock without being influenced by REM sleep.

Conclusions

The classification of the endocrine rhythms as endogenous or exogenous, as periodic or random, is extremely difficult due to the inherent biological variability and to the modulatory effect of factors such as sleep, meals and environmental changes.

A number of investigators have reported the persistence of a weak endogenous circadian rhythmicity in certain clearly defined sleep-related endocrine rhythms, such as those of prolactin and GH. The occurrence of a weak effect of sleep on hormones with clearly defined circadian rhythms – such as ACTH and cortisol – cannot justify the conclusion that an endogenous oscillator is the sole operative for these hormones. Finally, the observation that the renin–angiotensin system reflects sleep-stage alternation implies that processes linked to the REM–NREM sleep cycles play a prominent role in synchronizing and enhancing damped daytime fluctuations.

In summary, the findings so far are compatible with the existence of a multifactorial control of the 24-h secretory profiles. It is evident that many of these rhythms have little in common in terms of causal mechanisms and functional significance. A great variety of hormones listed as putative sleep factors support the view that several endogenous substances are capable of altering sleep, and vice versa, that sleep modifies endocrine secretion and that these interactions are biochemically linked to each other. Thus, the endocrine regulation during sleep and wakefulness has become an important facet in the science of sleep regulation.

References

Arendt J (1989) Melatonin: a new probe in psychiatric investigation. Br J Psychiatry 155:585–590
Bartter FC, Chan JCM, Simpson HW (1979) Chronobiological aspects of plasma renin activity, plasma aldosterone, and urinary electrolytes. In: Krieger DT (ed) Endocrine rhythms. Raven Press, New York, pp 225–245
Birkeland AJ (1982) Plasma melatonin levels and nocturnal transitions between sleep and wakefulness. Neuroendocrinology 34:126–131
Born J, Muth S, Fehm HL (1988) The significance of sleep onset and slow wave sleep for nocturnal release of growth hormone and cortisol. Psychoneuroendocrinology 13:233–243
Boyar R, Finkelstein J, Roffwarg H, Kapen S, Weitzman ED, Hellman L (1972) Synchronization of augmented luteinizing hormone secretion with sleep during puberty. N Engl J Med 287:582–586
Brabant G, Ranft U, Ocran K, Hesch RD, Von zur Mühlen A (1986) Thyrotropin – an episodically secreted hormone. Acta Endocrinol (Copenh) 112:315–322
Brabant G, Prank K, Ranft V et al. (1990) Physiological regulation of circadian and pulsatile thyrotropin secretion in normal man and woman. J Clin Endocrinol Metab 70:403–409
Brabant G, Prank K, Hoang-Vu C, Hesch RD, Von zur Mühlen A (1991) Hypothalamic regulation of pulsatile thyrotropin secretion. J Clin Endocrinol Metab 72:145–150
Brandenberger G, Follenius M, Muzet A, Ehrhart J, Schieber JP (1985) Ultradian oscillations in plasma renin activity: their relationships to meals and sleep stages. J Clin Endocrinol Metab 61:280–284
Brandenberger G, Simon C, Follenius M (1987) Night–day differences in the ultradian rhythmicity of plasma renin activity. Life Sci 40:2325–2330
Brandenberger G, Follenius M, Simon C, Ehrhart J, Libert JP (1988) Nocturnal oscillations in plasma renin activity and REM–NREM sleep cycles in humans: a common regulatory mechanism? Sleep 11:242–250
Brandenberger G, Krauth MO, Ehrhart J, Libert JP, Simon C, Follenius M (1990) Modulation of episodic renin release during sleep in humans. Hypertension 15:370–375
Claustrat B, Brun J, Garry P, Roussel B, Sassolas G (1986) A once repeated study of nocturnal plasma melatonin patterns and sleep recordings in six normal young men. J Pineal Res 3:301–310
Delitala G, Tomasi P, Virdis R (1987) Prolactin, growth hormone and thyrotropin-thyroid hormone secretion during stress states in man. Baillieres Clin Endocrinol Metab 1:391–414
Désir D, Van Cauter E, Fang V et al. (1981) Effect of "jet lag" on hormonal pattern. I. Procedures, variations in total plasma proteins and disruption of adrenocorticotropin-cortisol periodicity. J Clin Endocrinol Metab 55:628–641
Désir D, Van Cauter E, L'Hermite M et al. (1982) Effects of "jet lag" on hormonal patterns. III. Demonstration of an intrinsic circadian rhythmicity in plasma prolactin. J Clin Endocrinol Metab 55:849–857
Follenius M, Brandenberger G, Simeoni M, Reinhardt B (1982) Diurnal cortisol peaks and their relationships to meals. J Clin Endocrinol Metab 55:757–761
Follenius M, Simon C, Brandenberger G, Lenzi P (1987) Ultradian plasma corticotropin and cortisol rhythms: time-series analyses. J Endocrinol Invest 10:261–266
Follenius M, Brandenberger G, Simon C, Schlienger JL (1988) REM sleep in humans begins during decreased secretory activity of the anterior pituitary. Sleep 11:546–555
Follenius M, Krieger J, Krauth MO, Sforza F, Brandenberger G (1991) Obstructive sleep apnea treament: peripheral and central effects on plasma renin activity and aldosterone. Sleep 14:211–217

Frantz AG (1979) Rhythms in prolactin secretion. In: Krieger DT (ed) Endocrine rhythms. Raven Press, New York, pp 175–186

Greenspan SL, Klibanski A, Schoenfeld D, Ridgway EC (1986) Pulsatile secretion of thyrotropin in man. J Clin Endocrinol Metab 63:661–668

Jarrett DB, Greenhouse JB, Miewald JM, Fedorka IB, Kupfer DJ (1990) A reexamination of the relationship between growth hormone secretion and slow wave sleep using delta wave analysis. Biol Psychiatry 27:497–509

Judd HL, Parker DC, Rafkoff JS, Hopper BR, Yen SSC (1974) Elucidation of mechanism(s) of the nocturnal rise of testosterone in men. J Clin Endocrinol Metab 38:134–141

Kapen S, Boyar RM, Finkelstein JW, Hellman L, Weitzman ED (1974) Effect of sleep–wake cycle reversal on luteinizing hormone secretory pattern in puberty. J Clin Endocrinol Metab 39:293–299

Kapen S, Boyar RM, Hellman L, Weitzman ED (1976) The relationship of luteinizing hormone secretion to sleep in women during the early follicular phase: effects of sleep reversal and a prolonged three-hour sleep–wake schedule. J Clin Endocrinol Metab 42:1031–1039

Kleitman N (1963) Sleep and wakefulness. University of Chicago Press, Chicago, IL

Kleitman N (1982) Basic rest–activity cycle – 22 years later. Sleep 5:311–317

Knobil E, Hotchkiss J (1985) The circhoral gonadotropin releasing hormone (GnRH) pulse generator of the hypothalamus and its physiological significance. In: Schulz H, Lavie P (eds) Ultradian rhythms in physiology and behavior. Springer, Berlin Heidelberg New York, pp 32–40

Krauth MO, Saini J, Follenius M, Brandenberger G (1990) Nocturnal oscillations of plasma aldosterone in relation to sleep stages. J Endocrinol Invest 13:727–735

Lavie P, Kripke DF (1981) Ultradian circa $1\frac{1}{2}$ hour rhythms: a multioscillatory system. Life Sci 29:2445–2450

Lédée P, Brandenberger G, Follenius M (1990) Renin release stimulated by furosemide and perindopril is episodic both during sleep and wakefulness. Pharmacol. (Life Sci Adv) 9:459–465

Linkowski P, Mendlewicz J, Leclerq R et al. (1985) The 24-hour profile of adrenocorticotropin and cortisol in major depressive illness. J Clin Endocrinol Metab 61:429–438

MacLeod RM (1976) Regulation of prolactin secretion. In: Martini L, Ganong WF (eds) Frontiers in neuroendocrinology. Raven Press, New York, pp 169–194

Merriam GR, Wachter KW (1984) Measurement and analysis of episodic hormone secretion. In: Rodbard D, Forti G (eds) Computers in endocrinology. Raven Press, New York, pp 325–346

Miyatake A, Morimoto Y, Oishi T et al. (1980) Circadian rhythm of serum testosterone and its relation to sleep: comparison with the variation in serum luteinizing hormone, prolactin, and cortisol in normal men. J Clin Endocrinol Metab 51:1365–1371

Modlinger RS, Sharif-Zadeh K, Ertel NH, Gutkin M (1976) The circadian rhythm of renin. J Clin Endocrinol Metab 43:1276–1282

Mullen PE, James VHT, Lightman SL, Linsell C, Peart WS (1980) A relationship between plasma renin activity and the rapid eye movement phase of sleep in man. J Clin Endocrinol Metab 50:466–469

Parker DC, Rossman LG, Pekary AE, Hershman JM (1987) Effect of 64-hour sleep deprivation on the circadian waveform of thyrotropin (TSH): further evidence of sleep related inhibition of TSH release. J Clin Endocrinol Metab 64:157–161

Patel YC, Alford FP, Burger HG (1972) The 24 h plasma thyrotropin profile. Clin Sci 43:71–77

Quigley ME, Yen SSC (1979) A mid-day surge in cortisol levels. J Clin Endocrinol Metab 49:945–947

Quigley ME, Ropert JF, Yen SSC (1981) Acute prolactin release triggered by feeding. J Clin Endocrinol Metab 52:1043–1045

Rivest RW, Schutz P, Lustenberger S, Sizonenko PC (1989) Differences between circadian and ultradian organisation of cortisol and melatonin rhythms during activity and rest. J Clin Endocrinol Metab 68:721–729

Sassin JF, Parker C, Mace JW, Gotlin RW, Johnson LC, Rossman LG (1969) Human growth hormone release: relation to slow-wave sleep and slow-waking cycles. Science 165:513–515

Sassin JF, Frantz AG, Kapen S, Weitzman ED (1973) The nocturnal rise of human prolactin is dependent on sleep. J Clin Endocrinol Metab 37:436–440

Spratt DI, O'Dea L St, Schoenfeld D, Butler J, Rao N, Crowley WF (1988) Neuroendocrine–gonadal axis in men: frequent sampling of LH, FSH, and testosterone. Am J Physiol 254:E658–E666 (Endocrinol Metab 17)

Steiger A, Herth T, Holsboer F (1987) Sleep-electroencephalography and the secretion of cortisol and growth hormone in normal controls. Acta Endocrinol 116:36–42

Trinchard-Lugan I, Waldhauser F (1989) The short-term secretion pattern of human serum melatonin indicates apulsatile hormone release. J Clin Endocrinol Metab 69:663–669

Urban RY, Kaiser DL, Van Cauter E, Johnson ML, Veldhuis JO (1988) Comparative assessments of objective peak detection algorithms. II. Studies in men. Am J Physiol. 254:E113–E119 (Endocrinol Metab 17)

Van Cauter E, Honinckx E (1985) Pulsatility of pituitary hormones. In: Schulz H, Lavie P (eds) Ultradian rhythms in physiology and behavior. Springer, Berlin Heidelberg New York, pp 41–60

Van Cauter E, Refetoff S (1985) Multifactorial control of the 24-hour secretory profiles of pituitary hormones. J Endocrinol Invest 8:381–391

Vanhaelst L, Van Cauter E, Degaute JP, Golstein J (1972) Circadian variations of serum thyrotropin levels in man. J Clin Endocrinol Metab 35:479–482

Veldhuis JD, Iranmanesh A, Johnson ML, Lizarralde G (1991) Amplitude, but not frequency, modulation of adrenocorticotropin secretory burst gives rise to the nyctohemeral rhythms of the corticotropic axis in man. J Clin Endocrinol Metab 71:452–463

Waldhauser F, Steger H (1987) Physiology of melatonin secretion in man. In: Wagner TOF, Filicori M (eds) Episodic hormone secretion: from basic science to clinical application. TM-Verlag, Hameln, pp 105–112

Weeke J, Gundersen HJG (1978) Circadian and 30 minutes variations in serum TSH and thyroid hormones in normal subjects. Acta Endocrinol (Copenh) 89:659–665

Weitzman ED, Boyar RM, Kapen S, Hellman L (1975) The relationship of sleep and sleep stages to neuroendocrine secretion and biological rhythms in man. In: Greep R (ed) Recent progress in hormone research. Academic Press, New York, pp 399–446

Weitzman ED, Zimmerman JC, Czeisler CA, Ronda J (1983) Cortisol secretion is inhibited during sleep in normal man. J Clin Endocrinol Metab 56:352–358

Chapter 8

A Parsimonious Model of Amplitude and Frequency Modulation of Episodic Hormone Secretory Bursts as a Mechanism for Ultradian Signalling by Endocrine Glands

J. D. Veldhuis

Introduction

Organisms typically adapt to environmental cues and homeostatic perturbations by coordinated responses, which involve a time-ordered chain of biochemical, cellular, organ-specific and system-wide changes. A major challenge to adaptive and homeostatic responses is the temporal organization or sequential patterning of the response variables. Moreover, even under basal conditions when major exterior stresses are not imposed, coordinating signals within the organism must be regulated in relation to amplitude, frequency, duration and/or tonic output.

Although 24-h (circadian) variations in signalling amplitude and frequency and/or tonic output are relatively well-described phenomena in various organ systems, cell ensembles or biochemical reactions, there is considerably less information about: (a) ultradian endocrine signalling within an organism under physiological conditions as well as in specific pathophysiological states; (b) the contribution and relationship of the ultradian hormone signals to the robust circadian rhythm; (c) the specific impact of changes in ultradian secretory signal amplitude, frequency, or duration on individual target tissues; and (d) the mechanisms that generate episodic (apparently random) or rhythmic (approximately regularly occurring over time) ultradian endocrine signals. This chapter reviews and appraises critically these four issues regarding ultradian signalling by endocrine systems in general, and by the anterior pituitary gland in particular. The anterior pituitary gland is chosen as it is an endocrine organ manifesting ultradian changes in signal (glycoprotein hormone) output, highly coordinated feedback regulation, intimate association with neural mechanisms that generate time-dependent variations in the endocrine signal, and striking adaptive and homeostatic responses to external and internal environmental cues. Moreover, hormone secretion by the anterior pituitary gland exhibits distinctly ultradian variations that can be related to well-established circadian counterparts.

Neuroendocrine Physiology

Concept of Neural Regulation of the Anterior Pituitary Gland

Although limited descriptions of rhythmic or episodic hormone release by perifused anterior pituitary tissue in vitro have been reported (Stewart et al. 1985; Gambacciani et al. 1987), there is more compelling experimental evidence for minute-to-minute regulation of anterior pituitary hormone secretory activity in vivo by an array of blood and especially brain-derived effector molecules (e.g. steroids, prostaglandins, peptides and other neurotransmitters) secreted by nerve terminals originating in variably localized areas within the hypothalamus and contiguous structures (Daughaday 1981). For example, the secretion of growth hormone (GH) is subject to at least dual regulation by an inhibitory hypothalamic signal in the form of the tetradecapeptide somatostatin and a stimulatory hypothalamic peptide, GH-releasing hormone (GHRH), which has more than 40 amino acid residues (Plotsky and Vale 1985). On the other hand, the two primary gonadotrophic hormones, luteinizing hormone (LH) and follicle-stimulating hormone (FSH), are subject to hypothalamic stimulation by the decapeptide gonadotrophin-releasing hormone (GnRH) (Yen et al. 1975; Knobil 1980; Catt et al. 1983). In addition, although GnRH will stimulate the secretion of FSH under a variety of physiological and pathophysiological conditions in vivo and in vitro, other glycoprotein and/or polypeptide regulators such as gonadally derived inhibin, activin and follistatin, as well as male and female sex-steroid hormones, may play a role in regulating FSH synthesis and secretion by the anterior pituitary gland (Barraclough and Wise 1982; Grady et al. 1982). The stress-adaptive hormone adrenocorticotrophic hormone (ACTH) in some species is under dual hypothalamic stimulatory control by the octapeptide arginine-vasopressin and the 41 amino acid residue polypeptide ACTH-releasing hormone, or corticotrophin-releasing hormone (CRH) (Krieger 1979; Watanabe and Orth 1987). The last of these also triggers secretion of β-lipotropin and β-endorphin. On the other hand, although the tripeptide thyrotrophin-releasing hormone (TRH), releases both thyrotrophin (TSH) and prolactin in vitro and in vivo, TRH is not necessarily the sole agonist for the spontaneous release of TSH and prolactin (Leong et al. 1983). Indeed, hypothalamic dopamine is a predominant inhibitory modulator of prolactin release. In short, the anterior pituitary gland is under the direction of both hypothalamically and systemically derived regulators, as well as (see below) intrapituitary effectors.

Feedback Control

Hypothalamic releasing factors and/or inhibiting hormones are synthesized in cell bodies, which are variably dispersed within the hypothalamus and/or contiguous neural structures. Neurally derived effectors are released into the hypothalamo-pituitary portal microcirculation, by which they are transported to cells within the anterior pituitary gland (Daughaday 1981). The release of hormones is not unidirectional from hypothalamus to pituitary, since pituitary-derived hormones can also reach the brain. Thus, feedback regulation from pituitary to brain may occur, e.g. in the case of GH and prolactin.

Both hypothalamic regulatory neurons and pituitary cells are typically subject to inhibitory *feedback* regulation by hormones released by one or more remote target tissues stimulated by the anterior pituitary hormones, e.g. cortisol from the adrenal gland acts (or feeds back) negatively on brain regulatory centres as well as anterior pituitary ACTH-secreting cells; sex steroid hormones and the glycoprotein hormone inhibin derived from the ovaries and testes modulate neural and/or anterior pituitary governance of gonadotrophin secretion; and somatomedin C, an insulin-like growth factor produced by the liver (and other tissues) in response to GH action, can exert potent negative feedback effects on the hypothalamic control of the pituitary gland and on the GH-secreting cells per se within the anterior pituitary gland. In short, episodes of secretory activity enacted by the anterior pituitary gland trigger responses in remote target tissues that (with some appropriate time delay) tend to suppress further pituitary secretion by acting on the hypothalamus and/or anterior pituitary gland.

Intrapituitary Regulation of Hormone Secretion

In addition to the ability of remote target tissues stimulated by anterior pituitary hormones to secrete products that modulate the brain's stimulation or inhibition of the anterior pituitary gland and/or alter the activity of anterior pituitary secretory cells directly, there is increasing interest in, and evidence for, *intra*pituitary regulatory mechanisms, in which products released by one or more contiguous hypophysial cells act upon the cell of origin (*autocrine* feedback) or upon neighbouring cells (*paracrine* feedback). For example, in relation to the gonadotrophin axis, evidence exists for the intrapituitary synthesis and action of the glycoprotein hormones activin and inhibin, which respectively stimulate and inhibit FSH secretion by the anterior pituitary gland (Roberts et al. 1989). Moreover under some conditions, the hypothalamic peptide GnRH can activate not only the secretion of the gonadotrophins LH and FSH but also trigger release of prolactin from lactotrope cells (Denef and Andries 1983). The exact role played by such anterior pituitary autocrine and paracrine mechanisms in regulating time-variable episodic or rhythmic hormone secretion in vivo is not understood at present.

Episodicity of Hypothalamic Signalling of the Anterior Pituitary Gland

Direct catheterization studies of the hypothalamo-pituitary portal microcirculation in the conscious freely moving rat, sheep, horse and monkey indicate that hypothalamic releasing factors and inhibiting substances are secreted into the hypophysial portal blood in bursts that vary in amplitude and frequency under different endocrine conditions (Kasting et al. 1981; Clarke and Cummins 1982; Levine and Ramirez 1982; Levine et al. 1985; Plotsky and Vale 1985; Irvine and Alexander 1987). A spontaneous secretory episode of hypothalamic releasing factor typically encompasses one or more minutes, with a return of releasing factor secretion rates toward zero or basal levels thereafter. These neurally generated ultradian release episodes often occur at approximately hourly intervals (circahoral), although there is a wide range of variation in interburst intervals, viz. from several minutes to many hours.

The functional relevance of hypothalamic peptide and neurotransmitter release into the pituitary portal microcirculation is supported by the strong but not necessarily one-to-one concordance between such events and the nearly immediate secretion of the corresponding anterior pituitary hormone (Kasting et al. 1981; Clarke and Cummins 1982; Levine and Ramirez 1982; Levine et al. 1985; Plotsky and Vale 1985; Irvine and Alexander 1987). For example, hypothalamic GnRH release and pituitary LH secretion exhibit a strong temporal coincidence under many but not all experimental conditions (Clarke and Cummins 1982; Levine and Ramirez 1982). Appropriate electrical recordings, electrochemical stimulation, and neural lesioning procedures designed to monitor, stimulate or inhibit hypothalamic effector synthesis and/or release have supported further the correspondence between hypothalamic and anterior-pituitary hormone signalling patterns. Of interest, although the release of hypothalamic modulators is typically *ultradian* (occurring more often than once per day), the release patterns in general appear to be *episodic* (randomly distributed over time) rather than explicitly *rhythmic* (sinusoidally or otherwise regularly occurring). The term *pulsatile* has been employed to designate such episodic bursts of hormone release, which occur unpredictably and are characterized by an abrupt rise in secretory rate followed by a decline toward zero or baseline.

In summary, direct observations indicate that pituitary hormone secretion in vivo is characterized by episodic bursts or pulses of secretory activity, which are separated by intervals of low (or nearly zero) basal secretion at least under physiological conditions. In contrast, under pathological conditions, there may be significant basal or constitutive hormone release between major secretory volleys (Hartman et al. 1990a). In normal circumstances, the timing and amplitude of such secretory bursts are specified by hypothalamic stimulatory and/or inhibitory signals and the state of secretory responsiveness of the anterior pituitary cells.

Basal or Constitutive Hormone Secretion

Although high rates of basal or constitutive (unregulated) secretion of anterior pituitary hormones are well recognized in various pathological conditions (e.g. sustained GH secretion by pituitary adenomas in patients with acromegaly or gigantism (Hartman et al. 1990a); far less information is available about constitutive hormone synthesis by normal anterior pituitary cells in vivo. Pituitary fragments, cell aggregates, or dispersed pituitary cell preparations studied by in vitro perifusion can exhibit measurable amounts of basal hormone secretion (Stewart et al. 1985; Gambacciani et al. 1987). I suggest that significant basal secretion be defined as hormone release exceeding both the sensitivity and the experimental uncertainty in the assay, showing secretory independence from perifusion hydrodynamics (e.g. pressure and flow rates), and occurring in the absence of added known secretagogues. Such basal secretion may or may not be regulatable by prior exposure to hormone. Unregulatable basal secretion can be designated as constitutive. However, the exact extent to which either constitutive or regulatable basal hormone secretion in vitro might result from non-physiological effects of hypothalamo-pituitary disconnection, tissue removal, tissue dissociation, and other in vitro manipulations is not usually known.

Low rates of basal hormone synthesis are even more difficult to investigate in vivo under physiological circumstances. For example, significant basal GH secretion has been inferred in the rat, since administration of antiserum to the GH release-inhibiting hormone somatostatin increases mean serum GH concentrations between the prominent GH release episodes (pulses) (Tannenbaum and Ling 1984). Such experiments have suggested that under physiological conditions somatostatin restrains interpulse basal GH release. This interpretation has been complicated by recent clinical studies using an ultrasensitive GH assay, which have shown that even very low rates of puta- tively basal GH release between large GH pulses consist of pulsatile secretory episodes in normal men and women (Winer et al. 1990). In the case of LH and FSH secretion in humans and various experimental animals, administration of highly potent and selective peptide antagonists of GnRH will lower serum LH concentrations and virtually abolish the pulsatile mode of LH release (Pavlou et al. 1990; Urban et al. 1990). However, even under these circumstances low amplitude pulses of LH persist, which suggests that small pulses may be the mechanism of basal LH secretion (Pavlou et al. 1990), and/or that basal (like pulsatile) LH release is GnRH-dependent. These examples illustrate that the evaluation of basal (interburst) hormone secretion is difficult in normal physiology.

Secretory Behaviour of Endocrine Glands

Extensive physiological studies in vivo and in vitro support the general concept that endocrine glands signal their remote target tissues by way of an inter- mittent rather than a continuous (or time-invariant) mode of hormone secre- tion. This intermittency has been characterized as episodic, to denote that regular rhythmicity akin to sinusoidally explicit variation is not commonly identifiable, except in the case of certain hormones such as insulin where a strong periodicity is often evident (Lang et al. 1979; Polonsky et al. 1985). In addition, hormone release appears to occur predominantly in bursts or pulses consisting of a rapid and marked increase in the hormone secretion rates, which are sustained over an interval of seconds or minutes and then return toward baseline (Veldhuis et al. 1987a). The amount of basal hormone secretion estimated between such delimited secretory events appears to be relatively low and in some cases undetectable (Lang et al. 1979; Veldhuis et al. 1984, 1987a, 1989a, 1990a,b; Linkowski et al. 1985; Polonsky et al. 1985; Veldhuis and Johnson 1988a; Iranmanesh et al. 1989a; Hartman et al. 1990b; Sollenberger et al. 1990a; Urban and Veldhuis 1991). For example, even outside the pituitary gland, indirect investigations and direct catheteriza- tion studies have revealed episodic release of the ACTH-responsive adrenal steroid cortisol, with low levels of glucocorticoid secretion sustained between punctuated bursts (Nelson et al. 1955; Hellman et al. 1970; Veldhuis et al. 1989b). Similarly, catheterization of human testicular veins has disclosed sudden marked increases in steroid as well as glycoprotein hormone (e.g. inhibin) concentrations over time (Winters and Troen 1986; Winters 1990),

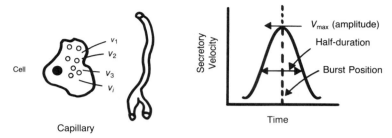

Cell

Capillary

Fig. 8.1. Schema of a hypothetical neuroendocrine secretory burst. In this model, a secretory *burst* results from the discharge of (neuro)hormone from a collection of cells contained within an endocrine gland or neuroendocrine ensemble. Within a burst, each molecule is released at its own theoretical secretory rate or velocity (v_i). The set of such secretory velocities constitutes the burst or pulse, which is characterized by some maximal value and some particular duration. In some cases, the set of secretory velocities in a burst may approach a Gaussian distribution. Indeed, the latter can be used to approximate many other distributions when i is large (i is the number of discharged molecules per burst). Such a Gaussian distribution of secretory rates can be defined fully by its mean, standard deviation and maximum. For the purposes of secretion modelling, a burst *half-duration* (which is the event duration at half-maximal amplitude) can be used, which equals the burst standard deviation multiplied by 2.354. The total mass of (homone) secreted in the burst is the area (integral) of the secretory pulse.

presumably reflecting the intermittency of pituitary LH and FSH release. Analysis of pituitary venous effluent shows a markedly pulsatile pattern of LH, FSH, ACTH, GH and prolactin release, e.g. as studied in the rat, sheep, monkey, or horse (Kasting et al. 1981; Clarke and Cummins 1982; Levine and Ramirez 1982; Levine et al. 1985; Irvine and Alexander 1987). In these studies, it is evident that the amount of basal hormone secretion between bursts is typically small compared to the mass of hormone secreted within a burst. Thus, as illustrated in Fig. 8.1, physiological endocrine glandular signalling on an ultradian time scale can be approximated by a model of intermittent burst-like release of hormone molecules into the bloodstream, with such events occurring as punctuated episodes distributed typically non-uniformly over time with or without a relatively low level of basal (interpulse) hormone secretion.

For the pituitary gland, intermittent bursts of hormone secretion occur at intervals that range between 30 min and 6 h (e.g. LH released half-hourly at prooestrus in the rat versus 6-hourly in the mid-luteal phase of young women). By definition, such interpulse intervals conform to the time structure of an ultradian rhythm: see Table 8.1 for examples of human pituitary hormone release. Indeed, by confining secretion to discontinuous ultradian events, only a fraction of the day is needed in which to achieve adequate total hormone release over a full 24 h (see Implications of Ultradian Endocrine Signalling, below).

In many but not all physiological circumstances evaluated to date, the timing of ultradian release episodes can be described as a *renewal process*, which designates that serial interburst intervals are statistically independent (Butler et al. 1986). In a renewal process, the system generating the bursts has no memory of the timing of the preceding events. Indeed, for LH, FSH, ACTH, β-endorphin, cortisol, and testosterone release in men, serial interburst intervals show no evident autocorrelation, i.e. the numerical values of consecutive

Table 8.1. Time characteristics of selected neuroendocrine secretory events as estimated by deconvolution analysis

Hormone	Approximate secretory burst half-duration[a] (min)	Number of bursts/day	Hours per day in which 95% of secretion occurs
LH (RIA)	6–10	21	5.0
LH (RICT)	12	24	6.5
GH (IRMA)	25	12	8.8
FSH (RIA)	6	13	2.0
Cortisol (RIA)	16	19	8.2
ACTH (IRMA)	19	40	21
Prolactin (RIA)	45	20	24[b]
TSH (RIA)	71	20	24[b]

RIA, radioimmunoassay; RICT, rat interstitial-cell testosterone bioassay; IRMA, immunoradiometric assay.
Data are mean estimates for six to eight normal men based on multiple-parameter deconvolution analysis. Adapted in part from Veldhuis et al. (1990a) with permission.
[a] Duration of the *secretory* burst at half-maximal amplitude.
[b] Consecutive secretory bursts are fused or confluent, which results in non-zero secretion at all times (i.e. interpeak secretion does not fall to zero but to the sum of the overlapping secretory peak edges, thus resembling basal hormone release).

intervals between secretory bursts are statistically independent (Veldhuis et al. 1987b, 1989b, 1990b; Veldhuis and Johnson 1988a; Iranmanesh et al. 1989a). An interesting exception to this concept of a randomly activated endocrine pulse generator is deconvolution-assessed pulsatile LH secretion in the mid-luteal phase of the normal human menstrual cycle (Sollenberger et al. 1990a), and pulsatile GH secretion in fed normal men (Hartman et al. 1990b). In both circumstances, serial interburst intervals exhibit significantly negative auto-correlations. Such statistical features denote that long and short interpulse intervals tend to alternate. A similar finding for episodic LH release in women but not men was reported earlier using conventional pulse analysis (Butler et al. 1986; Santoro et al. 1988). Mechanisms sustaining this interesting pituitary ultradian rhythmicity are not known, but may in principle involve time-delayed negative feedback by one or more hormones (e.g. intrapituitary paracrine or autocrine effectors, or gonadal products such as progesterone or oestradiol), and/or non-uniform intervals of pituitary refractoriness to hypothalamic releasing factor stimulation.

In summary, the sequential timing of the neural activity responsible for triggering episodes of anterior pituitary hormone release is not usually precisely rhythmic, and indeed often is apparently random. Such stochastic patterns have features of a statistical renewal process, rather than a simple oscillator. None the less, early studies of ultradian variations in anterior pituitary gland secretory activity attempted to approximate this time course with simple time series or Fourier analyses, which express the data in terms of one or more sinusoidal periodicities. This approach to quantifying and describing adenohypophysial secretion is handicapped by various special features of endocrine data (below).

Modelling and Analysis of Ultradian Hormone Secretory Bursts

Problems

The assessment of ultradian patterns of hormone release in vivo has been hampered by the episodic (apparently random) timing of release bursts, so that the data do not conform readily to classical time series analysis in the frequency domain (e.g. fitting the observations to a finite number of sine and cosine functions). Moreover, endocrine data series are typically sparse, since available numbers of observations range from several dozen to a few hundred per subject. Recent improvements in immunologically directed assay techniques have permitted the collection of more extensive clinical data, e.g. by assaying blood sampled every 5 min for 24 h or even every 30 s overnight (Veldhuis et al. 1988a; Hartman et al. 1990b; Holl et al. 1991). Moreover, hormone measurements are limited by assay sensitivity, and confounded to some degree by experimental uncertainty (so-called "noise"), which is introduced by sample collection, processing, assay and data reduction. Unfortunately, in many cases, the exact theoretical structure of the experimental uncertainty is not known, although occasionally this has been estimated, e.g. for LH and GH immunoassays (Hartman et al. 1990b; Veldhuis and Johnson 1991a). The idealized time course of the underlying secretory episode also is not known a priori. An exception is when direct catheterization of the veins draining the endocrine gland is used to assess the secretion event.

In the bloodstream, hormones undergo biotransformation, degradation, removal from the plasma compartment and/or association with other plasma macromolecules (e.g. GH binds to a specific high affinity plasma protein from which it dissociates only sparingly). The presence and affinity of binding proteins will influence hormone removal rates, and will vary not only among animal species but also in different pathophysiological conditions. Such post-secretory modification and/or sequestration can further alter the apparent kinetics of hormone distribution and clearance (Veldhuis and Johnson 1990a, 1991a).

In summary, the analysis of ultradian activity in endocrine data is challenging, because: (a) hormone release episodes are episodic rather than regularly distributed over time; (b) the data sets are sparse and noisy; (c) serum hormone concentrations may fall below assay sensitivity, i.e. into an undetectable range; (d) biotransformation, binding, and removal complicate hormone kinetics; and (e) the actual underlying structure of the secretory function is not necessarily known a priori. Despite these difficulties, several novel and effective approaches to analysing ultradian patterns of intermittent hormone secretion have been developed recently, and have yielded informative and provocative insights into the mechanisms that regulate ultradian endocrine glandular signalling.

Deconvolution as a Means of Investigating Secretory Events Based on Serial Measurements of Hormone Concentrations in Plasma

The plasma hormone concentration pulse generated by a burst of hormone release cannot be equated directly with the corresponding underlying hormone release episode itself, because secreted hormone molecules undergo distribution, biotransformation, binding, degradation and removal. However, adequate knowledge of the fate of the secreted molecules delivered into the bloodstream or other sampling compartment can be employed to gain quantitative insights into underlying secretory events. The class of mathematical procedures by which such quantitative features of secretory events are determined from the (plasma) hormone concentrations is referred to as deconvolution analysis (Veldhuis et al. 1988a; Veldhuis and Johnson 1990a,b, 1991a). Deconvolution in general designates the "unravelling" or "disentwining" of two or more contributing processes, given knowledge of the overall effect of the processes acting together. The specific processes that act jointly to specify an overall outcome are said to be convolved mathematically.

Deconvolution analysis entails estimating quantitatively the behaviour of one or more convolved functions, given information about their combined output and particular assumptions about the manner in which they interact. For example, deconvolution analysis of the profiles of plasma hormone concentrations over time consists of calculating the secretory activity of the endocrine gland that would be required to give rise to the observed serum hormone concentration profile, given pertinent assumptions about the expected kinetics of hormone entry into, and removal from, the blood (Veldhuis et al. 1988a; Veldhuis and Johnson 1990b). Under some circumstances, both the underlying secretory function *and* clearance (disappearance) kinetics can be evaluated from the plasma hormone concentration profile alone, for example when some particular structure for the secretory event is known or assumed (Veldhuis et al. 1988a; Veldhuis and Johnson 1990a,b, 1991a). This approach is particularly useful for hormones or substrates whose half-times of disappearance from plasma are likely to vary among different subjects or in different pathological conditions.

As reviewed elsewhere "in extenso" (Veldhuis et al. 1988a; Veldhuis and Johnson 1990a,b, 1991a,b), deconvolution analysis is a mathematical technique developed originally in the physical sciences to evaluate quantitatively the nature of a signal of interest based on remote measurements of that signal after several confounding influences have acted on it. For example, in seismology, if the intensity of a seismic shock wave is recorded at one or more points remote from its source, deconvolution analysis can be used to reconstruct the apparent tremor time course and/or intensity at the origin, assuming relevant dissipation kinetics acting on the shock wave as it travels to the points of observation. Similarly, the "source signal" of the endocrine gland, i.e. the time-specified secretory waveform, can be estimated quantitatively from the *resultant* hormone concentrations measured remotely from the endocrine gland, e.g. in the plasma compartment. This concept is illustrated schematically in Fig. 8.2.

Given reasonable assumptions about the temporal nature of the hormone secretory burst, which then can be represented by some relevant algebraic function to define the form of an idealized secretion rate over time, deconvolution analysis is concerned with calculating pertinent numerical values

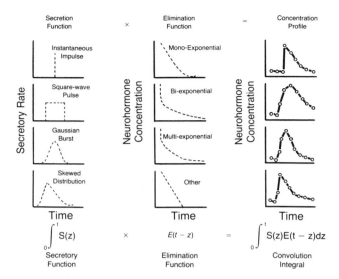

Fig. 8.2. Concept of *convolution integral*. A convolution integral is used to express the combined ("convolved", or "intertwined") effects of two distinct functions on an observed process. For example, an appropriate convolution integral containing secretion and clearance functions can define resultant changes in hormone concentrations over time. As illustrated, the *secretion function* could assume any one of several theoretical forms; e.g. an instantaneous impulse (zero duration secretory burst); a square-wave time course of secretion; a Gaussian waveform; a skewed distribution of secretion rates, etc. The clearance (*elimination*) *function* is most often represented as a mono or biexponential decay curve, but might in certain pathophysiological circumstances exhibit different properties (e.g. concentration-dependent decay constants). When several algebraic functions provide good mathematical descriptions of neuroendocrine data, independent knowledge of the system's biological behaviour is important to aid in choosing the most appropriate convolution functions. Note that the indicated combinations of functions are purely hypothetical. Biologically relevant secretion and elimination terms must be defined by physiological experiments.

to describe the secretory event quantitatively and in some cases simultaneously estimate hormone removal kinetics. For example, one particular deconvolution technique, a so-called multiple-parameter method, assumes that individual hormone secretory bursts can be approximated algebraically by a Gaussian distribution of molecular secretion rates, and that the combined effects of distribution, biotransformation, degradation and removal of the secreted hormone molecules can be approximated by a mono- or bi-exponential disappearance function (Veldhuis et al. 1987a). The inferred secretory burst therefore is characterized by some particular *amplitude* (maximal rate of secretion attained within the event), *half-duration* (duration of the secretory episode at half-maximal amplitude), and some approximate *centre* or location in time (burst position in the time series). Given these assumptions, adequate serial measurements of plasma hormone concentrations permit estimates of the number, duration, mass, amplitude and positions of statistically significant underlying hormone *secretory* bursts, and simultaneous calculations of the *half-life* of hormone removal from the blood.

The multiple-parameter deconvolution algorithm has been applied to serum concentration profiles of various anterior pituitary hormones (LH, FSH,

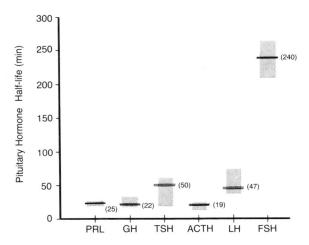

Fig. 8.3. Deconvolution predicted in vivo half-lives for various anterior pituitary hormones studied in the human. The vertical axis gives the computer-predicted half-time of disappearance from plasma of several distinct individual anterior pituitary hormones such as luteinizing hormone (*LH*), follicle-stimulating hormone (*FSH*), growth hormone (*GH*), adrenocorticotrophic hormone (*ACTH*), prolactin (*PRL*), and thyrotrophin (*TSH*). The double horizontal lines (and numerical values in parentheses) are nominal mean literature estimates of half-lives and the stippled boxes give the statistical confidence limits for deconvolution-based estimates. (Adapted, with permission, from Veldhuis et al. 1987a.)

GH, prolactin, TSH, ACTH and β-lipotropin/β-endorphin), and has yielded reasonable estimates of individual hormone half-lives, as shown in Fig. 8.3. In addition, the multiple-parameter deconvolution model provides good estimates of daily hormone secretory rates (Veldhuis et al. 1984, 1987a, 1988a, 1989b, 1990a,b; Veldhuis and Johnson 1988b, Iranmanesh et al. 1989a; Hartman et al. 1990b; Sollenberger et al. 1990a; Urban and Veldhuis 1991). Application of this methodology to ultradian GH release in a normal man is illustrated in Fig. 8.4.

A prediction of earlier studies using infusions of radiolabelled hormone to calculate the half-lives of various protein hormones, and ablative experiments to estimate the half-lives of endogenous hormones, is that the half-life of any given hormone can vary substantially among different subjects. For example, steady-state infusions of hormones in the 1970s disclosed GH metabolic clearance rates (litres of plasma irreversibly cleared of hormone every 24 h) of approximately 200, with a range among different healthy subjects of approximately four-fold (Thompson et al. 1972). This prediction of large interindividual variability in the kinetics of GH removal is confirmed by deconvolution analysis, when plasma hormone concentration profiles are used to calculate the half-life of *endogenous GH* in individual healthy subjects and/or patients with various pathophysiological conditions of interest: see Fig. 8.5a. Indeed, in both healthy ageing and obesity, a decrease in the half-life of endogenous GH has been inferred recently by deconvolution analysis (Iranmanesh et al. 1991a; Veldhuis et al. 1991). This inferred acceleration of GH removal in human obesity has been corroborated by steady-state metabolic clearance studies in obese orchidectomized rhesus monkeys, wherein infused recombinant human

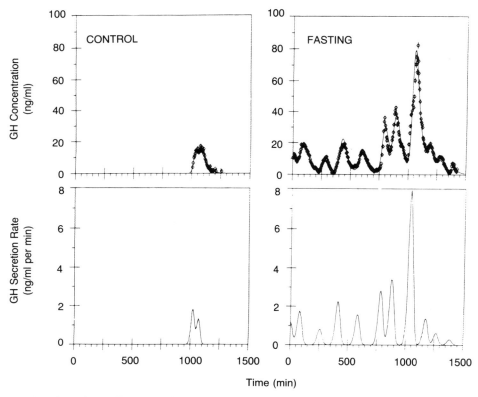

Fig. 8.4. Experimentally observed pulsatile profiles of serum growth hormone (GH) concentrations as assessed with the multiple-parameter convolution integral. Blood samples were withdrawn at 5-min intervals for 24 h in a healthy man in both the fed (left) and fasted (right) state. Top panels give mean sample immunoreactive GH concentrations (±range of duplicates) observed serially over time. Values below the detection limit for this assay (0.5 ng/ml) are not shown. Continuous curves through the observed data points represent calculated convolution fits predicted by the multiple-parameter model. Bottom panels show computer-resolved GH secretory bursts, whose amplitudes, durations, and temporal locations are defined with statistical confidence limits by deconvolution analysis. Zero time is 08:00 clocktime in these Figures. (Adapted, with permission, from Veldhuis and Johnson 1988b.)

GH was removed approximately one-third more rapidly than in lean control animals (Dubdey et al. 1988). Similarly, wide variations in the daily secretory rates of GH can be recognized among healthy human subjects (Fig. 8.5b).

The foregoing description of deconvolution analysis applies to model-specific algorithms, in which a particular *form* for individual secretion events (and the possible *existence* of discrete secretion events) is assumed. Although such téchniques have broad applicability to episodic hormone release, in some circumstances no plausible description of the secretion event is available and/or secretion may be generated in a time-invariant (non-pulsatile) manner. To evaluate such endocrine data, a *waveform-independent* deconvolution technique should be considered (Veldhuis and Johnson 1990b, 1991a,b).

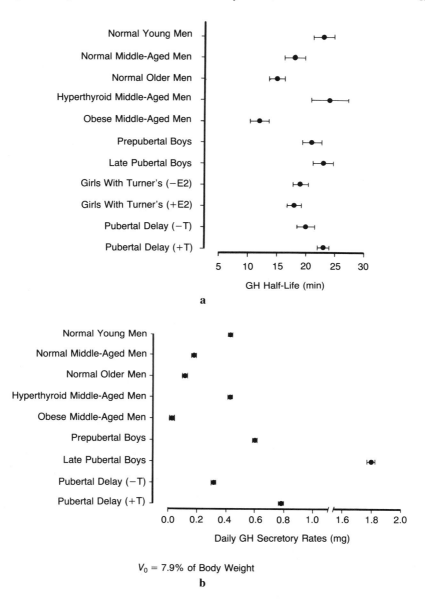

Fig. 8.5. a Spectrum of in vivo growth hormone (*GH*) half-lives and **b** daily GH secretion rates as estimated by deconvolution analysis in various clinical physiological and pathophysiological contexts. V_o, volume of distribution. (J. D. Veldhuis, A. Iranmanesh, M. O. Thorner, M. Hartman, W. S. Evans, A. D. Rogol, M. L. Johnson and G. Lizarralde, unpublished results).

Waveform-independent deconvolution analysis requires either some knowledge of the kinetics of hormone removal and/or an assumption that tonic (fixed basal) hormone secretion does not occur. A priori knowledge of the half-life is not obligatory if no basal secretion is present or the exact amount of basal secretion is already known. These constraints reflect the inherent statistical

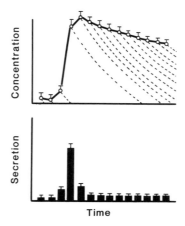

Fig. 8.6. Schematized model of the concept of a particular waveform-independent deconvolution technique. Serially observed sample hormone concentrations (upper panel) are considered to arise from the combined effects of individual sample secretion rates (lower panel) and clearance kinetics (e.g. exponential decay). Errors intrinsic to the hormone half-life estimate and dose-dependent experimental variance within the hormone assay are *both* used to estimate the statistical confidence limits for each of the calculated sample secretion values. In this model, some a priori knowledge of the half-life of hormone disappearance (decay) or assumptions about basal secretion must be available (Veldhuis et al. 1990a; Veldhuis and Johnson 1992).

difficulty of simultaneously evaluating multiple highly correlated parameters in a model of hormone release that has very little structure. Specifically, if there is no knowledge of a plausible waveform of secretion, estimated values of sample secretion rates will be strongly negatively correlated with simultaneous estimates of half-life and basal secretion. A more tractable numerical problem exists if the basal secretion term is zero or otherwise fixed (Linkowski et al. 1985; Veldhuis and Johnson 1990b, 1991b). Alternatively, assuming a priori knowledge of the hormone half-life, one may simultaneously estimate both basal and pulsatile secretion rates by a waveform-independent deconvolution technique, e.g. as illustrated in Fig. 8.6.

An application of waveform-independent deconvolution analysis to 24-h profiles of serum ACTH and cortisol concentrations is shown in Fig. 8.7. Here, we have assumed the literature-specified (two-component) half-lives of ACTH and cortisol and applied a waveform-independent technique to estimate their corresponding secretory rates at each sample observation. This waveform-independent methodology, like the multiple-parameter deconvolution approach discussed above, reveals that punctuated ACTH and cortisol secretory bursts occur on an ultradian time scale. Individual bursts of either hormone are episodic, inasmuch as knowledge of prior interburst intervals does not allow one to predict the next interburst interval. As discussed further below, waveform-independent deconvolution analysis has provided an interesting tool with which to relate ultradian bursts of hormone release to the well-established circadian variations in plasma hormone concentrations (Veldhuis et al. 1990a).

Waveform-independent deconvolution has also permitted an estimate of the percentage of hormone release that occurs within bursts versus the percentage that is "interburst" or basal (Veldhuis et al. 1990a). In the cases of GH, LH,

FSH, ACTH and β-endorphin (but not TSH or prolactin), a purely burst-like mode of hormone release without significantly measurable (interpeak) basal secretion can be inferred from available literature estimates of the hormone half-lives (Veldhuis et al. 1990a). Interestingly, the measurable amount of basal hormone secretion that can be inferred for prolactin is subject to apparent circadian variations in normal young men (Veldhuis et al. 1990a). The mechanisms that govern basal hormone secretion, specify whether it is constitutive (autonomous) or subject to long-term regulation, and/or impose 24-h variations on basal secretion are not yet defined in any detail.

Although no single deconvolution methodology is suited perfectly to the evaluation of diverse ultradian rhythms, several important and desirable features can be considered:

1. Estimate endogenous secretory and/or clearance rates by model-specific as well as waveform-independent methodologies.
2. Provide formal statistical error propagation to encompass experimental uncertainty in both the data measurements and the half-life/secretion estimates.
3. Avoid negative secretion and/or half-life calculations, and avoid "ringing" (oscillations of secretory rate about zero).
4. Compare and show fitted (predicted) function with observed data.
5. Operate robustly over a range of different sampling intensities, signal shapes (waveforms), frequencies and amplitudes as well as hormone half-lives, and signal-to-noise ratios.
6. Be validated by both computer simulations and in vivo biological data.
7. Allow for more complex systems behaviour (e.g. multiple pulse generators; multicompartmental kinetics, etc.).

I recommend that investigators consider the aim and design of the study and the nature of the data being collected when assessing which deconvolution technique to use in the quantitative examination of ultradian rhythms (see Veldhuis and Johnson 1992).

Rhythmic Modulation of Random Bursts in Normal Hypothalamo-pituitary Physiology

Considerable interest has focussed recently on the possible application of non-linear dynamic systems to various branches of biology and the physiological sciences (Swinney 1986; Pool 1989). The application of so-called deterministic chaos to endocrine data has been very limited to date, because far larger numbers of observations are required for the valid assessment of apparently chaotic systems than can be achieved readily in most physiological experiments; e.g. 100 000 observations or more can be required for chaos analysis, whereas approximately 100–300 samples are available typically to the experimental endocrinologist. None the less, some principles of stochastic analysis have been applied to hormone release episodes, as discussed above, to evaluate whether the timing of successive endocrine signals is statistically independent, i.e. the burst-generator system constitutes a renewal process.

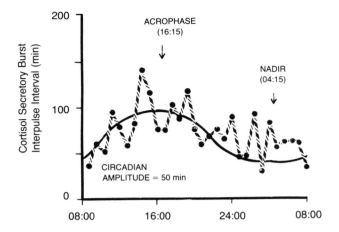

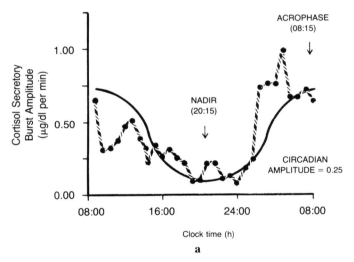

a

Fig. 8.7. Circadian and ultradian rhythms in deconvolution-resolved cortisol and ACTH secretion rates. **a** Multiple-parameter deconvolution analysis was applied to profiles of serum cortisol concentrations measured in blood collected at 10-min intervals for 24 h in six normal men. The hourly discretized mean amplitudes (maximal secretion rates within bursts) and interpulse intervals of the cortisol secretory profiles over 24 h were then analysed by Fourier transformation. This strategy allows assessment of circadian rhythms in secretory burst amplitude and frequency in a group of subjects. (Adapted, with permission, from Veldhuis et al. 1989b.) **b** Circadian rhythms in ACTH secretory features were also ascertained by cosinor analysis of the resultant time plots of waveform-independent deconvolution analysis to estimate peak ACTH secretory rates (amplitudes in pg/ml per min), and ACTH intersecretory burst intervals (min) in each of eight normal men. Data are shown for all eight men (rather than group means). Maximal ACTH secretory rates (amplitudes) but not interburst intervals varied significantly over 24 h. *NS*, not significant. (Adapted, with permission, from Veldhuis et al. 1990b.)

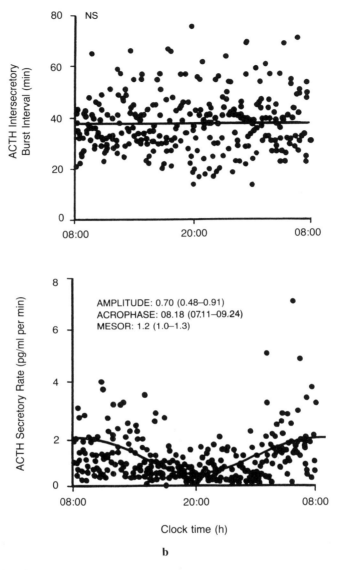

Fig. 8.7. *Continued*

Although anterior pituitary secretory activity manifests a predominantly stochastic nature, the patterns of variation of hormone pulse amplitudes and/or interpulse intervals over 24 h may exhibit deterministic trends. For example, GH secretory burst frequency appears to increase significantly at night during stage III and IV (delta or slow-wave) sleep (Holl et al. 1991). The decrease in GH interburst interval values in slow-wave sleep compared to that during wakeful hours contributes to a 24-h rhythm in GH secretory burst frequency

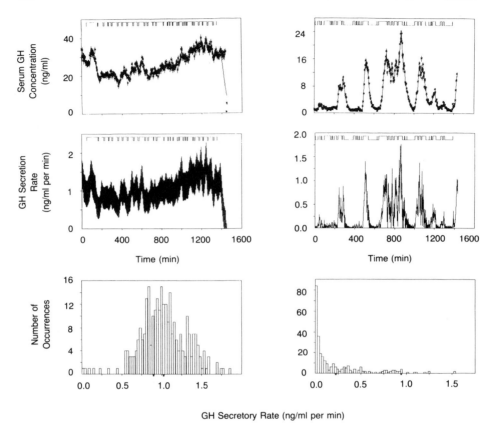

Fig. 8.8. Comparison of waveform-independent deconvolution analysis of 24-h serum growth hormone (*GH*) concentration profiles in a normal subject (right panels) and an acromegalic patient with a GH-secreting pituitary tumour (left panels). (Data were provided by Drs M. O. Thorner and M. Hartman.) (Adapted, with permission, from Veldhuis and Johnson 1991a.)

(Hartman et al. 1990b; Veldhuis et al. 1990a). Thus, ultradian bursts of GH release although they occur at seemingly random intervals can be subject to nycthemeral regulation, which includes sleep-associated changes. Similar in some respects is the approximately 2.2-fold circadian variation in cortisol secretory burst frequency in the human, which achieves a maximum shortly before the time of arising from sleep and a minimum in the late afternoon (Veldhuis et al. 1989b). Such findings indicate the influence of deterministic 24-h elements acting on what otherwise would appear to be a randomly ultradian-activated hypothalamic burst-generator system. Whether some or all of such deterministic nycthemeral variations arise from, or are coupled to, a circadian oscillator system in the suprachiasmatic nucleus or elsewhere is not known unambiguously for most hormones in humans.

When the secretion of various (anterior pituitary) hormones is estimated over 24 h by waveform-independent deconvolution analysis, significant nycthemeral variations in the *amplitude* of computer-resolved hormone secretory bursts can be demonstrated for GH, LH, TSH, prolactin, ACTH, β-endorphin and cortisol (Hartman et al. 1990b; Veldhuis et al. 1990a,b). Indeed, GH and ACTH secretory burst amplitudes vary by several fold over 24 h at one extreme, whereas FSH secretory burst amplitude does not vary significantly in a circadian manner. Some of these patterns are illustrated in Fig. 8.7. Notably, some hormones such as GH, TSH and β-endorphin exhibit both amplitude *and* frequency variations in their ultradian rhythms over 24 h (Hartman et al. 1990b; Veldhuis et al. 1990a). Other hormones, such as ACTH and LH (in certain conditions), show only amplitude modulation of ultradian pulses over 24 h.

The exact physiological mechanisms that specify 24-h variations in the *amplitudes* of otherwise randomly varying hormone secretory bursts are not fully understood, but may in principle involve differential hypothalamo-pituitary sensitivity to relevant secretagogues or antagonists and/or to negative feedback signals from the target tissue as a function of time of day, activity status, nutrient intake, sleep or other circadian cues. In addition, circadian-dependent variations in hormone secretory rates could arise from differential sensitivity of pituitary secretory cells to paracrine and autocrine factors that promote or oppose intraglandular hormone biosynthesis, storage, and secretion or degradation. Further neuroendocrine studies will be required to explain the 24-h variations that occur in ultradian secretory burst amplitudes as well as ultradian secretory burst frequencies.

Selected Pathophysiology of Disordered Ultradian Secretion of Anterior Pituitary Hormones

An increasing array of alterations in the episodic mode of ultradian pituitary hormone release has been recognized in clinical pathophysiology. Only a resumé of some of the more conspicuous disturbances can be given here. For convenience these are summarized in relation to several of the seven anterior pituitary hormones studied clinically.

GH

Alterations in the ultradian release of GH have been recognized in a vast number of pathophysiological conditions, as well as numerous physiological states (Veldhuis et al. 1985, 1988a, 1991, 1992a; Mauras et al. 1987, 1990; Ho et al. 1988; Asplin et al. 1989; Mauras and Veldhuis 1990; Ulloa-Aguirre et al. 1990; Iranmanesh et al. 1991a,b). For example, mean GH secretory burst amplitude increases approximately three-fold during the active stages of the human pubertal growth spurt, whereas there is no change in GH secretory burst frequency (Mauras et al. 1987; Ho et al. 1988). The amplification of GH secretory burst amplitude in midpubertal boys is specific, since the duration and number of GH secretory bursts do not change, nor does the half-life of endogenous GH. This increase in GH secretory burst amplitude results in an increased mass of GH secreted per burst, which can be induced in humans by androgen or oestrogen administration (Mauras and Veldhuis 1990; Ulloa-Aguirre et al. 1990). Indeed, small doses of oestrogens (100 ng/kg) given to prepubertal girls with Turner's syndrome for a week will increase the amount of GH secreted per burst by several fold (Mauras and Veldhuis 1990).

Various nutritional cues also markedly modify GH release. For example, catabolic states such as diabetes mellitus and fasting increase both the amplitude and detectable frequency of pulsatile GH release, whereas feeding rapidly suppresses pulsatile GH secretion (Ho et al. 1988; Asplin et al. 1989). Hyperthyroidism and short-term glucocorticoid excess also are accompanied by increased detectable GH secretory burst frequency with no change in GH half-life (Iranmanesh et al. 1991b; Veldhuis et al. 1992a). Conversely, obesity is associated with a decrease in the amplitude of GH secretory bursts (Veldhuis et al. 1991). Further studies in healthy individuals indicate that obesity and ageing in men is accompanied by progressive declines in GH secretory burst amplitude, frequency and half-life, which result in marked decreases in 24-h mean serum GH concentrations (Iranmanesh et al. 1991a). Further investigations are required to explicate the particular control mechanisms that amplify or suppress GH ultradian secretory burst frequency, amplitude, and/or GH half-life in various pathophysiological states and in healthy ageing.

Although tonic basal (non-pulsatile) secretion of GH cannot be demonstrated in normal children or adults, patients with GH-secreting pituitary tumours resulting in clinical acromegaly exhibit a marked amount of apparently non-pulsatile basal GH secretion (Hartman et al. 1990a; Fig. 8.8). The appearance of basal (non-pulsatile) hormone secretion may be a more general feature of endocrine tumours, since recently we have also observed augmented inter-pulse hormone secretion in patients with aldosterone-secreting adrenal tumours (Vieweg et al. 1991).

LH and FSH

Although relatively few data are available regarding the regulation of pulsatile FSH release on an ultradian time scale, one study in normal men has indicated that FSH secretory burst amplitude can be controlled negatively by dihydrotestosterone, whereas the half-life of endogenous FSH is negatively regulated by oestradiol (Urban and Veldhuis 1991).

The episodic secretion of LH has been evaluated more extensively than that of FSH in young and older men and in women of reproductive and post-menopausal ages. Both in vitro Leydig cell bioassay of LH and immunoassay (radioimmunoassay as well as immunoradiometric assay) have been employed in conjunction with deconvolution analysis to evaluate LH secretion in vivo (Veldhuis and Johnson 1988a; Veldhuis et al. 1989a). Bioassay and immunoassay have provided generally similar pictures of episodic LH release in young men, in whom the mean LH interburst interval is approximately circahoral. In both the bioassay and immunoassay, little evidence of tonic/basal LH secretion can be inferred in young individuals, although some element of basal secretion may appear with increasing age in healthy men (Urban et al. 1988; Veldhuis et al. 1992b). Interestingly, both the frequency and the half-duration (duration of the calculated secretory event at half-maximal amplitude) of LH secretory bursts manifest age-dependent increases, whereas LH secretory burst amplitude falls in older individuals (Veldhuis et al. 1992b).

Ultradian patterns of immunoreactive and bioactive LH release have also been quantified throughout the normal human menstrual cycle. Specific regulation of LH secretory burst half-duration, frequency, mass and amplitude can be demonstrated, whereas the total daily production rate of LH is independent of the stage of the menstrual cycle (see Fig. 8.9). Moreover, the estimated half-life associated with the irreversible metabolic removal of LH from plasma was stable across the entire menstrual cycle (Sollenberger et al. 1990a). Thus, the anterior pituitary gland secretes a similar total amount of LH each day at different stages of the menstrual cycle, but the *mechanisms* by which these equivalent daily LH secretion rates are achieved vary significantly.

Deconvolution studies of pituitary responsiveness to injections of synthetic GnRH have indicated that a significant contribution to the changing amplitude of spontaneous LH secretory bursts in the late follicular and mid-luteal phases of the menstrual cycle probably arises from the changing responsivity of LH-secreting cells to the GnRH stimulus (Sollenberger et al. 1990b). Such changes in gonadotrophic-cell secretory responsiveness presumably result from the previous pattern of endogenous GnRH release and the time and dose-dependent feedback effects of steroid hormones, such as oestradiol, on the hypothalamus and anterior pituitary gland.

Pathophysiological alterations in the episodicity of LH release have been inferred in various circumstances such as women long-distance runners, healthy ageing men, postpartum women, men with acute alcoholic abstinence syndrome, and chronic renal disease (Veldhuis et al. 1985b, 1989c, 1992b; Iranmanesh et al. 1988; Urban et al. 1988; Sollenberger et al. 1990b; Nunley et al. 1991). Further studies are required to evaluate the extent to which alterations in the ultradian pattern of LH release actually contribute to various clinical disorders of menstrual cyclicity, fertility or sexual function.

TSH and Prolactin

Marked increases in serum concentrations of TSH occur in primary thyroidal failure, but the exact secretory and clearance mechanisms generating such increases are not entirely clear. In addition, although increased serum prolactin concentrations are described in women with primary thyroidal failure, a recent study revealed the virtual absence of this occurrence in hypothyroid men (Iranmanesh et al. 1992). However, thyroid hormone replacement in such men temporarily suppressed the amplitude of ultradian prolactin secretory pulses, providing another instance of amplitude control (Iranmanesh et al. 1992).

Detailed studies of the ultradian dynamics of prolactin release in hyperprolactinaemia in the human are currently limited. Two reports of the mechanisms of ultradian prolactin release identified oestrogen- or suckling-induced amplitude modulation as the means by which variations in serum prolactin concentrations are achieved in response to steroid hormone or in postpartum lactation (Veldhuis et al. 1989c; Nunley et al. 1990). This is illustrated in Fig. 8.10. In contrast, a decrease in the amplitude of prolactin pulses has been recognized in diabetes mellitus (Veldhuis et al. 1989c), but the cause for this association is not evident.

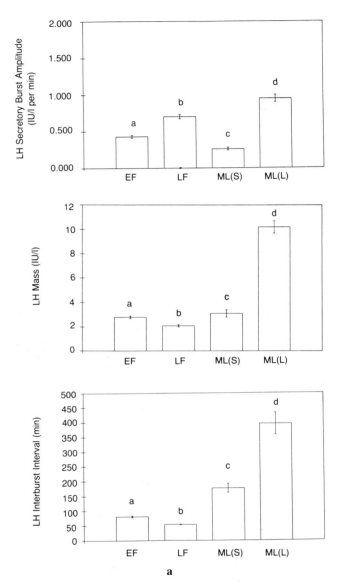

Fig. 8.9. Physiological variations in specific quantitative features of the episodic release of luteinizing hormone *(LH)* throughout the normal human menstrual cycle. **a** Mean secretory-event measures. *EF*, early follicular; *LF*, late follicular; *ML*, mid-luteal (*S*, small bursts; *L*, large). **b** Serum LH pulse profiles subjected to deconvolution analysis. Upper curves are predicted serum LH concentrations and lower curves are calculated LH secretory rates. Different letters in **a** denote significantly different means. [Adapted, with permission, from Sollenberger et al. 1990a.)

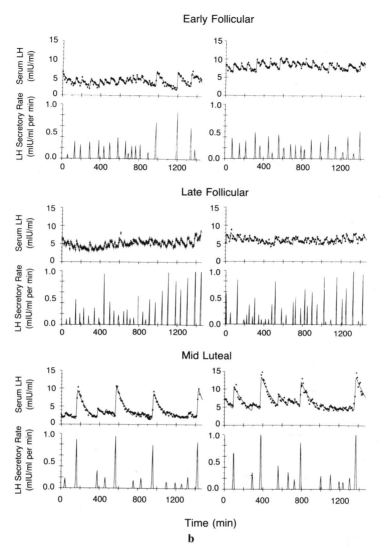

Fig. 8.9. *Continued*

ACTH and β-Lipotropin/β-Endorphin

The pulsatile patterns of plasma ACTH and β-endorphin/β-lipotropin concentrations have been described, and a strong correlation with concurrent or
10- and 20-min lagged serum cortisol concentrations demonstrated (Iranmanesh
et al. 1989a; Veldhuis et al. 1990b). In addition, all three hormones of the
corticotrophic axis are highly synchronized in the timing of their individual
release episodes in normal men (Veldhuis et al. 1990b). Deconvolution analysis

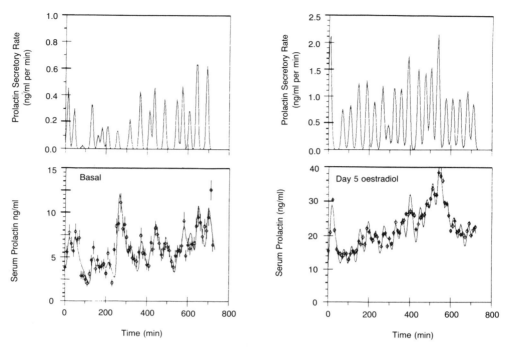

Fig. 8.10. Amplitude modulation of ultradian prolactin secretory bursts by oestrogen. Twelve-hour serum prolactin concentration profiles were obtained by sampling blood every 10 min in a post-menopausal woman before and after 5 days of oestrogen administration. The Figure gives deconvolution-determined prolactin secretory rates (upper panels) corresponding to the measured serum prolactin concentrations, and the deconvolution-specified fits of the data (lower panels). (Adapted, with permission, from Veldhuis et al. 1989d.)

of plasma ACTH time-series has demonstrated a predominantly burst-like mode of ACTH secretion in healthy men, and a prominent 24-h variation in ACTH secretory burst amplitude but not frequency (Veldhuis et al. 1990b). Thus, the well-established 24-h rhythm in blood ACTH concentrations can be accounted for by circadian control of the amplitude of apparently randomly dispersed ultradian ACTH secretory bursts.

Diurnal variations in both the frequency and amplitude of pulsatile cortisol secretion are recognized in normal physiology (Veldhuis et al. 1989b), with disturbances in clinical pathophysiology, such as the alcoholic abstinence syndrome (Iranmanesh et al. 1989b) or hypothyroidism (Iranmanesh et al. 1990). Other studies in patients with clinically excessive ACTH secretion have revealed increased frequency and/or increased amplitude of episodic ACTH /cortisol release (Linkowski et al. 1985). Because of the importance of this stress-responsive axis, further investigations of ultradian secretory rhythms of ACTH, β-endorphin and cortisol will be important to clarify the pathophysiology of a number of stressors such as transmeridian travel, neuropsychiatric disorders, Cushing's disease, depression, strenuous exercise, etc.

Implications of Episodic (Ultradian) Endocrine Signalling

The physiologically episodic mode of ultradian endocrine glandular signalling, rather than time-invariant hormone secretion, has several distinct implications. Some of these are summarized below.

1. Episodic exposure of the target tissues to the hormone (agonist) stimulus obviates down-regulation of tissue responses.
2. Brief bursts of secretion permit economy of glandular secretory activity over 24 h.
3. Selective amplitude and/or frequency modulation of ultradian secretory events can generate circadian rhythms.
4. Ultradian hormone rhythms reflect the transduction of phasic neural signalling by less rapidly responsive endocrine glands.

Plausible implications include (but are not limited to) the following. First, in many but not all endocrine systems, intermittency of the stimulatory input to an endocrine gland obviates down-regulation (progressively diminishing responses to the same stimulus), which otherwise can occur when a tissue is exposed to an unvarying amount of an otherwise effective secretagogue. This principle is illustrated particularly well in the gonadotrophic axis, inasmuch as down-regulation of pituitary LH and FSH secretion will occur if the pituitary gland is exposed to an unvarying GnRH stimulus, such as a constant infusion of GnRH (Yen et al. 1975; Knobil 1980; Catt et al. 1983). In contrast, pulsatile delivery of GnRH at an optimal frequency and amplitude can induce specific gene expression within gonadotrophic cells encoding biologically relevant subunits of LH and FSH glycoprotein hormones (Shupnik 1990). Indeed, the clinical use of long-acting GnRH agonist peptides in hormone-dependent tumours takes advantage of the physiological requirement of the anterior pituitary gland for intermittent ultradian exposure to GnRH, since long-acting potent GnRH agonists that exert a sustained unvarying effect on the anterior pituitary gland will profoundly depress gonadotrophin biosynthesis and release. This down-regulating property of an unvarying GnRH stimulus achieves reversible inactivation of the gonadotrophic axis, i.e. a castration effect (St Arnaud et al. 1986). Conversely, the intermittent (e.g. circhoral or sesquihoral) administration of GnRH in GnRH-deficient patients can establish a normal episodic pattern of gonadotrophic hormone release, obviate down-regulation of pituitary and gonadal target cells, and induce or restore normal sexual function and fertility (Valk et al. 1980; Hoffman and Crowley 1982).

Secondly, a burst-like mode of hormone release provides for economy of endocrine glandular secretory function. Indeed, for long-lived hormones, the parent gland spends only a fraction of the day in secretion (See Table 8.1). Intermittent stimulation provides an opportunity for the secretory gland to recover biosynthetic activity, process newly synthesized hormone, prepare exocytotic vesicles, and deliver secretory granules to the plasma membrane. This reasoning may explain the finding that pulses of GnRH evoke more biologically active LH secretion than that which occurs during the interburst interval (Veldhuis et al. 1987c). Note that a brief interval of secretory activity can achieve a sustained increase in circulating hormone concentrations, if the secreted molecules are subject to delayed biotransformation and/or retarded degradation (Veldhuis et al. 1987a). For example, the long half-life of FSH (approximately 4 h) for immunoreactive and bioactive hormone in humans (Urban et al. 1991) permits a sustained increase in circulating FSH concentrations despite small amounts of FSH secretion at relatively infrequent intervals (Urban and Veldhuis 1991). This intuitive inference can be verified by mathematical modelling (Veldhuis and Johnson 1988b). In contrast, the more rapid disappearance of ACTH from plasma (half-lives of about 3 and 25 min for first and second component decays) imposes a requirement for more frequent and/or higher-amplitude secretory bursts to sustain circulating ACTH concentrations (Veldhuis et al. 1990b). Accordingly, hormones with shorter half-lives tend to be secreted more frequently or over a more extended pulse duration (Table 8.1).

Prolongation of hormone half-life can also be achieved by the presence of one or more macromolecules in the blood that are capable of associating with the hormone with high affinity; e.g. sex-steroid-hormone-binding globulin, which binds testosterone and oestradiol; GH-binding protein, which can sequester the higher molecular weight protein hormone GH; or various insulin-like growth-factor-binding proteins. Such circulating receptor-like (glyco)proteins serve to extend the plasma residence time of a hormone, even when the hormone is discharged into the bloodstream infrequently in bursts of relatively short duration, e.g. minutes. Moreover, plasma macromolecules with specific and high affinity binding properties offer a mechanism to expand the apparent volume of hormone distribution by short-term retention of the hormone in a sparingly metabolized but circulating reservoir.

In addition to the presence of appropriate "capture" macromolecules in the circulation to delay the biotransformation or metabolic degradation of a hormone, specific structural features of secreted hormones may limit their removal to targeted metabolic sites. Examples of the latter specialization include the glycoprotein hormones LH, human chorionic gonadotrophin, FSH, and TSH, whose metabolic clearance is profoundly delayed by the post-translational addition of complex oligosaccharide residues to the polypeptide chain (Dufau and Veldhuis 1987). These carbohydrate residues direct metabolism of the glycoprotein hormones to selected tissues, e.g. in the reticuloendothelial system of the liver, etc. When hormone removal is delayed, the high energy processes of glandular hormone biosynthesis, post-translational modification, packaging in exocytotic granules, and active secretion can be restricted to a much shorter interval. For example, as shown in Table 8.1, we have estimated the times required for the secretion of 95% of the total daily amounts of the various anterior pituitary hormones. We note that secretion is not required over a full 24-h span, but can be compressed into substantially shorter intervals if frequency and/or amplitude-modulated bursts of secretion are combined with delayed metabolic removal. Indeed, the use of deconvolution modelling has highlighted this principle of secretory parsimony.

Recent analyses indicate that ultradian bursts of hormone release although apparently dispersed randomly over time can be subject to deterministic modulation in a circadian manner (above). In particular, the mean amplitude and/or the mean frequency of episodic hormone secretory bursts may be modulated over 24 h, which in the face of unvarying metabolic clearance rates yields a circadian variation in the mean plasma hormone concentration. Thus, a third plausible benefit of ultradian secretory bursts is to provide a structure with which to assemble the circadian rhythm of plasma hormone concentrations. Although the full physiological import of nycthemeral variations in multiple plasma hormone concentrations is not understood, changes in the amount of hormone bathing target tissues can produce prominent changes in the biochemical activity and genomic expression of target cells and thereby prepare the organism for optimal adjustments to homeostatic demands at relevant times in the sleep–wake cycle. Moreover, such adjustments in metabolism can be entrained by relevant environmental zeitgebers (Krieger and Aschoff 1979; Moore-Ede et al. 1983). Thus, the activities of numerous enzymes, ligands, metabolites, substrates, and gene expression itself can be made to vary as a function of time of day. Ultradian secretory episodes modulated in frequency and/or amplitude by a relevant interface with circadian oscillator systems can provide a mechanism for achieving such pertinent variations within the whole organism.

A fourth plausible implication of the ultradian intermittency of pituitary glandular secretory activity is the facilitation of neuronal-glandular communication. Since the biological functions of nervous tissue are carried out via intermittent electrical activation and quantal signalling at the synapse, suitable mechanisms for coupling this phasic operating mode to a less excitable endocrine gland would include episodic triggering of hormone-secreting cells by bursts of secretagogue released by the organized neuronal network. This strategy can achieve linkage of a dichotomous signal with what is ultimately a more nearly continuous graded response. Indeed, immediately after the hormone is released into the bloodstream, the pattern of circulating hormone concentrations continues to reflect in some measure the episodicity of the neural signal. After the hormone is distributed within (and in some cases outside) the vascular compartment into interstitial fluids, attenuation of its pulsatile nature occurs. In addition, hormones that have very prolonged plasma residence times and/or associate with high-affinity macro-molecules in blood achieve integration of the intermittent neuronal signal into a nearly smoothly varying endocrine output to target tissues.

Ultradian rhythms in the concentrations of hormones bathing the cell-surface receptors of target tissues can allow receptor internalization and recycling between successive stimuli; so that the target tissue is no longer refractory at the time the second stimulus is administered. This presumptive sequence may pertain for the insulin and GH receptors in vivo (Goodner et al. 1988; Baumann et al. 1989). Such recycling is economical of receptor synthesis, processing, and activation. In contrast, an unvarying amount of ligand presented to the receptor system would activate the tissue continuously (if it did not desensitize), but only at the expense of a larger receptor population and/or an increased efficiency of receptor recycling. According to this speculation, one would predict that receptors for hormones with a short half-life and a high frequency of pulsatile activity (e.g. ACTH, insulin and GH compared to FSH) would be recycled more rapidly and/or be less prone to desensitization.

Finally, ultradian episodes of hormone release allow for possible *interactions* between two (or among three or more) pulse-generator systems (Veldhuis and Johnson 1988b). Indeed, LH secretion in the normal luteal phase of the menstrual cycle might exemplify a two-pulse-generator system (Sollenberger et al. 1990a). Interactions between the output of distinct pulse generators

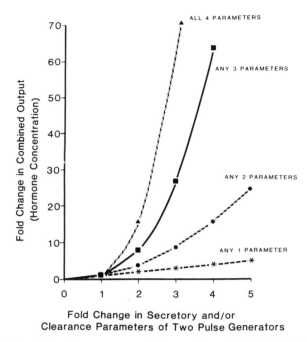

Fig. 8.11. Multiplicative interactions among multiple secretion and clearance parameters in controlling output (hormone concentrations). Note that in principle a marked increase in plasma hormone concentrations could occur by means of increased secretory burst frequency, amplitude, or duration and/or a prolonged half-life of hormone disposal. (Adapted, with permission, from Veldhuis et al. 1990c.)

provide one means of augmenting plasma hormone concentrations in a finely modulated manner (see Fig. 8.11).

Physiological relevance. Relatively few experimental studies are available that have investigated directly the physiological relevance of the episodic mode of hormone signalling to target tissues remote from the pituitary gland. Indeed, the importance of an episodic signal to avoid down-regulation is best demonstrated in the pituitary gland by hypothalamic GnRH stimulation of LH- and FSH-secreting cells (Valk et al. 1980; Hoffman and Crowley 1982; Catt et al. 1983; St Arnaud et al. 1986). The physiological relevance of intermittent hypothalamic signalling has been suggested also for stimulation by ACTH-releasing hormone (CRH) of the corticotrophic axis and possibly TRH stimulation of thyrotropes (Morley 1981; Rivier and Vale 1983). On the other hand, there is less desensitization of human somatotropes to sustained GH-releasing hormone exposure, possibly because intermittent somatostatin inhibition and withdrawal occurs (Vance et al. 1989).

The putative importance of an episodic endocrine signal at a remote extrapituitary site has been suggested by limited in vitro experiments in the rat and human ovary, demonstrating enhanced secretion of oestradiol and progesterone in response to ultradian pulses of LH (rather than continuous delivery of the same mass of hormone) in perifusion systems (Peluso et al. 1984; Weiss et al. 1989). In addition, pulsatile progesterone and LH release are significantly correlated in women (Veldhuis et al. 1988b). On the other hand, some in vivo experiments in the rat and ram indicate that testicular responses to pulsatile and continuous LH infusions can be similar (Chase et al. 1988; Gibson-Berry and Chase 1990), without a clear physiological benefit conferred by an experimentally pulsatile mode of gonadotrophin delivery.

A second endocrine system in which evidence exists for the importance of an intermittent hormone signal outside the pituitary gland is the somatotrophic axis. In the hypophysectomized rat, pulses of GH given in an ultradian fashion induce linear growth, and certain muscle and liver proteins and genes more effectively than the same dose of GH given continuously (Jansson et al. 1982, 1988; Isgaard et al. 1988). Equivalent data are not available for humans at present.

Investigations of the corticotrophic axis are less clear in defining the significance of an episodic ultradian ACTH signal to the adrenal glands. Intermittent ACTH-releasing hormone administration in the rat and sheep (Jansson et al. 1982; Rivier and Vale 1983) and pulsatile ACTH injections in the human (Manchester et al. 1983) often but not always favour continued secretory responsiveness. At the adrenal level, secretion of aldosterone is poorly sustained by continuous ACTH infusions but effectively maintained by pulsatile ACTH administration in normal men (Seely et al. 1989). Whether the actions of TSH, prolactin or β-lipotropin and β-endorphin on their respective extrapituitary target tissues are dependent upon, or enhanced by, the pulsatile mode of delivery of these hormones is not yet evident.

Acknowledgements

I thank Patsy Craig for preparation of the manuscript and Paula P. Azimi for the artwork. This work was supported in part by NIH grant RR 00847 to the General Clinical Research Center of the University of Virginia, RCDA 1 KO4 HD00634, Diabetes and Endocrinology Research Center Grant NIH DK-38942, NIH-supported Clinfo Data Reduction Systems, the Pratt Foundation, the University of Virginia Academic Enhancement Fund, and the NSF Science Center for Biological Timing (DIR-8920162).

References

Asplin CM, Faria ACS, Carlsen EC et al. (1989) Alterations in the pulsatile mode of growth hormone release in men and women with insulin-dependent diabetes mellitus. J Clin Endocrinol Metab 69:239–245

Barraclough CA, Wise PM (1982) The role of catecholamines in the regulation of pituitary LH and FSH secretion. Endocr Rev 3:91–119

Baumann G, Shaw MA, Buchanan TA (1989) In vivo kinetics of a covalent growth hormone-binding protein complex. Metabolism 38:330–333

Butler JP, Spratt DI, O'Dea LS, Crowley WF Jr (1986) Interpulse interval sequence of LH in normal men essentially constitutes a renewal process. Am J Physiol 250:E338–E340

Catt KJ, Loumaye E, Katikineni M, Hyde CL, Childs G, Amsterdam A, Naor Z (1983) Receptor and actions of gonadotropin releasing hormone (GnRH) on pituitary gonadotrophs. In: McCann

SM and Dhindsa DS (eds) Role of peptides and proteins in control of reproduction. Elsevier, Amsterdam, pp 33–68

Chase DJ, Schanbacher BD, Lunstra DD (1988) Effects of pulsatile and continuous luteinizing hormone (LH) infusion on testosterone responses to LH in rams actively immunized against gonadotropin-releasing hormone. Endocrinology 123:816

Clarke IJ, Cummins JT (1982) The temporal relationship between gonadotropin releasing hormone (GnRH) and luteinizing hormone (LH) secretion in ovariectomized ewes. Endocrinology 111:1737–1740

Daughaday WH (1981) Adenohypophysis. In: Williams RH (ed) Textbook of endocrinology. Saunders, Philadelphia, PA, pp 73–86

Denef C, Andries M (1983) Evidence for paracrine interaction between gonadotrophs and lactotrophs in pituitary cell aggregates. Endocrinology 112:813–818

Dubdey AK, Ahanukoglu A, Hansen BC, Kowarski AA (1988) Metabolic clearance rates of synthetic human growth hormone in lean and obese male rhesus monkeys. J Clin Endocrinol Metab 67:1064–1067

Dufau ML, Veldhuis JD (1987) Pathophysiological relationships between the biological and immunological activities of luteinizing hormone. In: Burger HG (ed) Bailliere's clinical endocrinology and metabolism, vol 1. WB Saunders, Philadelphia, PA, pp 153–176

Gambacciani M, Liu JH, Swartz WH, Tueros VS, Yens SC, Rasmussen DD (1987) Intrinsic pulsatility of LH release from the human pituitary in vitro. Neuroendocrinology 45:402–406

Gibson-Berry KL, Chase DJ (1990) Continuous and pulsatile infusions of luteinizing hormone have identical effects on steroidogenic capacity and sensitivity of Leydig cells in rats passively immunized against gonadotropin-releasing hormone. Endocrinology 126:3107–3115

Goodner CJ, Sweet IR, Courtenay-Harrison H (1988) Rapid reduction and return of surface insulin receptors after exposure to brief pulses of insulin in perifused rat hepatocytes. Diabetes 37:1316–1323

Grady RR, Charlesworth MC, Schwartz NB (1982) Characterization of the FSH-suppressing activity in follicular fluid. Recent Prog Horm Res 38:409–456

Hartman ML, Veldhuis JD, Vance ML, Faria ACS, Furlanetto RW, Thorner MO (1990a) Somatotropin pulse frequency and basal growth hormone concentrations are increased in acromegaly. J Clin Endocrinol Metab 70:1375–1384

Hartman ML, Faria ACS, Vance ML, Johnson ML, Thorner MO, Veldhuis JDV (1990b) Temporal structure of in vivo growth hormone secretory events in man. Am J Physiol 260: E101–E110

Hellman N, Nakada F, Curti J et al. (1970) Cortisol is secreted episodically in normal men. J Clin Endocrinol Metab 30:411–422

Ho KY, Veldhuis JD, Johnson ML et al. (1988) Fasting enhances growth hormone secretion and amplifies the complex rhythms of growth hormone secretion in man. J Clin Invest 81:968–975

Hoffman AR, Crowley WF (1982) Induction of puberty in man by long-term pulsatile administration of low-dose gonadotropin-releasing hormone. N Engl J Med 307:1237–1241

Holl R, Hartman ML, Veldhuis JD, Taylor WM, Thorner MO (1991) Thirty second-sampling of plasma growth hormone (GH) in man: correlation with sleep stages. J Clin Endocrinol Metab 72:854–861

Iranmanesh A, Veldhuis JD, Samojlik E, Rogol AD, Johnson ML, Lizarralde G (1988) Alterations in the pulsatile properties of gonadotropin secretion in alcoholic men. J Androl 9:207–214

Iranmanesh A, Lizarralde G, Johnson ML, Veldhuis JD (1989a) Circadian, ultradian and episodic release of beta endorphin in men, and its temporal coupling with cortisol. J Clin Endocrinol Metab 68:661–670

Iranmanesh A, Veldhuis JD, Johnson ML, Lizarralde G (1989b) Twenty-four hour pulsatile and circadian patterns of cortisol secretion in alcoholic men. J Androl 10:54–63

Iranmanesh A, Lizarralde G, Johnson ML, Veldhuis JD (1990) Dynamics of 24-hour endogenous cortisol secretion and clearance in primary hypothyroidism assessed before and after thyroid hormone replacement. J Clin Endocrinol Metab 70:155–161

Iranmanesh A, Lizarralde G, Veldhuis JD (1991a) Age and relative adiposity are specific negative determinants of the frequency and amplitude of GH secretory bursts and the half-life of endogenous GH in healthy men. J Clin Endocrinol Metab 73:1081–1088

Iranmanesh A, Lizarralde G, Johnson ML, Veldhuis JD (1991b) Nature of the altered growth hormone secretion in hyperthyroidism. J Clin Endocrinol Metab 72:108–115

Iranmanesh A, Lizarralde G, Veldhuis JD (1992) Robustness of the male lactotropic axis to the hyperprolactinemic stimulus of primary thyroid failure. J Clin Endocrinol Metab (in press)

Irvine CH, Alexander SL (1987) A novel technique for measuring hypothalamic and pituitary hormone secretion rates from collection of pituitary venous effluent in the normal horse. J Endocrinol 113:183–192

Isgaard J, Carlsson L, Isaksson OGP, Jansson JO (1988) Pulsatile intravenous growth hormone (GH) infusion to hypophysectomized rats increases insulin-like growth factor I messenger ribonucleic acid in skeletal tissues more effectively than continuous GH infusion. Endocrinology 123:2605–2610

Jansson JO, Albertsson-Wikland K, Eden S, Thorngren KG, Isaksson O (1982) Circumstantial evidence for a role of the secretory pattern of growth hormone in control of body growth. Acta Endocrinol (Copenh) 9:24–30

Jansson JO, Ekberg S, Hoath SB, Beamer WG, Frohman LA (1988) Growth hormone enhances hepatic epidermal growth factor receptor concentration in mice. J Clin Invest 82:1871–1876

Kasting NW, Martin JB, Arnold MA (1981) Pulsatile somatostatin release from the median eminence of the unanesthetized rat and its relationship to plasma growth hormone levels. Endocrinology 109:1739–1745

Knobil E (1980) Neuroendocrine control of the menstrual cycle. Recent Prog Horm Res 36:53–88

Krieger DT (1979) Rhythms in CRF, ACTH and corticosteroids. In: Krieger DT (ed) Endocrine rhythms. Raven Press, New York, pp 123–142

Krieger DT, Aschoff J (1979) Endocrine and other biological rhythms. In: DeGroott LJ (ed) Endocrinology, vol 2. Grune & Stratton, New York, pp 2079–2103

Lang DA, Matthews DR, Peto J, Turner RC (1979) Cyclic oscillations of basal plasma glucose and insulin concentrations in human beings. N Engl J Med 301:1023–1028

Leong DA, Rawley LS, Neill JD (1983) Neuroendocrine control of prolactin secretion. Ann Rev Physiol 45:109–131

Levine JE, Ramirez VD (1982) Luteinizing hormone-releasing hormone release during the rat estrous cycle and after ovariectomy, as estimated with push–pull cannulae. Endocrinology 111:1439–1445

Levine JE, Norman RL, Gliessman PM, Oyama TT, Bangsberg DR, Spies HG (1985) In vivo GnRH release and serum LH measurements in ovariectomized, estrogen-treated rhesus macaques. Endocrinology 117:711–721

Linkowski P, Mendlewicz J, Leclercq R et al. (1985) The 24-hour profile of adrenocorticotropin and cortisol in major depressive illness. J Clin Endocrinol Metab 61:429–438

Manchester EL, Lye SJ, Challis JRG (1983) Activation of ovine fetal adrenal function by pulsatile or continuous administration on adrenocorticotropin-(1-24). II. Effects of adrenal cell responses in vitro. Endocrinology 113:777–782

Mauras N, Veldhuis JD (1990) Increased hGH production rate after low-dose estrogen therapy in prepubertal girls with Turner's syndrome. Pediatr Res 28:626–630

Mauras N, Blizzard RM, Link K, Johnson ML, Rogol AD, Veldhuis JD (1987) Augmentation of growth hormone secretion during puberty: evidence for a pulse amplitude-modulated phenomenon. J Clin Endocrinol Metab 64:596–601

Mauras N, Rogol AD, Veldhuis JD (1989) Specific, time-dependent actions of low-dose estradiol administration on the episodic release of GH, FSH and LH in prepubertal girls with Turner's syndrome. J Clin Endocrinol Metab 69:1053–1058

Moore-Ede MC, Czeisler CA, Richardson GS (1983) Circadian time-keeping in health and disease. N Engl J Med 309:469–477

Morley JE (1981) Neuroendocrine control of thyrotropin secretion. Endocr Rev 2:396–436

Nelson DH, Egdahl RH, Hume DM (1955) Corticosteroid secretion in the adrenal vein of the non-stressed dog exposed to cold. Endocrinology 58:309–314

Nunley WC, Urban RJ, Kitchen JD, Bateman BG, Evans WS, Veldhuis JD (1990) Dynamics of pulsatile prolactin release during the postpartum lactational period. J Clin Endocrinol Metab 72:287–293

Nunley WC, Urban RJ, Evans WS, Veldhuis JD (1991) Preservation of pulsatile LH release during postpartum lactational amenorrhea. J Clin Endocrinol Metab 73:629–636

Pavlou SN, Veldhuis JD, Linner J, Souza KH, Urban RJ, Rivier JE, Vale WW, Stallard DJ (1990) Persistence of concordant LH, testosterone and alpha subunit pulses following LHRH antagonist administration in normal men. J Clin Endocrinol Metab 70:1472–1478

Peluso JJ, Downey MC, Gruenberg ML (1984) Role of LH pulse amplitude in controlling rat ovarian estradiol-17beta secretion in vitro. J Reprod Fertil 71:107–112

Plotsky PM, Vale W (1985) Patterns of growth hormone-releasing factor and somatostatin secretion into the hypophysial-portal circulation of the rat. Science 230:461–465

Polonsky KS, Given BD, Pugh W, Licinio-Paixao J, Thompson JE, Karrison T, Rubenstein H (1985) Calculation of the systemic delivery rate of insulin in normal man. J Clin Endocrinol Metab 63:113–120

Pool R (1989) Is it healthy to be chaotic? Science 243:604–607

Rivier C, Vale W (1983) Influence of the frequency of ovine corticotropin-releasing factor administration on adrenocorticotropin and corticosterone secretion in the rat. Endocrinology 113:1422–1426

Roberts V, Meunier H, Vaughan J et al. (1989) Production and regulation of inhibin subunits in pituitary gonadotropes. Endocrinology 124:552–557

St Arnaud R, La Chance R, DuPont A, Labrie F (1986) Serum luteinizing hormone (LH) biological activity in castrated patients with cancer of the prostate receiving a pure anti-androgen and in estrogen-pretreated patients with an LH-releasing hormone agonist and anti-androgen. J Clin Endocrinol Metab 63:297–306

Santoro N, Butler JP, Filicori M, Crowley WF Jr (1988) Alterations of the hypothalamic GnRH interpulse interval sequence over the normal menstrual cycle. Am J Physiol 225:E696–E701

Seely EW, Conlin PR, Brent GA, Dluhy RG (1989) Adrenocorticotropin stimulation of aldosterone: prolonged continuous versus pulsatile infusion. J Clin Endocrinol Metab 69: 1028–1032

Shupnik MA (1990) Effects of gonadotropin-releasing hormone on rat gonadotropin gene transcription in vitro: requirement for pulsatile administration for luteinizing hormone-beta gene stimulation. Mol Endocrinol 4:1444–1450

Sollenberger ML, Carlson EC, Johnson ML, Veldhuis JD, Evans WS (1990a). Specific physiological regulation of LH secretory events throughout the human menstrual cycle: new insights into the pulsatile mode of gonadotropin release. J Neuroendocrinol 2:845–852

Sollenberger MJ, Carlsen ES, Booth RA Jr, Johnson ML, Veldhuis JD, Evans WS (1990b) Nature of gonadotropin-releasing hormone self-priming of LH secretion during the normal menstrual cycle. Am J Obstet Gynecol 163:1529–1534

Stewart JK, Clifton DK, Koerker DJ, Rogol AD, Jaffe J, Goodner CJ (1985) Pulsatile release of growth hormone and prolactin from the primate pituitary in vitro. Endocrinology 116:1–5

Swinney HL (1986) Perspectives on studies of chaos in experiments. In: Shlesinger MF, Cawley R, Saenz AW, Zachary W (eds) Perspectives in nonlinear dynamics. World Scientific, Singapore, pp 18–23

Tannenbaum GS, Ling N (1984) The interrelationship of growth hormone (GH)-releasing factor and somatostatin in generation of the ultradian rhythm of GH secretion. Endocrinology 115:1952–1957

Thompson RG, Rodriguez A, Kowarski A, Blizzard RM (1972) Growth hormone: metabolic clearance rates, integrated concentrations, and production rates in normal adults and the effects of prednisone. J Clin Invest 51:3193–3199

Ulloa-Aguirre A, Blizzard RM, Garcia-Rubi E et al. (1990) Testosterone and oxandrolone, a non-aromatizable androgen, specifically amplify the mass and rate of growth hormone (GH) secreted per burst without altering GH secretory burst duration or frequency or the GH half-life. J Clin Endocrinol Metab 71:846–854

Urban RJ, Veldhuis JD (1991) Specific regulatory actions of dihydrotestosterone and estradiol on the dynamics of FSH secretion and clearance in man. J Androl 12:27–35

Urban RJ, Veldhuis JD, Blizzard RM, Dufau ML (1988) Attenuated release of biologically active luteinizing hormone in healthy aging men. J Clin Invest 81:1020–1029

Urban RJ, Pavlou SN, Rivier JE, Vale WW, Dufau ML, Veldhuis JD (1990) Suppressive actions of a gonadotropin-releasing hormone (GnRH) antagonist on LH, FSH, and prolactin release in estrogen-deficient postmenopausal women. Am J Obstet Gynecol 162:1255–1260

Urban RJ, Padmanabhan V, Beitins I, Veldhuis JD (1991) Metabolic clearance of human follicle-stimulating hormone in man as measured by radioimmunoassay, immunoradiometric assay, and in vitro Sertoli-cell bioassay. J Clin Endocrinol Metab 73:818–823

Valk TW, Corely KP, Kelch RP, Marshall JC (1980) Hypogonadotropic hypogonadism: hormonal responses to low dose pulsatile administration of gonadotropin-releasing hormone. J Clin Endocrinol Metab 51:730–738

Vance ML, Kaiser DL, Martha PM Jr, Furlanetto R, Rivier Vale W, Thorner MO (1989) Lack of in vivo somatotroph desensitization or depletion after 14 days of continuous growth hormone (GH)-releasing hormone administration in normal men and a GH-deficient boy. J Clin Endocrinol Metab 68:22–28

Veldhuis JD, Johnson ML (1988a) In vivo dynamics of luteinizing hormone secretion and clearance in man: assessment by deconvolution mechanics. J Clin Endocrinol Metab 66:1291–1300

Veldhuis JD, Johnson ML (1988b) A novel general biophysical model for simulating episodic endocrine gland signaling. Am J Physiol 255:E749–E759

Veldhuis JD, Johnson ML (1990a) New methodological aspects of evaluating episodic neuroendocrine signals. In: Yen SSC, Vale W (eds) Advances in neuroendocrine regulation of reproduction. Plenum Press, New York, pp 123–139

Veldhuis JD, Johnson ML (1990b) Returning to the roots of endocrinology: the challenge of evaluating in vivo glandular secretory activity. Endocrinology 127:2611–2617

Veldhuis JD, Johnson ML (1991a) Deconvolution analysis of hormone data. Methods Enzymol 210:539–575

Veldhuis JD, Johnson ML (1991b) A review and appraisal of deconvolution methods to evaluate in vivo neuroendocrine secretory events. J Neuroendocrinol 2:755–771

Veldhuis JD, Beitins IZ, Johnson ML, Serabian MA, Dufau ML (1984) Biologically active luteinizing hormone is secreted in episodic pulsations that vary in relation to stage of the menstrual cycle. J Clin Endocrinol Metab 58:1050–1058

Veldhuis JD, Evans WS, Demers LM, Thorner MO, Wakat D, Rogol AD (1985) Altered neuroendocrine regulation of gonadotropin secretion in women distance runners. J Clin Endocrinol Metab 61:557–563

Veldhuis JD, Carlson ML, Johnson ML (1987a) The pituitary gland secretes in bursts: appraising the nature of glandular secretory impulses by simultaneous multiple-parameter deconvolution of plasma hormone concentrations. Proc Natl Acad Sci USA 84:7686–7690

Veldhuis JD, King JC, Urban RJ et al. (1987b) Operating characteristics of the male hypothalamo-pituitary-gonadal axis: pulsatile release of testosterone and follicle-stimulating hormone and their temporal coupling with luteinizing hormone. J Clin Endocrinol Metab 65:929–941

Veldhuis JD, Johnson ML, Dufau ML (1987c) Preferential release of bioactive luteinizing hormone in response to endogenous and low-dose exogenous gonadotropin releasing hormone (GnRH) pulses in man. J Clin Endocrinol Metab 64:1275–1282

Veldhuis JD, Faria A, Vance ML, Evans WS, Thorner ML, Johnson ML (1988a) Contemporary tools for the analysis of episodic growth hormone secretion and clearance in vivo. Acta Paediatr Scand 347:63–82

Veldhuis JD, Evans WS, Kolp LA, Rogol AD, Johnson ML (1988b) Physiological profiles of episodic progesterone release during the mid-luteal phase of the human menstrual cycle: analysis of circadian and ultradian rhythms, discrete pulse properties, and correlations with simultaneous LH release. J Clin Endocrinol Metab 66:414–421

Veldhuis JD, Johnson ML, Dufau ML (1989a) Physiological attributes of endogenous bioactive luteinizing hormone secretory bursts in man: assessment by deconvolution analysis and in vitro bioassay of LH. Am J Physiol 256:E199–E207

Veldhuis JD, Iranmanesh A, Lizarralde G, Johnson ML (1989b) Amplitude modulation of a burst-like mode of cortisol secretion gives rise to the nyctohemeral glucocorticoid rhythm in man. Am J Physiol 257:E6–E14

Veldhuis JD, Urban RJ, Beitins I, Blizzard RM, Johnson ML, Dufau ML (1989c) Pathophysiological features of the pulsatile secretion of biologically active luteinizing hormone in man. J Steroid Biochem 33:739–750

Veldhuis JD, Evans WS, Stumpf P (1989d) Mechanisms subserving estradiol's induction of increased prolactin concentrations in man: evidence for amplitude modulation of spontaneous prolactin secretory bursts. Am J Obstet Gynecol 161:1149–1158

Veldhuis JD, Iranmanesh A, Johnson ML, Lizarralde G (1990a) Twenty-four hour rhythms in plasma concentration of adenohypophyseal hormones are generated by distinct amplitude and/or frequency modulation of underlying pituitary secretory bursts. J Clin Endocrinol Metab 71:1616–1623

Veldhuis JD, Iranmanesh A, Johnson ML, Lizarralde G (1990b) Amplitude, but not frequency, modulation of ACTH secretory bursts gives rise to the nyctohemeral rhythm of the corticotropic axis in man. J Clin Endocrinol Metab 71:452–563

Veldhuis JD, Lassiter AB, Johnson ML (1990c) Operating behavior of dual or multiple endocrine pulse generators. Am J Physiol 259:E351–E361

Veldhuis JD, Iranmanesh A, Ho KKY, Lizarralde G, Waters MJ, Johnson ML (1991) Dual defects in pulsatile growth hormone secretion and clearance subserve the hyposomatotropism of obesity in man. J Clin Endocrinol Metab 72:51–59

Veldhuis JD, Lizarralde G, Iranmanesh A (1992a) Divergent effects of short-term glucocorticoid excess on the gonadotropic and somatotropic axes in normal men. J Clin Endocrinol Metab 74:96–102

Veldhuis JD, Urban RJ, Lizarralde G, Johnson ML, Iranmanesh A (1992b) Attenuation of luteinizing hormone secretory burst amplitude as a proximate basis for the hypoandrogenism of healthy aging men. J Clin Invest (in press)

Vieweg VW, Veldhuis JD, Carey RM (1992) Amplitude modulation of pulsatile aldosterone and renin release. Am J Physiol (in press)

Watanabe T, Orth DN (1987) Detailed kinetic analysis of adrenocorticotropin secretion by dispersed anterior pituitary cells in a microperifusion system: effects of ovine corticotropin-releasing factor and arginine vasopressin. Endocrinology 121:1133–1145

Weiss TJ, Steele PA, Umapathysivam PK (1989) Perifusion of human granulosa-luteal cells: response to LH stimulation. Clin Endocrinol (Oxf) 31:285–294

Winer LM, Shaw MA, Baumann G (1990) Basal plasma growth hormone levels in man: new evidence for rhythmicity of growth hormone secretion. J Clin Endocrinol Metab 70:1678–1680

Winters SJ (1990) Inhibin is released together with testosterone by the human testis. J Clin Endocrinol Metab 70:548–550

Winters SJ, Troen PE (1986) Testosterone and estradiol are co-secreted episodically by the human testis. J Clin Invest 78:870–872

Yen SSC, Lasley BL, Wang CF, Leblanc H, Siler TM (1975) The operating characteristics of the hypothalamic–pituitary system during the menstrual cycle and observations on biological action of somatostatin. Recent Prog Horm Res 31:331–363

Chapter 9

Ontogenesis of Human Ultradian Rhythms

Toke Hoppenbrouwers

Introduction

Ultradian cycles constitute a broad spectrum in the hour, minute, second and millisecond domain (Hildebrandt 1986). This chapter is restricted to those in the hour and minute domain and, with few exceptions, the data reviewed and presented were derived from human infants. There are a number of issues which have been a part of the debate about rhythms in ontogeny. The first issue is unique to the immature organism and deals with functional equivalence. Quite simply, at what point when observing cyclic behaviours in utero reminiscent of behaviours in the newborn is one justified in asserting that these behaviours are functionally equivalent? It leads to such questions as: "is the foetus awake?" "when can the basic rest–activity cycle (BRAC) be first identified?" and "what is the true birthday of the circadian rhythm?" These questions currently have no definitive answers, although attempts have been made to specify the criteria by which preliminary conclusions are reached (Nijhuis et al. 1982).

The second issue is not restricted to the ontogenetic literature: can one speak of real rhythms (Stratton 1982)? Many studies have described apparent cyclic manifestations during development, but a number of investigators also have submitted their data to a variety of statistical treatments to bolster their arguments for the presence of real periodicities (Sterman and Hoppenbrouwers 1971; Hellbrügge 1974; Hoppenbrouwers and Sterman 1975; Hoppenbrouwers et al. 1978b; Meier-Koll et al. 1978; Harper et al. 1981; Robertson 1987; Thoman and McDowell 1989). Because of the emerging nature of ultradian cycles during development, assumptions of some techniques, such as stationarity, may not always be fulfilled. More basically, however, the nature of these ultradian rhythms at *any* age, may render the use of classical techniques for detecting hidden periodicities, such as fast Fourier transform, akin to fitting a round peg into a square hole. One of the objectives of this chapter and, for that matter the entire volume, is to enquire whether the study of ultradian rhythms can benefit from strategies derived from non-linear systems or chaos theory.

A third issue deals with the question of whether it is fruitful in our search for the manifestations, origins and mechanisms of ultradian rhythms to use the circadian rhythm as a reference. For instance, is it lucrative to search for single/dual oscillators analogous to putative circadian oscillators (Wever 1979; Daan et al. 1984; Kronauer 1987; Edmunds 1988)? Can the alternating states of active sleep (AS) and quiet sleep (QS), which together constitute the BRAC during sleep, be considered analogous to sleep and wakefulness as a circadian cycle? Amplitude changes can be observed in the multitude of hands of the circadian clock such as neurotransmitter levels, autonomic and metabolic activity. Is there reason to speak of the hands of the BRAC as comprising different amplitudes in heart rate variability, in rapid eye movement (REM) levels or in electroencephalographic (EEG) activity sigma bursts?. To what extent is coupling in the BRAC and the circadian cycle comparable? And, is there evidence that classical zeitgebers influence any ultradian rhythms?

In this chapter, rhythms in the minute and hour domain are reviewed, including their early history, their manifestations after birth and the factors that seemingly influence these rhythms during ontogeny. Special emphasis is placed on the BRAC and the prandial cycle, their relationship with each other and with the circadian rhythm. The radical change in environmental input which marks the transition from intra- to extrauterine life is relevant for biological rhythms. In some respects, birth represents a discontinuity in development. The data on ultradian rhythms presented here, however, favour continuity at least from the last two months of gestation to throughout the first two months of life. In the third month a transition appears with significant changes especially in the cyclic domain. These unique maturational events are discussed for their potential practical consequences.

History of Ultradian Rhythms

Rhythms in the Minute Domain

Motility cycles with a frequency of between twice a minute and once every 4 to 5 min (0.13–1.91 cycles/min) have been described in the foetus as early as 21 weeks of gestation (Robertson 1987). Strain gauges on the maternal abdomen were used to measure foetal movements, a relatively insensitive method which makes it plausible that the actual birthday of this cyclic motor activity is earlier. Fourier transform of the autocorrelation revealed peaks in the movement spectra which differed significantly from the spectral estimates of random noise. The spectral peaks had broad bases, suggesting wide variability in cycle times. The same range of frequencies were found in these subjects as in newborns.

Heart rate and breathing periodicities have much higher frequencies than these movement cycles (Harper et al. 1976; Hoppenbrouwers et al. 1978a). One heart rate rhythm, however, described in both the foetus and the young infant seems to parallel the faster frequency components of this motor rhythm. This heart rate rhythm is sometimes seen in isolation and sometimes as background oscillation for respiratory sinus arrhythmia (RSA), the influence of breathing on heart rate (Fig. 9.1). The absence of this 2–6 cycle/min modulation

Cardiac intervals
RSA and 2-6 cycles/min rhythms

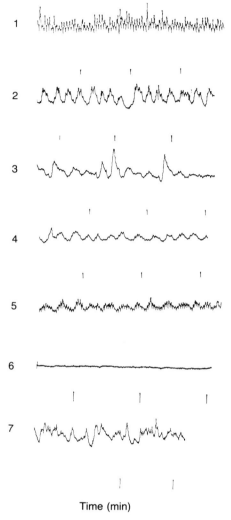

Time (min)

Fig. 9.1. Beat-to-beat heart rate intervals with evidence of respiratory sinus arrhythmia (RSA) and 2–6 cycles/min heart rate rhythms. Each tracing is approximately 4 min long.

1. Tracing from 3-month-old infant who experienced an apparent life threatening event. Extremely prominent RSA rides on 3 cycles/min rhythm.
2. Tracing from 3-month-old subsequent sibling of SIDS during QS. RSA riding on prominent 5 cycles/min rhythm, accompanied by sucking.
3. Minimal evidence of RSA but 5 cycles/min HR rhythm present in normal 3-month-old infant.
4. Same infant as in (3), above. Rhythmic mouth movements during QS.
5. Prominent RSA superimposed on 4 cycles/min rhythm. Infant was sucking on pacifier.
6. Absence of both RSA and 4–6 cycles/min rhythms in infant with apparent life threatening event.
7. Tracing from a 6-month-old infant during AS when RSA and 2–6 cycles/min rhythms tend to be less pronounced.

has been associated with poor prognosis in both the foetus and the premature infant (Hon and Lee 1963; Urbach et al. 1965). The burst–pause pattern in non-nutritive sucking follows a similar frequency, and indeed these 2–6 cycle/min oscillations in heart rate were sometimes associated with regular sucking and mouthing during QS or with episodes of periodic breathing, but neither motility pattern needs to be present for these rhythms to be observed. Movement rhythms resemble in their frequency vasomotor waves and thermoregulatory cycles, thus far described only in adults (Golenhofen and Hildebrandt 1961; Lovett Doust 1979). Of course, similarity in frequency does not indicate that these rhythms are related or are manifestations of a common oscillator, and to date little is known about their underlying mechanisms or their function. In immature vertebrates, Corner (1990) discovered neuronal firing patterns in this frequency range. These were observed in spinal preparations, in brainstem cells and in cortical cells in vivo and in vitro, across all stages of wakefulness and sleep, suggesting a diffuse, primitive, endogenous bursting system. On the basis of his animal studies, Corner (1990) made a strong case for considering these rhythmic movement patterns forerunners of phasic REM activity. In the human foetus the progressive emergence of the stable foetal BRAC, discussed below, seems to involve inhibition or clustering of movements, allowing for periodic episodes of quiescence (Nijhuis et al. 1982). The question is: how do originally independent movement generators integrate to produce stable patterns at a BRAC frequency which is much lower than the frequencies of its putative constituents?

Rhythms in the Hour Domain

BRAC

Regular waxing and waning of foetal activity is observed as early as 21 weeks of gestation (Sterman and Hoppenbrouwers 1971), but may well be present before that time. The BRAC is not well established at this early age. State-defining variables such as body–eye movements and heart rate patterns begin to be coupled only between 32 and 36 weeks of gestation, when patterns unique to AS and QS recur every 40–60 min (Junge 1980; Nijhuis et al. 1982; Visser et al. 1987). The BRAC continues to undergo changes in both the degree of coupling of state-defining variables and in the regularity of its cycle time. The BRAC has shown stability in the face of drastic changes in the circadian cycle (Siffre 1986), but it can be manipulated experimentally; for instance, cycle times have been lengthened and shortened by treatment with agonists and antagonists of acetylcholine (Sitaram et al. 1978).

Fig. 9.2. Foetal, maternal and neonatal heart rates during the night. Heart rate data are grouped into 10-min intervals. Top: maternal and infant sleep states. Note the increase in foetal heart rate after maternal sleep onset and the similarity between foetal and neonatal patterns. Maternal heart rate patterns are rhythmic but the coherence between the heart rate peaks and her *REM* cycle is poor. *BPM*, beats/min; *A* or *AW*, awake; *AS*, active sleep; *QS*, quiet sleep; *IN*, indeterminate. (Reproduced, with permission, from Hoppenbrouwers, 1990.)

SUBJECT 6–38 WEEKS

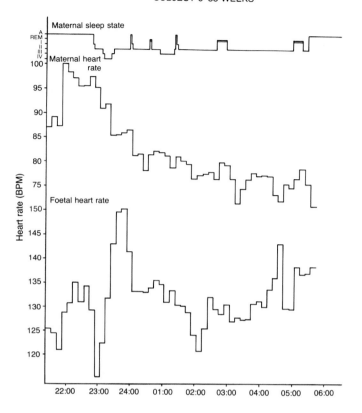

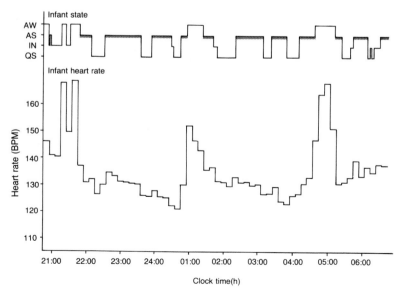

Clock time(h)

Complex demodulation

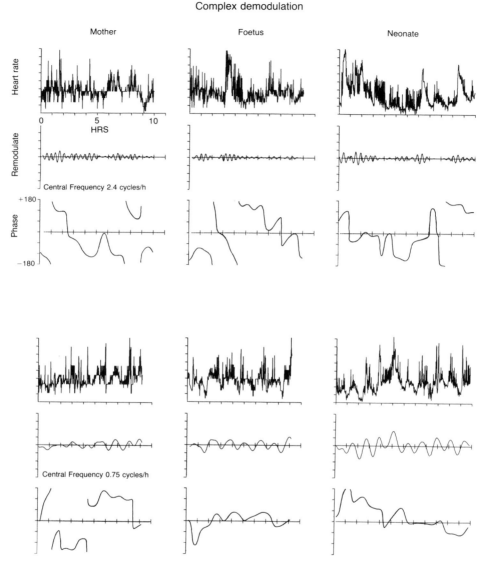

Fig. 9.3. Minute-by-minute maternal, foetal and neonatal heart rates on top with the amplitude and phase plot (remodulate) below. In the top portion, a centre frequency of 30 min was selected and both amplitude and phase plot revealed the presence of that cycle time during a portion of the night. In the lower part, heart rate values of another subject pair are shown; the centre frequency of 80 min appeared to be a good estimate of a dominant periodicity during either a portion of the night or the entire night. (Reproduced, with permission, from Hoppenbrouwers et al. 1978b.)

Simultaneous recordings of maternal sleep states indicated that the foetal 40–60 min BRAC was infrequently in phase with the maternal 90–110 min AS–QS cycle (Fig. 9.2). Complex demodulation of maternal, foetal and infant heart rates, however, revealed in *both* the presence of oscillations in the 30–80 min range, suggesting at least the possibility of some maternal influence on the foetus (Hoppenbrouwers et al. 1978b; Fig. 9.3). The BRAC therefore appears during the second trimester in utero, while coupling awaits the third trimester. The postnatal BRAC does not substantially differ from that in late prenatal life except for its tighter coupling and regularity. A maternal influence supporting development of stable states in utero is quite plausible, but whether this influence is mediated through an independent foetal oscillator or represents maternal driving of some foetal variables has not been established.

Feeding Cycle

A 3–4 h postnatal feeding cycle dominates early postnatal life and constitutes the boundaries of a polyphasic sleep–wakefulness cycle (Kleitman and Engelmann 1953; Parmelee 1961). Feeding is established after birth, but there is some indirect evidence that preparation for this timing takes place prior to birth (Hoppenbrouwers et al. 1978b). Elevated foetal heart rate levels in utero, very similar to those of wakefulness after birth, suggest that the foetus exhibits "wakefulness" in utero. Such "wakefulness" episodes typically occur at 3–4 h intervals and may be related to maternal glucose levels (Figs. 9.2 and 9.4). This 3–4 h cycle is present immediately after birth prior to the first feed, again supporting its origin in utero (Emde et al. 1975). Moreover, a not uncommon phenomenon documented by pregnant women and polygraphic recordings alike is increased foetal movement after maternal sleep onset. Maternal sleep onset could thus also function as a device for the foetus to set its putative, internal arousal/feeding clock (Hoppenbrouwers et al. 1978b).

In one infant on a self-demand feeding schedule, Morath (1974) found a free-running feeding cycle early after birth, with entrainment becoming obvious during the second month (Fig. 9.5). A phenomenon, akin to phase-trapping described by Kronauer and Gander (1989) was seen in this feeding record as well. With the advent of an overt circadian influence on sleep and waking, infants begin to sleep through the night and at least one night-time feeding typically disappears (Kleitman and Engelmann 1953; Morath 1974; Meier-Koll et al. 1978; Hoppenbrouwers et al. 1988). Nowadays most hospitals and caregivers do not enforce a strict feeding regimen, but rather reinforce an infant's spontaneous feeding cycle. Approximately 30 years ago, however, a strict feeding routine was usually imposed and adhered to rigidly in maternity hospitals. Hellbrügge (1960) capitalized on this condition and his data show how hunger might affect sleep and wakefulness. In his clinical setting, newborns were not fed between 21:00 and 06:00 h. This resulted in enhanced wakefulness in the nursery (Fig. 9.6). At subsequent ages a nightly feeding was interspersed in his study sample causing a significant increase in sleep. After the first year the night-time feeding was deleted again and now infants managed to remain asleep throughout the night. This subordination of the night-time feeding to the circadian cycle begins during the second and third months. A seemingly endogenous feeding rhythm can be entrained by an external zeitgeber.

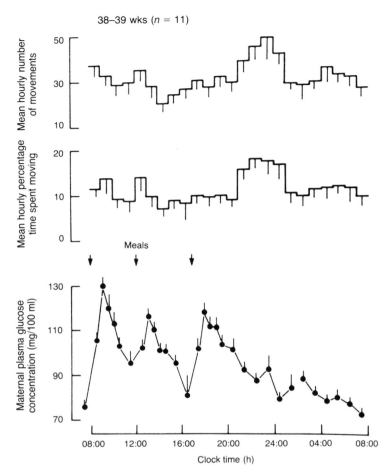

Fig. 9.4. Foetal movement patterns during a 24-h period in 11 foetuses. Note the increases at 3–4 h intervals and their daytime relationship with maternal plasma glucose concentrations and the transient rise in movements prior to midnight. (Reproduced, with permission, from Patrick et al. (1982) Am J Obstet Gynecol 142:363.)

That this is largely a cultural zeitgeber is also suggested by preliminary studies of feeding in Trobriand infants (Papua New Guinea) who were breast fed at irregular intervals (Siegmund et al. 1990). Because food intake is essential for survival, it is intriguing that the feeding cycle should have a frequency in the same range as that of the putative, primitive solar day during prebiotic evolution (Klevecz et al. 1984).

In summary, the 3–4 h feeding cycle may have its origin in foetal life, where maternal influences could prepare the infant for gratification of postnatal metabolic needs. After birth, internal gastric and metabolic influences may help to sustain the cycle. The rhythm can be entrained by caregivers, although the data from the baby on a self-demand feeding schedule may point to other zeitgebers as well. Zeitgeber influence would favour the categorization of the

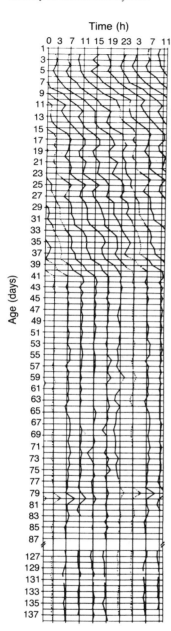

Fig. 9.5. Feeding patterns during the first 137 days in one infant. Two periods of free-running interspersed with what appears to be phase-trapping (Kronauer and Gander 1989) are followed by an entrained 4-h feeding cycle. The dotted lines toward the end of the monitoring period refer to missed feedings. (Reproduced, with permission, from Morath 1974.)

feeding cycle as a forerunner of the circadian cycle. Perhaps, the circadian cycle is an ontogenetically late integration of six to eight shorter units which stem from early evolutionary time. The question raised in the section on motility rhythms in the minute domain applies here as well: if the 4-h cycle represents a primitive clock mechanism, what would be the mechanism that integrates this clock into a 24-h circadian rhythm?

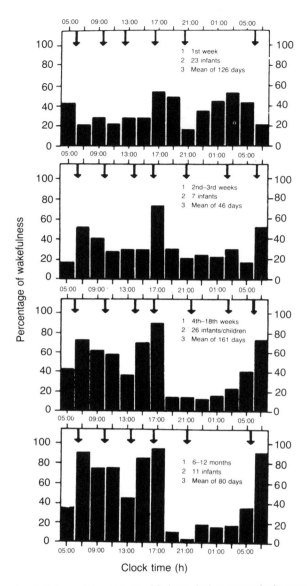

Fig. 9.6. Sleep and wakefulness from groups of infants during respectively, week 1 (top), weeks 2–3, 4–18 and during the first year of life. Arrows indicate feedings. The absence of feeding between 21:00 and 06:00 h during the first week of life seemed to induce wakefulness. Addition of a feeding during subsequent weeks promoted sleep. Omission of feeding in the bottom trace during the first year of life did not affect the sleeping pattern anymore (Reproduced, with permission, from Hellbrügge 1960.)

Circadian Cycle

Overt manifestations of an endogenous circadian modulation of temperature, sleep, wakefulness and the manifold hands of the clock appear at various times. For instance, a circadian modulation of cell mitosis is present at birth (Cooper 1939), whereas a circadian modulation of sleep and wakefulness, temperature and such autonomic variables as heart and respiratory rate appears after birth, with the most obvious signs evolving around two months of age (Hellbrügge 1960; Hoppenbrouwers et al. 1980; Minors and Waterhouse 1981). Like the BRAC, at the time of first appearance, circadian rhythms do not exhibit fully adult characteristics. After birth, sleep and wakefulness have occasionally been shown to be free-running as if the maturation of the clock precedes the ability of the infant to integrate zeitgeber information with the clock (Kleitman and Engelmann 1953; Parmelee 1961). This is, however, not the only immature pattern that has been observed. A recent study (Tomioka and Tomioka 1991) described sleep and wakefulness behaviour in three infants on a self-demand feeding schedule, followed from birth until 8–12 weeks of age. Two of the infants began with a free-running sleep–wakefulness pattern, but duration of free-running varied from 1 to 6 weeks. In the third infant, short sleep episodes occurred throughout the 24-h day, but this infant developed a biphasic sleeping pattern during the first 7 weeks, where sleep episodes clustered around the afternoon and night. This same pattern was followed by the second infant after his short free-running period ended. In both infants this 12-h cycle converted to a 24-h cycle around 12–14 weeks. The authors stressed the possible significance of social zeitgebers in explaining these two immature patterns, a free-running 24-h cycle and a biphasic 12-h cycle.

Studies of premature infants raised in temperature controlled incubators with relatively constant ambient light have shown that circadian and ultradian modulation of heart rates is weak at 25 weeks of gestation but rises during the next 10 weeks (Minors et al. 1990). The primacy of the role of maturation over the influence of zeitgebers or vice versa on the development of overt circadian rhythms has yet to be firmly established and an interaction between these two variables is most likely. In one study it was found that premature newborns developed a circadian kidney rhythm between 7 and 12 weeks postnatally, at the same time as term infants once their postconceptional age was equated (De Kraker 1978), suggesting a prominent role for maturation. McMillen et al. (1991), however, recently compared sleep–wakefulness rhythms in term and preterm infants. They found that preterm infants adjusted for postconceptional age, once at home and exposed to natural zeitgebers, showed entrainment sooner than term infants, suggesting to the authors a more prominent role for zeitgeber influence than had been previously believed. It is doubtful that neonatal intensive care units contain no zeitgeber information; day and night levels of illumination, for instance, varied at between 6210 and 1566 lux. Therefore, the duration of the exposure of the two groups of infants to zeitgebers was probably not identical.

Other aspects of immaturity involve phase and amplitude of the circadian rhythm. With increasing age, troughs in functioning appear typically at progressively later times in the night and amplitudes rise, often as a result of the attainment of lower minimal values during the night (Hellbrügge 1960; Abe et al. 1978). The circadian temperature rhythm, for instance, does not reach an adult configuration until the child is 7 years of age. The question of coupling between the temperature and sleep–wakefulness cycle or internal synchronization has not been thoroughly explored in infants, although Kleitman et al. (1937) provided some evidence that the temperature cycle was phase advanced relative to the circadian sleep–wake cycle in infants beyond the first year of life. The transition period, when phase relationships and amplitudes are changing, appears to be much longer in the circadian system than in the BRAC, since in some cases an adult configuration has to await years rather than months.

During both the development of the BRAC and the circadian cycle an interesting phenomenon has been observed consisting of an overshoot. At three months of age the BRAC exhibited unprecedented regularity and coupling. A good example is heart rate, which is modestly coupled with EEG and other sleep state variables in the foetus, and becomes tightly coupled at 3 months of age in the infant. Subsequently it begins to decouple again, as demonstrated by the mothers of our infants, who during their sleep exhibited 80-min heart rate oscillations that were not tightly coupled to their REM-QS cycle (Fig. 9.2). In a similar vein, the amplitude of the circadian temperature cycle was found to be transiently enhanced during the second year, compared to either earlier and later ages (Kleitman et al. 1937; Abe et al. 1978).

Although it appears that the birthday of most overt circadian rhythms must be sought post partum this does not mean the infant has not received circadian influences in utero. As early as 20–22 weeks of gestation, peaks in foetal motility are found between 20:00 and 23:00 h (Fig 9.4) and troughs between 09:00 and 12:00 h. Such movements are accompanied by peaks and troughs in foetal heart rate variability (Patrick et al. 1982; Visser 1985). Romanini et al. (1985) demonstrated the concomitant changes in maternal circadian plasma cortisol levels. More pertinent are the data from Selinger and Levitz (1969), who measured maternal plasma oestriol, presumably of foetal placental origin. They found a fall between 08:30 and 16:30 h followed by a gradual rise until 21:30 h.

Postnatal maturation is marked by a recruitment of more and more "hands" into the circadian system in much the same way as the BRAC seemed to recruit gradually more variables into a system of alternating stable states. It is not unreasonable to expect recruitment to be a function of maturation. For example, the circadian modulation of water excretion, dependent on maturation of the glomeruli, appeared at an earlier age than that of mineral excretion, which depends on the function of the tubules (Hellbrügge 1960). On the basis of information from other mammals, one should expect not only maturation of the output system to be critical but also that of the putative clock and its input, for instance the retino-hypothalamic pathway. Does the entrainability of foetal heart rates, motility and even hormone levels indicate a prenatal maturation of the putative oscillator? In the rat, Reppert et al. (1987) have demonstrated that in utero the suprachiasmatic nucleus of the hypothalamus seems to be entrainable, at an age when only the most rudimentary innervation and synaptic contacts can be seen and prior to maturation of the retino-hypothalamic pathway. In the rat the maternal circadian system thus seems to act on a functioning foetal clock rather than just by driving foetal oscillations. To what extent a similar mechanism is involved in humans has yet to be determined.

Relationships Among Rhythms

Numerous examples can be cited, suggesting some mutual dependence between the cycles in the minute and hour domain and the circadian cycle both during development and in adulthood. This dependence can be manifest in both internal coupling *and* phase relationships between rhythms with different frequencies. For instance, in the foetus and young infant, respiration and heart rate show a tighter coupling expressed in coherence or amplitude of RSA, during QS when breathing is regular, than during AS or wakefulness (Fig. 9.1; Harper et al. 1978). The 2–6 cycle heart rate rhythm tends to be more regular during QS in infants (T. Hoppenbrouwers, unpublished results), while a state relationship in the foetus has not been established. Furthermore, spectral peaks of fast motility rhythms tended to be lower in AS (Robertson 1987). In adults, coupling in ultradian rhythms is also affected by circadian rhythms (Hildebrandt 1986).

Ultradian rhythms of alertness were found to be more rapid and pronounced in the morning compared to the afternoon (Lavie 1985). Moreover, the BRAC has a propensity to emerge during a specific interval of the 24-h day, which led Lavie (1985) to identify "gates" and what he called "forbidden zones" for the BRAC. In newborns, AS dominates during the first part of the night and the shift toward the late night peak begins to occur at 3 months (Hoppenbrouwers et al. 1982). In young infants, Schulz et al. (1983) revealed that "gates" for entering a REM episode after wakefulness have a bimodal distribution. After 3 months of age, a significant circadian influence on these REM latencies was found, with the longest ones (mean: 20 min) occurring between 12:00 and 16:00 h and the shortest ones between 04:00 and 08:00 h. Because the onset of sleep affects the phase of the first REM episode and awakening tends to originate from REM sleep, Aschoff and Gerkema (1985) proposed a bi-directional coupling between the REM cycle and the circadian cycle in adults. In infants we observed further evidence of bi-directional coupling: circadian modulation in respiratory rate in infants appeared in AS at 2 months of age before it appeared in QS (Hoppenbrouwers et al. 1980).

In adults, a recurring and unanswered question is whether the BRAC is sleep dependent and the term REM cycle is more appropriate or whether the BRAC is a separate rhythm which continues during the entire 24-h day as proposed by Kleitman (1963). In infants who sleep throughout the day, the BRAC can be easily observed and seems autonomous. Furthermore, Aserinsky and Kleitman (1955) presented some evidence that the BRAC's phase influenced the feeding cycle. Meier-Koll's (1979) longitudinal data in one infant provides additional evidence for the continuation of the BRAC throughout the 24-h day. The infant showed an enhanced occurrence of wakefulness 1 *and* 4 h after awakening and frequencies of REM episodes prior to and after awakening followed a 50-min cycle. Awakenings occurred predominantly at a predicted REM sleep time and the end of the waking period was highly correlated with the 50-min cycle as well.

A study by Harper et al. (1977) showed that feeding altered the regularity of the BRAC, although it was not determined whether tighter coupling was responsible for this effect. A case can thus be made for bi-directionality not just between the BRAC and the circadian cycle but also between the BRAC and the feeding cycle. Martin du Pan (1974) submitted two fullterm infants to constant light from the eighth day of life, for 72 days. These infants were on a self-demand feeding schedule and their feedings were carefully plotted. By 2 months of age they were typically awake between 18:00 and 06:00 h and asked for their feedings during the night hours when their interfeeding intervals were less than 2.5 h. They were typically asleep between 06:00 and 18:00 h, with their excretion of urine and urinary electrolytes also following this inverted pattern. Nine days after exposure to the natural rhythm of day and night, they established their normal circadian sleep–wakefulness cycle. The inverted sleep–wakefulness cycle in these two infants increased the frequency of their feeding cycle, again suggesting bi-directional coupling. This idea is further supported by a study in newborn infants who underwent two treatments: half were raised in a communal nursery with feedings at 4-h intervals, the other half in adjacent, single rooms where they received feedings on demand from foster mothers. Among the girls, the ones who had been submitted to the 4-h feeding routine established their circadian sleep–wakefulness cycle sooner than those who had been on a demand feeding schedule (Sander et al. 1972).

On the basis of their systematic observations of one infant during the first 6 months of life, Meier-Koll et al. (1978) postulated a frequency modulation of the feeding cycle by the circadian cycle. They devised a mathematical model in which the amplitude of the circadian sleep–wakefulness cycle was progressively increased, representative of the first 11–12 weeks of life. The ultradian feeding cycle was represented by a sine wave with a cycle time of 4 h and 20 min. In the simulated data, complete phase-locking between these two cycles appeared around 11 weeks. The simulated data were quite similar to those from their infant and suggested that gradual coupling between these two rhythms may be the actual mechanism underlying sleep–waking behaviour in the infant.

Lastly, there is the intriguing frequency relationship between these rhythms supplemented by the ontogenetic story (Hildebrandt 1986; Broughton 1985; Hoppenbrouwers 1989; Table 9.1). The immature BRAC will ultimately double in duration from 45 to 90 min and a further doubling makes 180 min, the typical duration of the feeding cycle. Broughton (1989) has emphasized the close resemblance of some of these frequencies, especially the slower ones, with environmental rhythms. Synchronization with the present, and possibly the primitive solar day, the mean tidal period, and the gyro compass, as well as time compartmentalization confer evolutionary advantages (Broughton 1989). Also, foetal and maternal heart and respiratory rates tend to follow a 3:1 ratio. According to Hildebrandt (1986) an organism with oscillator subsystems at constant integer ratios would have the ability more easily to maintain consistent phase relationships between them.

In the hour domain, a striking phenomenon in both the immature and mature human is the simultaneous presence of several frequencies. In the foetus, the neonate and the pregnant woman we found both immature and mature cycle times (Fig. 9.3), a finding also observed by Gagel (1972) and Ullner (1974) in the neonate. Lavie (1985) found two ultradian falling asleep rhythms in adults at 1.5 h and 3.5–4 h, while Klein and Armitage (1979) found three peaks in performance on verbal and spatial matching tasks at 37, 96 min and 4 h, respectively. It is unclear whether these multiple peaks in the ultradian domain are simply artefacts of analyses or whether, as Aschoff and Gerkema (1985) have suggested, "The structural and functional units of the organism are

Table 9.1. Frequency relationships between rhythms

Cycle time in min (h)	Biological oscillation
45	Foetal BRAC
90 (1.5)	Adult BRAC
180 (3)	Prandial cycle
360 (6)	
720 (12)	Immature sleep–wakefulness cycle
1440 (24)	Circadian cycle
2880 (48)	Free-running circadian cycle

Adapted from Broughton 1985.

arranged in a way that they get 'in resonance' to the circadian rhythm and its submultiples more easily than to any other frequency." Support for the independent influence of the BRAC and the feeding cycle on performance derives from a two factorial research design. In the 1950s, an era when feeding and sleeping regimens were "de rigueur", Koch (1968) carried out an extraordinary study in 5-month-old infants. He managed to manipulate feeding and sleeping times so that he could assess their separate contribution to the latency of a conditioned orienting response to sound. Both digestion and waking affected latencies, which were minimal 40 min after awakening and 145 min after feeding.

Unique Developmental Events

The period of development between 1 and 3 months, the age of highest risk for sudden infant death syndrome (SIDS), is one of reintegration of a multitude of physiological functions (Hoppenbrouwers and Hodgman 1982). A few of these, such as breathing, cardiac and EEG activity and movement, are depicted in Fig. 9.7. Each exhibits oscillatory behaviour in the minute or second domain and undergoes monotonic and non-monotonic changes during this age period. Simultaneously, changes occur in the rhythms in the hour domain. As the overt circadian sleep–wakefulness cycle emerges, feedings continue during the day but disappear during the night. The BRAC is gradually becoming more periodic. Inevitably, shifting phase relationships between these cycles of various frequencies are the rule.

The end of this period of reintegration around 3 months of age seems to be marked by *transient order and periodicity* (Hoppenbrouwers 1989). The somatic and autonomic variables described above level off and attain a more stable relationship with one another; within the BRAC they are tightly coupled. The BRAC attains an unprecedented regularity (Harper et al. 1981; Hoppenbrouwers et al. 1988) which may be a result of both tight coupling and the emergence of the circadian rhythm with the simultaneous disappearance of night-time feedings permitting uninterrupted sleep. A similar regularity was observed in the tracings of kittens at around 3 to 4 weeks, an age comparable

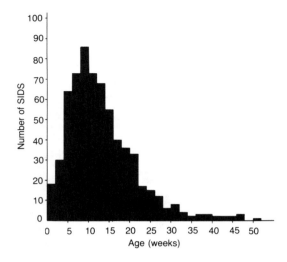

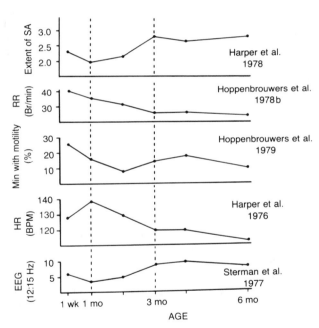

Fig. 9.7. Top: incidence of SIDS in Los Angeles County (1974–77). Bottom: schematic representation of the ontogenesis of a number of physiological variables measured during quiet sleep. *SA*,
respiratory sinus arrhythmia; *RR*, respiratory rate (breaths/min); *HR*, heart rate; *BPM*, beats/min.
(Reproduced, with permission, from Hoppenbrouwers and Hodgman 1982, 1990.)

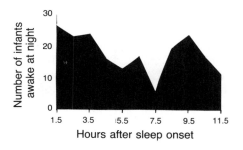

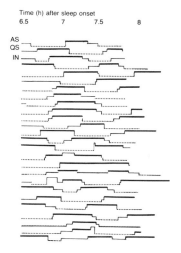

Fig. 9.8. Data from three-month-old infants. Top: number of infants awake during the night. Note "a forbidden zone" for awakening between 6.5 and 8h after sleep onset. Plots of their BRAC (bottom) showed great regularity. Mean duration of AS 21.3 min, mean duration of AS 21.7 min, compared to 18.8 and 24.3 when calculations were based on the entire night. *AS*, active sleep; *QS*, quiet sleep; *IN*, indeterminate.

to the human infant at 3 months (Hoppenbrouwers and Sterman 1975). An ultradian coefficient in heart rates as reported by Ullner (1974) reached its maximum around this time as well. AS, which dominated during the first months of life, has decreased compared to QS and the mean durations of these two states are nearly the same at this age. In addition, at 3 months of age *certain times of the night begin to show unique patterns* which previously were not as obvious: I have described what might be a "forbidden zone" for arousal during night sleep (Hoppenbrouwers 1989). Infants slept in the laboratory between 18:00 and 07:00h while they were on a self-demand feeding schedule. In the 3-month-old group, the only ones surveyed for this purpose, infants were least likely to be awake 6.5–8h after sleep onset. A plot of the sleep states during that time revealed an extremely regular BRAC compared to prior to or after that interval (Fig. 9.8), while respiratory rates during QS reached their lowest levels (T. Hoppenbrouwers, unpublished results). Interestingly in 1953, Kleitman and Engelmann had already pointed to the time of night between 01:00 and 03:00 h as one in which their infants were most likely to be asleep, especially after 3 months of age. These are clearly manifestations of a circadian influence. Lastly, although sleep state data tend to be fairly variable, the age of 3 months is marked by a minimal intersubject variability (Hoppenbrouwers et al. 1988).

It is postulated that there is a brief corridor of time or a fault line in development. An initial period of instability culminates in a transient period of relative stability and order that seems likely to be an overshoot in the system akin to that seen in the development of the circadian system at a later date (Kleitman et al. 1937; Abe et al. 1978). It is manifest as enhanced periodicity and internal synchronization. An elevated threshold for arousal is accompanied by a trough in autonomic functioning during the early morning hours. In other words, the central nervous system during sleep seems more prone to mode locking or synchronization and less responsive to exogenous stimulation, including that leading to arousal.

Potential Mechanisms

Aschoff and Gerkema (1985) have pointed out that the diversity in the frequencies of ultradian rhythms argues against the presence of a common mechanism or even self-sustaining master oscillators. As has been shown, this diversity is of early ontogenetic origin. Also the bi-directionality in coupling between the ultradian BRAC, the feeding cycle and the circadian cycle emerges early. In the 1960s, Wever (1965) stressed the usefulness of distinguishing two types of oscillators: relaxation and pendulum. In biology, completely stable conditions rarely prevail, but under such conditions both oscillations should show constant amplitude and frequency. A number of characteristics distinguish these two types, especially their purpose. Relaxation-type oscillators foster energy transfer and conversion from constant energy to discrete bursts of energy, while the pendulum type serves a timing function, especially in the absence of timing feedback. Also, in the case of the pendulum type one expects different amplitudes but always the same frequency, unless there are structural changes, and, conversely, with relaxation-type oscillations different frequencies accompany the same amplitude, again unless structural changes take place. Wever (1965) speculated that because of their shorter periods, ultradian rhythms tend to be more like relaxation-type oscillators, whereas the strong, slow, fundamental circadian temperature cycle has pendulum-type characteristics. Many biological rhythms exhibit characteristics of both: for instance, the EEG and respiration where both amplitude and frequency change. Stratton (1982) adapted this idea and speculated that maturation involves a decrease in relaxation oscillations and an increase in pendulum characteristics. With the growth in the size of the organs, cycle times increase and a shift toward pendulum-type characteristics is expected. In addition, Stratton argues, the feature of energy exchange is particularly needed in the immature organism, which frequently operates with limited reserves. In the case of breathing, for instance, infants are less able to increase their breathing amplitude and typically adjust by increasing their frequency which is a characteristic of relaxation-type oscillators. Moreover, relaxation oscillators have the ability to synchronize with a wide range of frequencies including multiples and submultiples of the basic frequency. According to Stratton, this adaptability should be advantageous in early stages of development when frequencies are typically variable. Perhaps this interesting idea deserves further exploration. Development is characterized by many structural changes, one would therefore expect a fluid situation where rhythms exhibit both relaxation-type and pendulum-type characteristics even those that ultimately will have exclusively pendulum-type characteristics. But it seems that variability in frequencies during early development as claimed by Stratton is not universal. In utero, for instance, the foetal ECG begins with a low fixed rate which gradually rises and increases in variability. After birth, variability in heart rate during wakefulness and REM rises and attains a peak at 2 months of age (Harper et al. 1976). In young infants, lack of variability in heart rates is seen under two conditions: first, during wakefulness with crying when the rate is fixed at very high heart rate levels. Second, as an antecedent to death when a fixed rate is seen at a low heart rate level when the modulating effects from the brainstem have ceased to influence the intrinsic rhythm of the heart (Hodgman et al. 1987).

All oscillatory systems contain the seeds of chaos. When positive feedback processes are present the system can be forced into a chaotic regime depending on the size of the positive feedback parameters (Pool 1989). Investigators have begun to scrutinize some of the better known oscillations such as EEG and electrocardiographic (ECG) for transitions to chaotic behaviour. Some argue, for instance, that epileptic seizures and cardiac fibrillations can be explained by chaotic principles (Babloyantz and Destexhe 1986; Goldberger and West 1987). Wever (1985) has offered a model for ultradian rhythms in relation to the circadian cycle which includes a reference to chaotic behaviour. According to this model a threshold divides the sleep–wakefulness cycle into two states: awake above the threshold, asleep below the threshold (see Wever, Chap. 15). In this model, the ultradian rhythmicity is considered to be self-sustained and the circadian amplitude is the most significant determinant of the system. The model predicts that when the threshold for awakening is slightly elevated and the amplitude of the circadian rhythm is small, the system has a high likelihood of falling below the threshold during wakefulness. The ultradian cycle may thus start during a waking episode, as for instance in narcolepsy. "Under these conditions the system shows a highly sensitive dependence on initial conditions or random fluctuations. In other words, reducing the circadian amplitude leads from rhythmic order to chaos in the ultradian domain" (Wever 1985). In young infants the circadian amplitude is originally very small and rises only gradually. In line with this model, infants have a polyphasic sleep–wakefulness cycle and, like narcoleptics, they initiate sleep by entering REM sleep. According to this model it should be advantageous to survey ultradian rhythms during ontogeny to determine the applicability of non-linear strategies. Should it turn out that, during certain stages of development, non-linear models apply, questions of mechanisms can be viewed from a new perspective. The search for master clocks would have to be amended to include less hierarchical models. This task has barely begun. The study of Szeto et al. (1990) stands virtually alone at this time. They have shown in lambs that clusters of foetal breathing show self-similarity over different time scales and, interestingly, the records from younger lambs exhibited stronger fractal properties than those from older foetuses.

Speculations

There are practical considerations for studying ultradian oscillatory behaviour during development. The timing of an individual's rhythms may exert a profound effect upon social behaviour and, in the case of infants and children, upon learning, a theme well developed by Thoman (1975). The recognition of such rhythms as the BRAC is of tremendous importance in the interpretation of data. No longer does is suffice to stipulate what treatments were given or what variables were surveyed, unless it is also stipulated in what behavioural state, or in what phases of the BRAC the infants were during the study.

The data can also be reviewed from a much more conjectural perspective. It is plausible that, as a result of evolutionary changes, human uterine life has been curtailed and infants are born about 8 to 12 weeks prematurely (Prechtl 1984). Among other evolutionary pressures, the theory goes, the transition from a four-legged to an upright posture narrowed the birthcanal, preventing passage of the foetal head and necessitating an earlier delivery. Since birth is a typical time of high risk for mortality, we have elsewhere asked whether the peak death from SIDS at 3 months of age could be a manifestation of this primitive birthday (Hoppenbrouwers and Hodgman 1988). In utero, foetal temperature has been fully dependent on maternal temperature, while after birth infants must regulate their own. A prolonged intra-uterine life would have entailed prolonged dependence on maternal temperature regulation and a need to entrain to external zeitgebers only 8 to 12 weeks later than the current age of delivery. It is rather striking that the overt circadian temperature cycle emerges around this time. Undoubtedly, biological rhythmicity is a very primitive organizational principle. If the second and third postnatal months have, in an evolutionary sense, special significance, it is reasonable to expect some unique events in the functional domain in general, and, specifically, in the rhythmic domain. This is precisely what seems to occur. In the rhythmic domain a signature of development at 3 months of age is the subordination of the feeding cycle to the circadian sleep–wakefulness cycle. The BRAC shows unprecedented regularity and tight coupling. Breathing, cardiac and EEG activity have reached a stable, relatively low activity, with stable interrelationships. EEG sleep spindles have become typical of QS and breathing and cardiac activity have begun to follow a circadian rhythm with minima during the early morning hours. It is the tight coupling between the components of the BRAC, its regularity, and its seeming lack of susceptibility to external and internal feedback at 3 months of age that raise questions about its dynamic underpinnings. Three months of age seems truly an age of transition.

Lastly, there are the questions relating to non-linear models in the development of biological rhythms. Is the BRAC pushed in and out of the chaotic domain during various states of development, dependent on its relationship with the changing feeding and circadian cycles? Could there be an age or ages during development that the BRAC makes a relative switch from chaos to periodicity or from more toward less fractal organization? Could there be an advantage in terms of energy transfer and environmental adaptation if the BRAC, perhaps a multiple of fast ultradian rhythms, and a submultiple of the circadian sleep–wakefulness cycle were obeying chaotic principles? Or, would there instead be liabilities? Are there zones in the 24 h when liabilities are enhanced? Are there peculiar characteristics of maturation that "seduce" oscillatory systems at faster frequencies, which themselves are prone to occur in certain phases of the BRAC, into chaotic regimes with maladaptive or fatal outcomes: for instance enhanced changes for periodic breathing, cardiac fibrillations or epileptic seizures? Could external stimuli, such as maternal cocaine use, known to be associated with SIDS, or abrupt changes in sleep and waking or temperature, constitute stimuli that drive a healthy chaotic regime into a maladaptive regime? These questions are extremely speculative but perhaps worth pursuing.

Summary

In this chapter the ontogenesis of ultradian biological rhythms in the minute and hour domain is reviewed. This includes the basic rest activity cycle (BRAC) and the prandial cycle. Ultradian rhythms during foetal life, early infancy and adults are characterized by an instability in cycle lengths which has raised questions about the presence of "real" periodicities. The multiplicity of periodicities and their intrinsic variability seems to argue against a simple hierachical model involving a few discrete oscillators, as has been proposed for the circadian cycle. Their mutual dependence and their interactions with the emerging circadian rhythm suggest bi-directionality on several levels. In human infants between 1 and 3 months of age, central and autonomic nervous system changes are abundant, and ultradian and circadian rhythms undergo significant changes in their amplitude and phase relationships. This period of transition culminates in a period of pronounced stability and periodicity. Perhaps the study of these ultradian rhythms can benefit from strategies derived from non-linear dynamics. Should fractal or chaotic priciples during certain phases of development be discovered, alternating with more periodic behaviour at other times, insight might be gained about underlying mechanisms during development.

Acknowledgements

The studies from our own laboratory presented here were carried out in collaboration with Drs J. E. Hodgman, R. M. Harper, M. B. Sterman, M. Durand and L. Cabal, under National Institutes of Health, NICHD contracts N01-HD-22777, HD4-2810 and grant HD-13689. The financial support of the Arthur Zimtbaum Foundation of New York, the Orange County Guild for Infant Survival and the Los Angeles Chapter of the National SIDS Foundation is greatly appreciated. I thank Dr Joan Hodgman for her valuable feedback.

References

Abe K, Sasaki H, Takebayashi K, Fukui S, Nambu H (1978) The development of circadian rhythm of human body temperature. J Interdiscipl Cycle Res 9:211–216

Aschoff J, Gerkema M (1985) On diversity and uniformity of ultradian rhythms. Exp Brain Res 12 [Suppl]:321–334

Aserinsky E, Kleitman N (1955) A motility cycle in sleeping infants as manifested by ocular and gross bodily activity. J Appl Physiol 8:11–18

Babloyantz A, Destexhe A (1986) Low-dimensional chaos in an instance of epilepsy. Proc Natl Acad Sci USA 83:3513–3517

Broughton RJ (1985) Three central issues concerning ultradian rhythms. Exp Brain Res 12 [Suppl]:217–231

Broughton RJ (1989) Chronobiological aspects and models of sleep and napping. In: Dinges DF, Broughton RJ (eds) Sleep and alertness: chronobiological, behavioral and medical aspects of napping. Raven Press, New York, pp 71–97

Cooper ZK (1939) Mitotic rhythm in human epidermis. J Invest Dermatol 2:289–300

Corner MA (1990) Brainstem control of behavior: ontogenetic aspects. In: Klemm WR, Vertes RP (eds) Brainstem mechanisms of behavior. John Wiley & Sons, Inc., New York, pp 239–267

Daan S, Beersma DGM, Borbely AA (1984) Timing of human sleep: recovery process gated by a circadian pacemaker. Am J Physiol 246:R161–R178

De Kraker J (1978) Development of circadian rhythm in the kidney. A study in premature infants. Doctoral dissertation, University of Amsterdam

Edmunds LN Jr (1988) Cellular and molecular basis of biological clocks. Springer, Berlin Heidelberg New York, pp 368–395

Emde RN, Swedberg J, Suzuki B (1975) Human wakefulness and biological rhythms after birth. Arch Gen Psychiatry 32:780–783

Gagel J (1972) Untersuchung zum zeitlichen Verlauf von Herz und Atemtätigkeit im Schlaf reifer und gesunder neugeborener Kinder. Doctoral dissertation, University of Dusseldorf

Goldberger AL, West BJ (1987) Applications of nonlinear dynamics to clinical cardiology. Ann NY Acad Sci 504:195–213

Golenhofen K, Hildebrandt G (1961) Zur relativen Koordination von Atmung und Blutdrückwellen dritter Ordnung. Z Biol 122:451–458

Harper RM, Hoppenbrouwers T, Sterman MB, McGinty DJ, Hodgman J (1976) Polygraphic studies of normal infants during the first six months of life. I. Heart rate and variability as a function of state. Pediatr Res 10:945–951

Harper RM, Hoppenbrouwers T, Bannett D, Hodgman J, Sterman MB, McGinty DJ (1977) Effects of feeding on state and cardiac regulation in the infant. Dev Psychobiol 10:507–517

Harper RM, Leake WB, Hoffman HJ et al. (1978) Development of sinus arrythmia during sleeping and waking states in normal infants. Sleep 1:33–48

Harper RM, Leake B, Miyahara L, Hoppenbrouwers T, Sterman MB, Hodgman J (1981) Development of ultradian periodicity and coalescence at 1 cycle per hour in electroencephalographic activity. Exp Neurol 73:127–143

Hellbrügge T (1960) The development of circadian rhythms in infants. Cold Spring Harbor Symp Quant Biol 25:311–323

Hellbrügge T (1974) The development of circadian and ultradian rhythms of premature and full-term infants. In: Scheving LE, Halberg F, Pauly JE (eds) Chronobiology. Igaku Shoin Ltd, Tokyo, pp 339–341

Hildebrandt G (1986) Functional significance of ultradian rhythms and reactive periodicity. J Interdiscipl Cycle Res 17:307–319

Hodgman JE, Pavlova Z, Hoppenbrouwers T et al. (1987) Clinical predictors of imminent brain death. Clin Res 35:234A

Hon EF, Lee ST (1963) Electronic evaluation of the fetal heart rate. VIII. Patterns preceding fetal death, further observations. Am J Obstet Gynecol 87:814–826

Hoppenbrouwers T (1989) Sudden infant death syndrome (SIDS) and sleep. Proc IEEE Conf Eng Med Biol 11:310–312

Hoppenbrouwers T (1990) Polysomnography in the newborn and young infant In: Buela-Casal G, Navarro Humanes JF (eds) Advances en la investigation del sueno y sus trastornos. Siglo Veintiuno Editores, Mexico, Spain

Hoppenbrouwers T, Hodgman JE (1982) Sudden infant death syndrome (SIDS) (1982) an integration of ontogenetic, pathologic, physiologic and epidemiologic factors. Neuropaediatrie 13:36–51

Hoppenbrouwers T, Hodgman JE (1988) An alternative evolutionary and physiologic model of SIDS. Med Anthropol 10:61–64

Hoppenbrouwers T, Hodgman JE (1990) Sudden infant death syndrome. In: Timiras PS, Meisami E (eds) Handbook of human growth and developmental biology, vol 3, pt 2. CRC Press, Boca Raton, FL, pp 181–207

Hoppenbrouwers T, Sterman MB (1975) Development of sleep state patterns in the kitten. Exp Neurol 49:822–838

Hoppenbrouwers T, Harper RM, Hodgman JE, Sterman MB, McGinty DJ (1978a) Polygraphic studies of normal infants during the first six months of life. II. Respiratory rate and variability as a function of state. Pediatr Res 12:120–125

Hoppenbrouwers T, Ugartechea JC, Combs D, Hodgman JE, Harper RM, Sterman MB (1978b) Studies of maternal–fetal interaction during the third trimester of pregnancy. I. Ontogenesis of the basic rest–activity cycle. Exp Neurol 61:136–153

Hoppenbrouwers T, Hodgman JE, Harper RM, Sterman MB (1979) Motility patterns as a function of age and time of night. Sleep Res 8:124

Hoppenbrouwers T, Jensen D, Hodgman JE, Harper RM, Sterman MB (1980) The emergence of a circadian pattern in respiratory rates: comparison between infants at low and increased risk for SIDS. Pediatr Res 14:345–351

Hoppenbrouwers T, Hodgman JE, Sterman MB, Harper RM (1982) Temporal distribution of sleep states, somatic and autonomic activity during the first half year of life. Sleep 5:131–144

Hoppenbrouwers T, Hodgman J, Arakawa K, Geidel SA, Sterman MB (1988) Sleep and waking states in infancy: normative studies. Sleep 11:387–401

Junge HD (1980) Behavioral states and state-related heart rate and motor activity patterns in the newborn infant and the fetus antepartum: a comparative study. III. Analysis of sleep state-related motor activity patterns. Eur J Obstet Gynecol Reprod Biol 10:239–246

Klein R, Armitage R (1979) Rhythms in human performance: $1\frac{1}{2}$-hour oscillations in cognitive style. Science 204:1326–1327

Kleitman N (1963) Sleep and wakefulness. University of Chicago Press, Chicago, IL

Kleitman N, Engelmann TG (1953) Sleep characteristics of infants. J Appl Physiol 6:269–282

Kleitman N, Titelbaum S, Hoffmann H (1937) The establishment of the diurnal temperature cycle. Am J Physiol 119:48–54

Klevecz MS, Kauffman SA, Skymks RM (1984) Cellular clocks and oscillators. Int Rev Cytol 86:97–128

Koch J (1968) The change of conditioned orienting reactions in 5 month old infants through phase shift of partial biorhythms. Hum Dev 11:124–137

Kronauer RE (1987) Temporal subdivision of the circadian cycle. Lect Math Life Sci 19:63–120

Kronauer RE, Gander PH (1989) Commentary on article of Daan et al. (Timing of human sleep: recovery process gated by a circadian pacemaker). Am J Physiol 246:R178–R183

Lavie P (1985) Ultradian rhythms: gates of sleep and wakefulness. Exp Brain Res 12 [Suppl]: 148–164

Lovett Doust JW (1979) Periodic homeostatic fluctuations of skin temperature in the sleeping and waking state. Neuropsychobiology 5:340–347

Martin du Pan R (1974) Some clinical applications of our knowledge of the evolution of the circadian rhythm in infants. In: Scheving LE, Halberg F, Pauly JE (eds) Chronobiology. Igaku Shoin Ltd, Tokyo, pp 342–347

McMillen IC, Kok JSM, Adamson TM, Deayton JM, Nowak R (1991) Development of circadian sleep–wake rhythms in preterm and full-term infants. Pediatr Res 29:381–384

Meier-Koll A (1979) Interactions of endogenous rhythms during postnatal development: observations of behaviour and polygraphic studies in one normal infant. Int J Chronobiol 6:179–189

Meier-Koll A, Hall U, Hellwig U, Kott G, Meier-Koll V (1978) A biological oscillator system and the development of sleep–waking behavior during early infancy. Chronobiologia 5:425–440

Minors DS, Waterhouse JM (1981) Development of circadian rhythms in infancy. In: Davis JA Dobbing J (eds) Scientific foundations of paediatrics. Heinemann Medical Books, London, pp 980–997

Minors DS, Tenreiro S, D'Souza T, Waterhouse JM (1990) Ultradian and circadian rhythms in heart rate in premature babies maintained in a constant environment. J Interdiscipl Cycle Res 21:220–222

Morath M (1974) The four-hour feeding rhythm of the baby as a free running endogenously regulated rhythm. Int J Chronobiol 2:39–45

Nijhuis JG, Prechtl HFR, Martin CB Jr, Bots RSGM (1982) Are there behavioural states in the human fetus? Early Hum Dev 6:177–195

Parmelee AH (1961) A study of one infant from birth to eight months of age. Acta Paediatr Scand 50:160–170

Patrick J, Campbell K, Carmichael L, Natale R, Richardson B (1982) Patterns of gross fetal body movements over 24-hour observation intervals during the last 10 weeks of pregnancy. Am J Obstet Gynecol 142:363–371

Pool R (1989) Ecologists flirt with chaos. Science 243:310–313

Prechtl HFR (1984) Continuity and change in early neural development. In: Prechtl HFR (ed) Continuity of neural functions from prenatal to postnatal life. Spastics International Medical Publications, London, pp 1–15

Reppert SM, Duncan MJ, Weaver DR (1987) Maternal influences on the developing circadian system. In: Krasnegor NA et al. (eds) Perinatal development: a psychobiological perspective. Academic Press, Orlando, FL, pp 343–356

Robertson SS (1987) Human cyclic motility: fetal–newborn continuities and newborn state differences. Dev Psychobiol 20:425–442

Romanini C, Arduini D, Valensise H et al. (1985) Circadian variation of fetal quiet/activity cycle: correlation with endocrine maternal parameters. Poster presented at International Workshop on Developmental Neurology of the Fetus and Preterm Infant, Groningen, The Netherlands

Sander LW, Julia HL, Stechler G, Burns P (1972) Continuous 24-hour interactional monitoring in infants raised in two contrasting environments. Psychosom Med 34:270–282

Schulz H, Salzarulo P, Fagioli I, Massetani R (1983) REM latency: development in the first year of life. Electroencephalogr Clin Neurophysiol 56:316–322

Selinger M, Levitz M (1969) Diurnal variation of total plasma estriol levels in late pregnancy. J Clin Endocrinol Metab 29:995–997

Siegmund R, Biermann K, Schiefenhovel W (1990) Ontogenic development of time patterns in food intake – a study of German infants and preliminary data from Trobriand infants (Papua New Guinea). J Interdiscipl Cycle Res 21:246–248

Siffre M (1986) Desynchronization of human biological rhythms during long-term temporal isolation studies in the constant environment of caves. J Interdiscipl Cycle Res 17:155

Sitaram N, Moore AM, Gillin JC (1978) Experimental acceleration and slowing of REM sleep ultradian rhythm by cholinergic agonist and antagonist. Nature 274:490–492

Sterman MB, Hoppenbrouwers T (1971) Development of sleep–waking and rest activity patterns from fetus to adult in man. In: Sterman MB, McGinty DJ, Adinolfi A (eds) Brain development and behavior. Academic Press, New York, pp 203–227

Sterman MB, Harper RM, Havens B, Hoppenbrouwers T, McGinty DJ, Hodgman JE (1977) Quantitative analysis of central cortical EEG activity during quiet sleep in infants. Electroencephalogr Clin Neurophysiol 43:371–385

Stratton P (1982) Rhythmic functions in the newborn. In: Stratton P (ed) Psychobiology of the human newborn. John Wiley & Sons Ltd, New York, pp 119–145

Szeto H, Dwyer G, Cheng P, Decena J (1990) Fractal scaling properties of fetal breathing dynamics. Proc Conf IEEE Eng Med Biol 12:1390–1391

Thoman EB (1975) Sleep and wake behaviors in neonates: consistencies and consequences. Merrill-Palmer Q 21:295–341

Thoman EB, McDowell K (1989) Sleep cyclicity in infants during the earliest postnatal weeks. Physiol Behav 45:517–522

Tomioka K, Tomioka F (1991) Development of circadian sleep–wakefulness rhythmicity of three infants. J Interdiscipl Cycle Res 22:71–80

Ullner RE (1974) On the development of ultradian rhythms: the rapid eye movement activity in premature children. In: Scheving LE, Halberg F, Pauly JE (eds) Chronobiology. Igaku Shoin Ltd, Tokyo, pp 478–481

Urbach JR, Phuvichit B, Zweizig H et al. (1965) Instantaneous heart-rate patterns in newborn infants. Am J Obstet Gynecol 93:965–974

Visser GHA (1985) Circadian rhythms in the fetus. Paper presented at the International Workshop on Developmental Neurology of the Fetus and Preterm Infant, Groningen, The Netherlands

Visser GHA, Poelmann-Weesjes G, Cohen TMN, Bekedam DJ (1987) Fetal behavior at 30 to 32 weeks of gestation. Pediatr Res 22:655–658

Wever RA (1965) Pendulum versus relaxation oscillations. In: Aschoff J (ed) Circadian clocks. North-Holland, Amsterdam, pp 79–83

Wever RA (1979) The circadian system of man. Springer-Verlag, Berlin Heidelberg New York

Wever RA (1985) Modes of interaction between ultradian and circadian rhythms. Exp Brain Res 12 [Suppl]:309–317

Chapter 10

Phase Plots of Temporal Oscillations

A. Garfinkel and R. Abraham

Oscillator Theory in Qualitative Dynamics

The last decade has witnessed an important development in biology, both theoretical and applied. It has come to be recognized that oscillatory processes play a central role in physiological systems (Chance et al. 1969; Winfree 1980; Yates 1982; Iberall and Soodak 1987). This is true at many scales, from the level of organ systems, such as the rhythmic behaviours of the heart, lung and endocrine systems, to the finer scales of cellular metabolism. An image has emerged of the typical physiological system as a system of loosely coupled oscillators cooperating to produce marginally stable collective behaviour modes. This image seems appropriate to the human circadian system.

It is remarkable that, over the same period, mathematics has been developing a language, and a set of techniques, for dealing with precisely the same concepts: coupled oscillators with multiple stable modes. We think that these mathematical techniques can give real insights into physiological systems. The mathematical techniques are drawn from qualitative dynamics, an approach to dynamic systems that stresses the importance of understanding the qualitative forms of behaviour that are possible in various dynamic systems, and the factors that can bring about qualitative changes in a system from one form of behaviour to another (see Abraham and Marsden 1979; Abraham and Shaw 1982; Garfinkel 1983). Qualitative dynamics gives definiteness to the idea of "form of motion" through the concept of an *attractor*. An attractor is a subset to which a large number of trajectories tend over time. (For definitions, see Hirsch and Smale 1974; Abraham and Marsden 1979.) We may think of it as the qualitative behaviour of the system after transients die away. The simplest attractor is a *point attractor*. Orbits that approach a point attractor represent an approach to a steady equilibrium. The next simplest type of attractor is a closed cycle, and orbits that approach such a *periodic attractor* represent periodic or oscillatory processes. Other attractors, called *chaotic attractors* represent aperiodic processes, ranging from nearly periodic oscillations with a slight wobble to wildly turbulent motions.

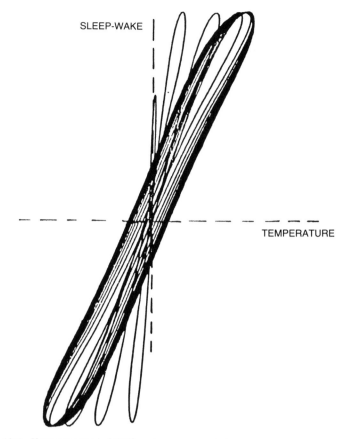

SLEEP-WAKE

TEMPERATURE

Fig. 10.1. Kronauer et al. (1982) system: approach to attractor.

Phase Plots

Qualitative dynamics suggests that we look at the behaviour of systems, in terms of their attractors. In order to do this we must plot their behaviours in a certain way: instead of plotting each of the variables as a function of time and then trying to compare them, we plot the variables of interest *against each other*; that is, as a trajectory in the space defined by the variables. In this method, time is eliminated as an explicit variable; it is suppressed by comparing state variables directly. Thus, rather than plotting $X(t)$ versus t and $Y(t)$ versus t, we plot an orbit, the curve through $X-Y$ space of the point $(X(t), Y(t))$ moving with time t.

We call such trajectories *phase plots*. They have several virtues, the most obvious of which is that they give a graphic representation of the covariation of the variables. For example, we can study how the temperature and activity oscillations in the Kronauer et al. (1982) model covary by plotting the trajectory of the model as it projects on to the subspace of the X and Y variables. If the system is tracking an attractor, then we may expect to see a projection of that attractor on to the subspace we have chosen. This we call a *projected attractor*, and its shape gives us information about the attractor itself. For example, an oscillation (a limit cycle attractor) will be indicated by a projection of a cycle: a circle, figure of eight, or other closed curve.

For example, if we start the Kronauer system at arbitrary initial conditions of body temperature and sleep–wake activity, the state of the system moves to an attractor. Looking at the trace of this motion, after a number of "days" have passed, we obtain the phase plot shown in Fig. 10.1. In this case the phase plot of the projected attractor is the cigar-shaped object lying diagonally across the first and third quadrants, and its shape tells us a number of useful things about the relation between the X and Y oscillators.

First, the X and Y oscillators have become *frequency entrained*. This can be read off the phase plot from the fact that there is no further precession of the orbit. In the transient phase, on the other hand, they are not frequency entrained, and the precession of the orbit from a nearly vertical orientation to the final diagonal marks the process of frequency entrainment. Second, once at the attractor, the two oscillators are *phase entrained*: they reach their maxima, like their minima, nearly simultaneously. Moreover, they stay highly correlated throughout their cycle, as can be read from the thinness of the cigar-shaped projected attractor.

We can see in this example some of the advantages of this style of presentation, especially in giving graphic and immediately readable representations of important notions. Imagine, for example, being given parallel time series for two or three variables and being asked for their frequency- or phase-entrainment tendencies, or even a simple question such as: is it ever the case that one is low while another is high? The phase plot answers such a question at a glance.

Further, the phase plot serves to suggest causal hypotheses about the inter-relations of X and Y. We see in the example of Fig. 10.1 that the activity oscillator has its full amplitude of oscillation almost immediately, while the temperature oscillator is barely active, being progressively "pumped up" by the activity oscillator as it comes more and more into phase coherence with it.

Finally, models of physiological processes enable us to simulate the effects of experiments, pathologies, or environmental perturbations. Phase plots of the results of these simulated experiments can then serve as predictions. As an example, we simulated the effects of a sudden phase shift, the equivalent of an instantaneous flight from Los Angeles to London or an isolation bunker experiment in which the day/night cycle is shifted by 8 h. The results of the simulation are shown in Fig. 10.2. The subject is originally in the cigar-shaped mode, stably entrained to the sun. At the arrow, we phase-shift the entraining oscillator, and the stable mode becomes destabilized temporarily. Over a number of days of "jet-lag", the subject becomes gradually re-entrained.

Forms of Desynchronization (Raster Presentation)

Kronauer et al. (1982) were particularly concerned to study desynchronization between subsystems of the circadian system. In some classic experiments,

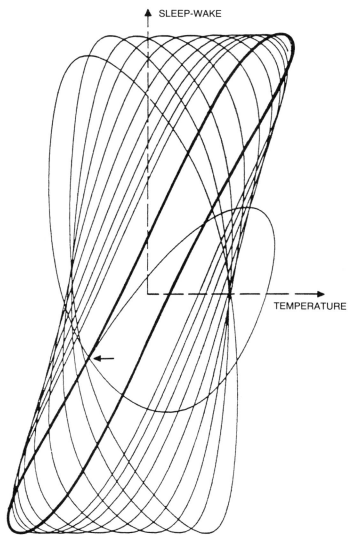

Fig. 10.2. Jet-lag: model of resynchronization after phase reversal.

volunteers were placed in environments which were completely cut off from the effects of the sun. In these "free-run" conditions, various forms of desynchronization between the body's core temperature cycle and sleep–wake cycle were observed. Kronauer et al. modelled these experiments by a simulation in which the sinusoidal driving force is set to zero on day 5 and the model is allowed to free-run until day 100. In order to match observed data, they impose an additional assumption in their simulation: from day 5 until day 100, they impose a uniform decline in the native frequency of the Y-oscillator (the quantity they call $\hat{\omega}_y$) from 0.920 to 0.778. We followed this additional

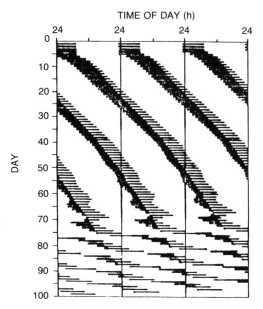

Fig. 10.3. Raster plot from Kronauer et al. (1982).

assumption in our initial simulations (but see the final section of this chapter for a discussion of this condition).

They present the results of their simulation as a raster plot, a style of presentation in which a time series is taken for one epoch (in this case 1 day) and then the time series for the next epoch is placed immediately below it, the following epoch immediately below that, and so on. In order to ease visualization, the data in such cases are usually binarized; that is, converted to a yes−no scale (temperature ⩾ or < average, sleep or awake), which is then represented by white versus black bars. In addition, several copies of each epoch are placed alongside each other for better visualization. Kronauer et al.'s description of the scenario, drawn from this presentation, is as follows.

Days 5−9. "Transient adjustment of relative phase between sleep and temperature", to a new relationship.

Days 9−40. The secular decline in $\hat{\omega}_y$ brings about a "progressive shift of the relative phase between sleep and temperature".

Days 40−75 ("phase trapping"). "The sleep−wake cycle and the temperature cycle have on average the same periods. Yet there are distinct modulations of activity period that grow in amplitude as the progressive lengthening of $\hat{\tau}_y$ [i.e. shortening of $\hat{\omega}_y$] separates it further from the intrinsic period of the X oscillator, $\hat{\tau}_x$." These properties cannot be read off the raster plot of Fig. 10.3, and require a distinct plot of period versus time over the 100 days (their Fig. 4B).

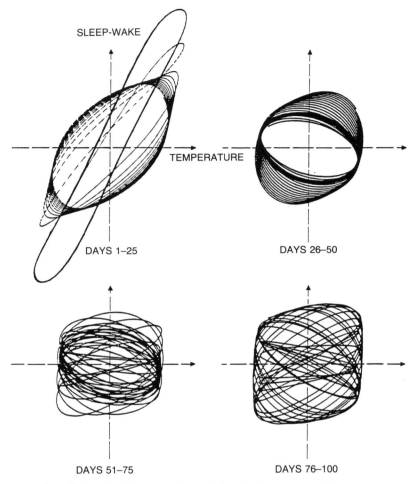

Fig. 10.4. Circadian oscillation model under free-run conditions.

They remark that this "phase trapping" is similar to what Wever calls "relative coordination".

Days 75–100. "The *X* and *Y* oscillators are desynchronized" (albeit with "considerable relative coordination").

Forms of Desynchronization (Phase Presentation)

As an example, we reproduced their simulation and plotted our results as phase plots in the *X*–*Y* plane, obtaining Fig. 10.4. This representation suggests a rich story that supports their interpretation and also supplements it in several ways. We can see definite developments as the orbit evolves.

Days 1–25. The temperature (X) and sleep–wake activity (Y) cycles begin in their normal synchronized mode, moving along the cigar-shaped attractor. But on day 5, when the entraining effects of the sun are removed, the synchronized mode is no longer stable, and the system moves into a new mode of behaviour: the cigar shape is quickly replaced by the oval shape, and a steady progression is begun in which the oval gets shorter and fatter over the next 20 days. This progression vividly shows a gradual loss of phase entrainment and a progression of relative phase, as reported by Kronauer et al.

Days 26–50. The tendency to oblateness continues. By day 50, the trajectory has achieved a new stable mode, an attractor pointing orthogonally to the orientation of the cigar shape. In this mode, we can see that the temperature and activity oscillators are directly out of phase with each other.

Days 51–75. In this period, the continuing decline of the native frequency of the sleep oscillator $(\hat{\omega}_y)$ ultimately destabilizes the new attractor, and the spaghetti-look of the orbit signifies what appears to be an instability in the relative coordination of the two cycles, but is in fact a prolonged transient; the orbit is tracking periodic attractors of long Lissajous (braided) type, through a rapid sequence of bifurcations brought about by the imposed secular decline of $\hat{\omega}_y$.

Days 76–100. Finally, the trajectory becomes less disordered, and the orbit, while quite complex, has a definite structure. The actual attractor of the coupled system has ceased to change, and the prolonged transient has died away. The final attractor, shown here, resembles a roll of twine, and this appearance is significant. A roll of twine is a line wrapped around a torus, and the look of the orbit suggests that it has stabilized itself onto an attractor on a 2-torus, whose projection onto the X–Y axis we are seeing.

This last observation suggests a useful application of phase plots of real-world data. The 2-torus, T^2, is the natural state space for two coupled oscillators, using a cyclic coordinate around the outside (long) circumference for oscillator 1 and a second cyclic coordinate around the hollow tube for oscillator 2. (Topologically, $T^2 = S^1 \times S^1$.) Consequently, when we find a toroidal orbit in a plot, it suggests the possibility that the underlying system producing it has the structure of two coupled oscillators. In other words, we would have some justification for inferring that this circadian system was two coupled oscillators even if we did not already know it. Thus, a plot of this kind of (real-world) data would suggest a hypothesis for a possible model.

The Decline of $\hat{\omega}_y$

The specific model of Kronauer et al. (1982) has the property, as we mentioned, that an extrinsic feature is added for the simulation of desynchronization: the uniform decline of the native frequency of the Y-oscillator

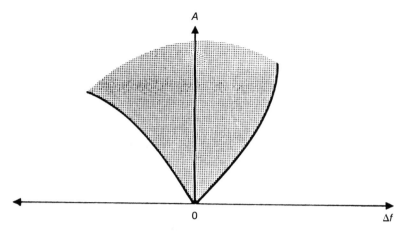

Fig. 10.5. Region of entrainment. Δf = difference in native frequency $(\hat{\omega}_x - \hat{\omega}_y)$; A = coupling strength.

(the quantity $\hat{\omega}_y$) from 0.920 to 0.778 from day 5 to day 100. There is no physiological justification for this feature; it was added solely to model the observed data.

It seems possible to model this phenomenon in a more intrinsic way. It is a general property of coupled Van der Pol oscillators that their regions of entrainment depend on their relative frequencies and amplitudes: two Van der Pol oscillators with distinct native periods will entrain to a common frequency if the strength of the coupling is sufficiently large relative to the frequency difference. A partial diagram of the region of entrainment is shown in Fig. 10.5. The shaded area represents a typical region of entrainment. This general property suggests that a secular decline in the observed period of the Y oscillator could be modelled as a transient response, after the external drive corresponding to daylight has been removed, to the final attractor shown in Fig. 10.4.

To test this idea, we repeated the simulation of Kronauer et al. (1982), but with a different approach to modelling the decline of the native frequency of the sleep–wake oscillator. The point was to see whether the observed data could be modelled as a loss of 1:1 entrainment (the cigar shape) followed by a transient relaxation, over 95 days, to a new attractor representing a more complex entrainment. So instead of externally imposing a secular decline in $\hat{\omega}_y$, we set it equal to 0.78, fixed from the start.

In this new simulation, with external forcing corresponding to daylight, the orbit tracks a cigar-like attractor, exactly as in Fig. 10.1. (In other words, the system is inside the shaded region of entrainment in Fig. 10.5.) After the suppression of daylight, the system relaxes to the Lissajous shown in Fig. 10.6, giving good qualitative agreement with observed data. (The quantitative agreement is less good: the relaxation in this simulation is faster than in the original simulation. We presume that this transient could be slowed, to better fit the observed data, by weakening the coupling between the two oscillators.)

We can summarize the difference between our simulation and that of Kronauer et al. by means of another form of representation (one that we think

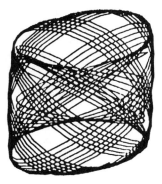

Fig. 10.6. Free-run simulation with native frequency of sleep–wake oscillator held constant at low value: final 85 days.

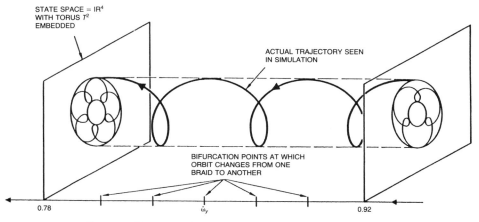

STATE SPACE = IR⁴
WITH TORUS T²
EMBEDDED

ACTUAL TRAJECTORY SEEN
IN SIMULATION

BIFURCATION POINTS AT WHICH
ORBIT CHANGES FROM ONE
BRAID TO ANOTHER

0.78 $\hat{\omega}_y$ 0.92

Fig. 10.7. Response diagram of Kronauer et al. (1982) simulation.

is generally useful) called a *response diagram*. In a response diagram, we represent a series of phase plots corresponding to different values of a chosen parameter. In this case, we let the horizontal axis represent different values of $\hat{\omega}_y$: corresponding to each value, there is a phase plot of the attractor of the system with that parameter value. Fig. 10.7 shows the schematic representation of the simulation of Kronauer et al., an orbit tracking a changing attractor, which passes through a number of bifurcations as $\hat{\omega}_y$ is exogenously lowered. In contrast, the orbit seen in Fig. 10.6 is an approach to a single attractor.

Summary

Phase plots, in which variables are plotted as trajectories through state spaces, give useful insights into the behaviour of systems, both model and real. We show here that a number of significant facts about systems behaviour can be seen in these plots, and illustrate our claims with applications to a model of human circadian oscillations. We advocate a method of presenting data, both model and real, that is especially useful for studying oscillatory processes. We think that the methods can be widely used and apply them to the circadian oscillator model of Kronauer et al. (1982). Kronauer et al. represent human circadian oscillations by two coupled Van der Pol oscillators, representing the (core) body temperature cycle and the sleep–wake activity cycle, respectively. In addition, there is a sinusoidal oscillatory driver, representing the sun, acting on the activity oscillator.

References

Abraham R, Marsden J (1979) Foundations of mechanics. Addison-Wesley, Reading, MA

Abraham R, Shaw C (1982) Dynamics, the geometry of behavior. Aerial Press, Santa Cruz, CA

Chance B, Ghosh A, Pye B, Hess B (1969) Biological oscillators. Academic Press, New York

Garfinkel A (1983) A mathematics for physiology. Am J Physiol Reg Int Comp Physiol 14: R455–R466

Hirsch M, Smale S (1974) Differential equations, dynamical systems, and linear algebra. Academic Press, New York

Iberall AS, Soodak H (1987) A physics for complex systems. In: Yates FE (ed) Self-organizing systems: the emergence of order. Plenum Press, New York

Kronauer RE, Czeisler CA, Pilato SF, Moore-Ede MC, Weitzman ED (1982) Mathematical model of the human circadian system with two interacting oscillators. Am J Physiol Reg Int Comp Physiol 11:R3-R17

Winfree AT (1980) The geometry of biological time. Springer, Berlin Heidelberg New York

Yates FE (1982) Outline of a physical theory of physiological systems. Can J Physiol Pharmacol 60:217–248

Chapter 11

Metabolic and Behavioural Long Period Ultradian Rhythms in Endotherms

M. Stupfel

Nous sommes tous l'heure qui sonne.

Blaise Cendrars, *Aujourd'hui.*

To ultradian Ina.

Introduction

We humans are much aware of a daily rhythm of behaviour. We wake up, wash, dress, breakfast, go to work or to accomplish a predetermined morning programme, then have lunch, perhaps go back home; after that comes dinner time and later on we go to bed and fall asleep. Of course this is not always at strictly regular times. Certain days there are variations in the schedule of occupation, weekends and holidays have many individualities. Seasons, weather, ageing and retirement modify ways of life. Some people are strict with time keeping, others are not. But the sequence of bed, breakfast, activity, lunch, activity, dinner, activity and bed are for many unavoidable, though, of course, at not regularly fixed times, but rather dependent on external current events. Furthermore, in a single day, we feel ourselves to be sometimes in a happy mood, for instance after good food, and sometimes depressed as a consequence of bad news. Even so we could have the sensation of rhythmic daily changes in humour, even independently of our psychological environment. Moreover, we know or rather we feel that, in a day, there are some kinds of hourly rhythms ruled by the clock; these have become societal cues in our busy lives. These cues have, for a long time, replaced the astronomical and solar light and dark signals that were originally the determinants of the activities of our forefathers. Time life tables of so-called "primitive" ethnic Australian and African people are still regulated by their temporal surroundings. In those cases diurnal and/or nocturnal food-seeking and protection against natural risks (for instance camp-fires) are primary determinants.

This chapter refers to rhythms of periods shorter than 24 h (ultradian) – in fact it will be limited to those falling in a range 40 min to 6 h, because in endotherms (or homeotherms) such periods separate different behavioural and metabolic episodes.

Examples will be largely confined to the comparative physiology of endothermic vertebrates (mammals and birds), which in their ordinary thermic surroundings, maintain their internal body temperatures in the range 34–42 °C.

Animal and human behaviour has, for the last 30 years, become the subject of the new science ethology. The behaviour of an endotherm is complex, comprising respiration, locomotion, feeding, search and exploration, herding and dispersion, sleep, reproduction, social relations, care of body surface, nest-building and more. To investigate the chronobiology of behaviour it is necessary to identify sequences of movements which are uniform and hence highly predictable within a species. This corresponds to what Barnett (1963) called "fixed action patterns".

In fact knowledge is limited by the means of measurement and recording. These must be performed with the use of non-invasive techniques. Amongst them, video recording necessitates a time consuming and difficult separation of behavioural types. Motility and displacement have long been the subjects of several measuring processes (light-beam cutting, inductance, capacitance variations, suspended cages, cages with springs); however these methods are practically impossible to calibrate. On the other hand, respiratory exchanges may be quantitatively assessed using expired/inspired gas analysis and feeding by recording food and water consumptions. These experimental techniques have been the most commonly used, but the list is not exhaustive, for it includes, for instance, dust-bathing in birds. Most of these measurements need sophisticated equipment and therefore must be performed by animals kept in laboratories, where their usual behaviour is modified by captivity.

However, for several years, field measurements have been performed using radiotelemetry. Originally this was limited to electrocardiograph and temperature recordings. More recently transmitters, designed either with a broadcasting antenna forming a collar or as a back and breast harness, with a whip antenna extending along the back, enable the recording of activity over long distances. These techniques are generally used for animal research.

Investigations of human rhythms are performed with sensors taped at several parts of the body (wrist, ankle, head) to investigate local displacements. Rhythm research in "sleep laboratories" uses beds equipped to measure global motility, electroencephalography (EEG), and electromyography to record the different phases of sleep, as well as other polygraphic recorders (electrocardiograph, respiratory movements, respiratory gases, temperatures at several body places). Processes that automatically analyse the sleep phases of EEG, such as those proposed by Schulz (1988), would be a great advance in the ultradian rhythms measurements of long-term records of brain electrical activity in sleeping individuals as well as during wakefulness.

According to Yates (1982) metabolism is a broad term used by biologists to designate the sum of all the free-energy-requiring or -generating processes of transport (diffusion, convection, wave propagation) and transformation of compounds (energy or information), that occur in a single living organism, ranging in complexity from virus to man. Yates also defines metabolism as the total inflow and outflow of matter and energy and its intermediary transformation within a living organism. Thus determination of the metabolic rate should give thermodynamic exchanges between an organism and its surroundings. This overall metabolic rate may be quantitatively measured by the emission of heat in a calorimeter, or more simply by the respiratory gaseous exchanges (oxygen consumption ($\dot{V}O_2$) and carbon dioxide emission ($\dot{V}CO_2$)) of a living organism placed in an environment of known physical parameters (temperature, humidity, air velocity). In fact metabolic deter-

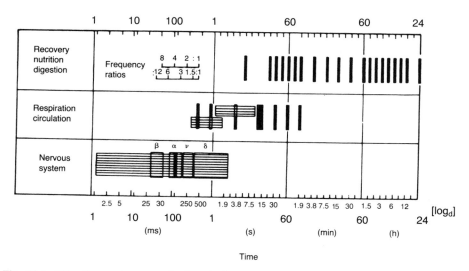

Fig. 11.1. Ultradian endogenous rhythms in humans: recovery (rest), nutrition (food intake), digestion, respiration, cardiovascular function and nervous system. Abscissa: periods (log scales). Black vertical bars indicate preferred frequency bands, horizontal hatching indicates the range of the frequency modulation. $\log_d$ is log duration. (Reproduced, with permission, Hildebrandt 1988.)

minations reveal the thermodynamic exchanges between the organism and its actual surroundings. The great interest of a continuous determination of metabolism is that, as it depends on every behavioural episode (respiration, displacement, muscular movement, feeding, excretion, etc.), it integrates most behavioural changes (e.g. feeding, foraging, displacement towards a thermo-neutral site) of an organism aimed at maintaining its metabolic integrity.

Ultradian rhythms have periods shorter than 24 h, which means that they may be defined as repetitions of the same physiological event several times per day, with (theoretically) no limit to the number of repetitions. Therefore, there would be an infinite number of biological ultradian rhythms down to the atomic level if our measurements were not technically limited. Figure 11.1 shows the scale of ultradian rhythms in humans. Until now the span of frequencies investigated have been mostly in the rapid range, e.g. in humans: electroencephalograph (0.5–30 Hz), electromyograph (0.3–8.0 Hz), heart rate (0.5–6 Hz) and respiratory rate (0.3 Hz).

This chapter discusses slow repetitions of behaviour during a 24-h span: eating, drinking, moving, resting, sleeping, awaking, naps, playing, toileting, etc. It also could have been extended to changes in attention and mindness, vigilance, in so far as these influence motivation and behaviour.

These various events may repeat themselves several times in a day, either separated, as sequences, or as bursts of the same behaviour. Everyone familiar with pet animals can witness how a polycylic alternation of motion, rest, feeding, licking, preening, etc. occurs several times in the day. The pattern is not the same in the cat, dog or bird. At first sight these various behavioural events seem impossible to compute when one considers the difficulty of separating them (e.g. feeding and displacement), and of identifying the "fixed

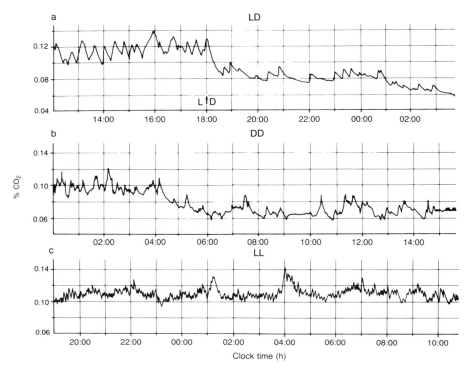

Fig. 11.2. Carbon dioxide concentration versus time in respiratory chamber containing one Japanese quail: **a** in light–darkness alternation (*LD*); **b** in continuous dark (*DD*); **c** in continuous 100 lux light (*LL*). In **a** the mean CO_2 level is higher and the ultradian oscillations are more frequent during light (L: from 12:00 to 18:00 h) than during darkness (D: after 18:00 h). In **b** (continuous darkness) the differences between ultradian oscillations are not so marked. In **c** (continuous light) there are no circadian differences in mean CO_2 concentration levels, and ultradian oscillations are more frequent than in **a** and **b**. Note the irregular occurrence of ultradian oscillations in the three tracings. (Reproduced, with permission, from Stupfel et al. 1990.)

action patterns" proposed by Barnett (1963). It would be interesting to investigate whether or not these behavioural sequences are periodic (Aschoff 1981), changing with the moment of the day and also whether these patterns are characteristic of different individuals as well as of species. This evidently necessitates statistical computation of many hundreds of data points collected over a long time of continuous survey. A visual examination of continuous recordings of movements, of respiratory gaseous exchanges (Fig. 11.2) or of an electroencephalograph reveals a succession of irregular and regular oscillations, often of several different periods, and sometimes there are no oscillations at all. It is difficult to know how best to compute their amplitudes and their frequencies. Most of the research performed in analyses of ultradian rhythms utilizes those classic analytical processes which should be used only for strictly periodic phenomena.

A periodic phenomenon is a mathematical entity which reproduces itself at constant intervals of time, called periods. The mathematical characterization of periodicity requires that it must be the same over a great number of periods. In other words, the investigated periodic function should be in an equilibrium or stable state independent of time. This is what is called a stationary state.

This stationary state condition is practically never found for a long span of time in the biological rhythms. It must be said that the same lack of stationarity is also observed in physical systems (astronomy provides some obvious examples). Thus, the oscillations or time variations in the biological literature which are called "rhythms", "cycles", "periodicity", "frequencies" are, in the long term (several minutes, a few hours), practically never really periodic. The best example is the heart rate, which shows, in any individual, a great number of variable episodes. For instance, in the same day, a healthy man has a heart rate of between 60 and 80 beats/min. These changes in rate result from respiratory inspiration and expiration, body movements, emotional excitement, digestion or sleep. Many years ago physicians described sinusal and nodal heart rhythms which originate from endogenous processes of cardiac neural centres and nerves, and underlie heart contraction irregularities. According to Jalife (1990), there is not for the moment an adequate mathematical method which can completely analyse heart periodicity, although new procedures (e.g. chaotic dynamics, fractal analysis) have been proposed. Ever since the beginning of medicine, pulse has been considered as somewhat "periodic" by physicians and physicists; Galileo (1564–1642) measured his pulse rate to demonstrate the periodicity of the pendulum.

Data

Short periods in animals' movements were first measured about 70 years ago. Szymanski (1918) reported that rabbits are typical polyphasic animals since they have 16–21 regular periodic alternations of activity–rest per 24-h cycle. In 1927, Richter demonstrated in rats the existence of rhythms of activity with periods of 1–2 h, e.g. of drinking with a period of 2–3 h, eating with a period of 3–4 h, defecating with a period of 3–5 h, and urinating with a period of 2–3 h). Apart from these activity measurements, even now, there are rather few observations of the various types of ultradian change. Almost all the available data concern circadian periods.

Activity and Rest

In 1922, Richter was able to describe bursts of activity with periods of between 1 and 2 h by placing rats in spring suspended cages. Much later, Wollnik and Döhler (1984) using sophisticated methods of measurements and computation, found ultradian mobility rhythms of periods of 12 h, 6 h, 4.8 h and 4 h in several strains of rat. These investigators also described ultradian rhythms of activity of 1–2 h for guinea pigs. In a British shrew (one of the smallest of all mammal species), Crowcroft (1954) reported many ultradian activity–rest alternations, the longer ones of which were 2–3 h. In 1957, Aschoff distinguished in mice several series of activity rhythms: short ones corresponding to grooming and scratching, and superimposed longer ones of 2–3 h periods. In isolated and grouped mice, Del Pozo et al. (1978) reported major peaks of activity with periods of near 3 h. Jilge et al. (1986) measured simultaneously several behavioural rhythms or variations in rabbits: e.g. locomotor activity, food and water intake, hard faeces as well as urine emission, which they found to be tightly coupled together, with periods of 1–6 h, 2–8 h and 3–12 h, respectively for activity, faeces, water and food intake, and urine emission. Several studies have been performed in monkeys. In the years 1973–78, Delgado-Garcia's group investigated the mobility as well as several types of behaviour (observed through a window) in 11 rhesus monkeys (see Delgado-Garcia et al. 1978). Each monkey was caged and carrying an FM (frequency modulated) transmitter sensor. A total of 328 recordings during 24 h of the mobility episodes showed dominant periods of 0.8 h, 1–2 h and 1–7 h. Furthermore, spontaneous behavioural non-stationary sequences of 70 min were observed during the day. Bowden et al. (1978) also reported, in seven rhesus monkeys, behavioural ultradian rhythms of 1.33–2.0 h periods, but they did not simultaneously measure any other physiological parameters.

In humans, several types of motility sensor have been used. Globus et al. (1973) reported a rhythm of gross body movements with a period of about 2.4 h. Later, in 1984, Okudaira et al. using small motility sensors placed on the wrists, heads and the ankles of seven men and three women, were unable to find the expected ultradian rhythms of the 90–100 min basic rest–activity cycle (BRAC).

An interspecies comparative study of these movements is very difficult. These measurements (including displacement, motility, muscular activity) cannot be related to a specified standard unit, and cannot consequently be quantified. Moreover, their rates are not the same in quadrupeds (e.g. rodents) as in bipeds (e.g. birds).

Food Intake

Location and gathering of food are only possible when endotherms are active. Therefore both normal food intake and drinking are coupled to rest–activity alternations, for they are not possible during sleep. These correlations were noted and documented in laboratory rats by Richter, as early as 1927. In the field, unlike in laboratory experiments, food is not always available. Food availability is a major determinant of whether an endotherm species is nocturnal or diurnal. As explained by Moore-Ede et al. (1982), in the laboratory, the allocation of feeding between light and darkness depends on the species: for instance rats eat 75% of their food in the dark, while squirrel monkeys eat 90% of their food in the light (Sulzman et al. 1978).

In rodents, under laboratory conditions, the intervals of food intake diminish with decreasing size of animals (e.g. rats 4 h, field mice (*Microtus*) 2.5 h, *Apodemus* and *Peromyscus* mice 2 h, domestic mice 0.75–1.5 h). Voles (diurnal rodents), when kept in cages exposed to natural meteorological conditions have, according to Hoogenboom et al. (1984), a periodicity of 2 h of feeding activity per day.

As has been pointed out by Daan and Aschoff (1975), discontinuous feeding enables ruminants to proceed to an efficient breakdown of cellulose with the help of microbes in their gastrointestinal tract. Since these grazers have to be vigilant for predators they usually stay within a herd to provide a collective awareness against this risk. Hughes and Reid (1951) and many other investigators have demonstrated that the herding tendencies of these herbivores synchronize their ultradian rhythmic behaviour.

Digestion

The occurrence of ultradian contractions in the small intestine has been recognized for many years, but only more recently recorded (e.g. by the use of implanted electrodes, in conscious dogs, rabbits and sheep). These recordings show that segmental contractions begin in the duodenum and have a frequency of 15–20 cycles/24 h. In the dog, these segmental contractions maintain gut muscles in condition during the interdigestive episodes, whereas in herbivores they assist the mixing of food and digestive secretions and facilitate the exposure of food mixture to the absorptive mucosa. According to Hiatt and Kripke's (1975) experiments, in both fasting humans and animals, motor complexes issued from the stomach propagate themselves towards the duodenum and ileum at time intervals of between 1.5 and 2 h. After feeding, the cycling activity of these motor complexes is disrupted for several hours. Bueno (1986) mentioned several experimental arguments that support the hypothesis that the frequency of these motor complexes, as well as their food-induced disruption, are under the control of brain neuropeptides. This would provide a chemical basis for the "psychological" participation, even in ruminants, in the appreciation of a good meal.

Temperature Regulation

In 1979, Lovett Doust reported 15–50 min variations in cutaneous temperatures (dorsum of the hand and axilla) of 23 fully clothed healthy men. But the measurements have been made for only 1–2 h. More extensive investigations have been performed in cows, by Lefcourt (1990), who reported that core body temperature rhythms of a 90-min period are superimposed on a 24-h rhythm. Moore-Ede et al. (1982) noted that squirrel monkeys, in continuous light, may show ultradian rhythms of core body temperature while they still have a circadian rhythm of skin temperature.

Cardiovascular Function

Eleven dogs implanted with intra-aortic recorders showed, according to Shimada and Marsh's (1979) investigations, heart and blood pressure oscillations with periods of about 1.5 h that were in phase together. Livnat et al. (1984) reported ultradian variations of both blood pressure and heart rates that were in phase and had 1–2-h periods. These results were obtained in nine unrestrained dogs; the blood pressure and the heart rates were followed by radiotelemetry during time spans of between 14 and 24 h. In dogs carrying telemetric sensors, Broten and Zehr (1989) found ultradian 0.6–1-h periods for both heart rate and blood pressure. They remarked that surgical suppression of the barosensitive nerves, the carotid sinuses, and the aortic cross nerves modify the amplitudes (but not the frequencies) of the blood pressure oscillations, without changing the ultradian oscillations of the heart rate. Blinowska and Marsh (1985) reported, in 11 dogs, ultradian oscillations of mean arterial blood pressure (MABP), with periods of between 1 and 4 h, and also ultradian fluctuations of electromyographic activity (which were not coupled with the ultradian oscillation of MABP).

Wilson et al. (1977) were not able to assess 90–100-min periods for blood pressure and heart rate measured from ten surgical intensive care patients. In two groups, one of young and old men and one of women, Benton et al. (1990) using ambulatory monitoring, reported ultradian rhythms of blood pressure with broad peaks centred at about 3 h, 6 h and 9 h, but which appeared to be not so marked as those of blood pressure circadian rhythms. An alternating congestion and decongestion of the right and left nasal airways has been known since 1895 when it was discovered by Kayser. Its periodicity varies very much according to the subjects, from 30 min to 6 h. In 1975, Eccles and Maynard reported a similar nasal cycle in pigs. In 1983, Werntz et al., in a study of 19 human volunteers, related this nasal vascular rhythmicity to an alternating brain electroencephalographic activity.

Urine Excretion

In humans, micturitions are more frequent during the day than at night, but differ between individuals and are greatly affected by ambient temperature and liquid volume ingestion. Though the circadian variations in urine volume and also in its constituents have been investigated for more than a century (Roberts 1860; Weigelin 1868) data on ultradian measurements are lacking. This may be because they are difficult to make. However, it has been reported that the responsiveness of diuretic mechanisms varies with the time of the day; for instance Stanbury and Thompson (1951) showed that, in humans, 1 litre of water ingested at 10:00 h gives a peak urine flow rate of 16 ml/min, whereas if the same challenge occurs at 17:00 h the peak flow does not exceed 8 ml/min.

For 25 male students drinking water and urinating (as far as it was possible) every 10 min over 10 consecutive hours, Lavie and Kripke (1977) described periods of 80–133 min of urine flow out of phase with the osmolality and electrolyte concentration. Gordon and Lavie (1985) reported that, in six dogs, ultradian rhythms of urine flow and osmolality showed periods of about 200 min and Jilge et al. (1986) that, in rabbits drinking water ad libitum, the frequency of urine excretion is seven to eight times per day.

Care and Body Cleaning

Grooming, scratching, licking in rats and mice, and feather preening and dust-bathing in birds are repeated behaviours of body surface care and cleaning which show ultradian rhythmicities. Aschoff (1957) reported periods of 2–3 h for such rhythms in mice; this activity is increased and disturbed by surface parasites.

Sensory Systems

Though they are not directly related to behaviour, it must be assumed that rhythms in sensory perception could modify behavioural reactions.

Long-term temporal variations of the responsiveness of sensory systems are difficult to identify in animals, unless inferred indirectly. Among several psychophysiological measurements performed in humans (for reviews, see Lavie and Kripke 1981, Kripke 1982; Rossi 1982, 1986), only two will be quoted here. In 1978, Lovett Doust et al. reported that, in 53 healthy men, the reaction times to determine the momentary stopping of the second hand of a clock show a periodicity of 4–15 min. As the length of their experiments on each subject was only 30 min they could not in fact discriminate periods longer than 15 min. It must be added that tests of this kind are difficult to replicate with accuracy as a long repetition may induce habituation. More convincing were the experimentation of Pöllman and Pöllman (1988), who determined pain in the front teeth of human subjects, on the application of cold. They found that the minimum time to provoke pain showed cycles with periods which ranged between 85 and 130 min.

Central Nervous System

Long-term ultradian data relating to the central nervous system are more difficult to collect than those concerning activity–rest, food intake, tempera-ture, urine excretion or metabolism. Nevertheless, ultradian patterns of motivation, which are susceptible to behavioural influences, have been demon-strated in three kinds of animal experiments. In 1972, Sterman et al. reported, in 20 cats, a mean periodicity of electrical self-stimulation of 21 min related to food availability. In 1980, Katz implanted Sprague-Dawley rats with intra-cranial electrodes and trained them to self-administration of intracranial stimulation in order to obtain food. He reported that, during the night, the numbers of stimulations that the rats dispensed to themselves, according to the individuals, were separated by periods of 12–30 min. In four rhesus monkeys, Maxim and Storrie (1979) implanted stimulating electrodes in two predeter-mined places in the brain. They found ultradian rhythms with periods of 40 min, 45 min and 85 min for rewarding brain stimulation.

Sleep Phases

During sleep in humans, electroencephalographic and electromyographic ultradian rhythms of 190 min (range 30 min–2 h) occur (Aserinsky and Kleitman 1955), one of rapid eye movement (REM, also called paradoxical sleep) often accompanied by erections of the penis or clitoris, and the other of non-rapid eye moments (NREM). Kleitman (1963) surmised that these 90-min rhythms found during sleep could also exist in humans during wakefulness, i.e. during all the active part of the day. He called this rhythm of 90 min the basic rest–activity cycle (BRAC). These observations have provoked numerous investigations relative to all kinds of physiological, psychological, psychiatric ultradian rhythms with periods that range from 30 min to a few hours.

Several investigators have looked for ultradian electroencephalographic (EEG) rhythms of low frequencies in humans during wakefulness. Thus, Ortega and Cabrera (1990) recording EEG activity in eight awake volunteers, during the morning and the afternoon, reported slow EEG oscillations of periods of between 3 and 4 h.

Tobler (1984) has studied the comparative physiological data of sleep phases in several birds and endothermic mammals; he found that the time interval between the beginning and the end of a sleep episode depends of the species. For instance it is 10 min in mice, 12 min in rats and quail, 19 min in cats, 90 min in humans and 125 min in elephants. Zepelin and Rechschaffen (1974) correlating the length of time of the REM–NREM sleep cycle with several physiological parameters in 53 mammalian species, found coefficients of +0.92 for brain weight, +0.81 for life span, −0.83 for metabolic rate and +0.79 for gestation period. Allison and Cichetti (1976) looked for interrelations between sleep phases, and physiological and ecological parameters in 39 mammalian species. They found that slow wave sleep (SWS) is negatively correlated with brain weight and they concluded that long durations of SWS are disadvantageous for the survival of large species. They also demonstrated that the duration of REM sleep is associated with the risk of exposure to predators.

In humans, regardless of genetic factors, it is well known that the timing of sleep and its periodicity (e.g. the number of naps or of food intake) are greatly dependent of culture, climate or working habits. For instance, hunting and fishing determine rest–activity and feeding times per day in primitive societies. Monastic rules, such as those of the Benedictines, have imposed strict ultradian societal rhythms, dividing up the 24-h day for religious practices.

The development of "sleep laboratories" has motivated intensive research on the alternation of sleep–wakefulness episodes in humans (for a review, see Schulz 1988). Classic sleep, EEG and electromyographic investigations have been associated with simultaneous recordings of body movements and other physiological parameters (temperatures, heart and respiratory rates, and endocrine determinations). For instance Gardner and Grossmann (1976) have estimated that the total number of sleep movements in humans per night ranged from 70 to 160; this corresponds roughly to time intervals of between 6 and 3 min. It has also been shown that major irregularities in heart and respiratory rates and also in blood pressure coincided with those body movements. However, to date, nothing is known about the temporal interrelations of these physiological parameters in humans or in animals, except for some hormone secretion processes.

Endocrinology

Endocrine hormones play important roles in behaviour. For example: stimulation of certain portions of the hypothalamus may set off sham rage; in humans hypothalamic lesions secondary to encephalitis not uncommonly result in hypersomnia; and in men hyperthyroidism is characterized by a considerable elevation of the basic metabolism, excitability and intolerance to heat. Addison's disease may lead to deficiency of all adrenocortical hormones and produce anorexia and aesthenia in humans. Hormones regulate and control the basic behaviour of the organism, i.e. body temperature, hunger, thirst, aggressivity and sexual drives.

Endocrine secretion takes some time. It is now about 15 years since it was shown that, in humans, all the pituitary hormones and many peripheral ones enter the blood in episodic pulses rather than continuously (Weitzman et al. 1974). As a result of many studies on humans, most endocrinological estimations may be now made on very small samples of blood; it is now possible to measure hormonal concentrations at short time intervals by sampling through a catheter that remains in a vein for a long time. The results have demonstrated that secretions of hypophysial hormones show peaks with ultradian periodicities.

Tannenbaum and Martin (1976) reported a 3.3-h ultradian periodicity of growth hormone which has been confirmed by Mori et al. (1986).

Many investigations concern the secretion rhythmicity of the luteinizing hormone (LH) by the anterior lobe of the hypophysis, both in humans and in all the studied endothermic mammalian species. Dierschke et al. (1970) demonstrated in ovariectomized rhesus monkeys a series of oscillations in circulating LH of periods of about 1 h; these were not observed in normal cycling females. Yen et al. (1972) found a pulsatile period of 1–2 h in LH in blood of women during their menstrual cycles. Rasmussen et al. (1981) and Ellis et al. (1983) reported in castrated rats a 20-min secretion of LH which corresponds to a 20-min REM–NREM rhythm in this species. Johnson and Gay (1981) showed a pulsatile secretion of LH in ovariectomized cats. Pelletier and Thiery (1986) reported, in sheep, periods of LH pulses varying between 2.4 and 24 h, depending on the individuals and also on environmental factors, such as daylength, nutritional levels or social surrounding. Odell and Griffin (1987), measuring LH and gonadotrophin (hCG) in eight postmenopausal women, reported rhythms of 0.54 pulses/h for LH and of 0.56 pulses/h for hCG. Rasmussen (1986) has suggested a species-dependent correlation between the periods of the pulsatile secretion of LH and the periods of Kleitman's (1963) hypothetical BRAC as a result of the relationships between both rhythms in humans, monkeys, rats and cows. Morin (1986) has established for LH a correlation between the periods of its secretion (0.3–4.0 h) and the body weights of ovariectomized rats, macaque monkeys, sheep and women, which, according to him, may correspond to a single cycle of blood circulation. Brandenberger et al. (1987) showed, in humans, ultradian rhythms of adrenocorticotrophic hormone with a period of about 100 min. Other important observations were that, in humans, hypophysial hormones (LH, prolactin, thyroid-stimulating hormone) exhibit secretion pulses synchronized with rest–activity alternations of sleep phases. Angeli and Carandente (1988) surmised that "superpulses" of hypophysial hormones synchronized with rest–activity or sleep–wake schedules occur in a circadian fashion.

Pineal glands of all mammals, whether they are diurnally or nocturnally active, produce melatonin abundantly at night and minimally during the day. The secretion of melatonin exhibits ultradian patterns: peaks at intervals of 10 min in humans (Vaughan et al. 1979), 5 min in sheep (Cozzi et al. 1988) and 25–30 min in rats (Pang and Yip 1988).

For peripheral hormones, Weitzman and Hellman (1974) reported ultradian pulses of cortisol with circadian components, in the plasma of seven human subjects. Levin et al. (1979) measuring plasma noradrenaline levels in blood

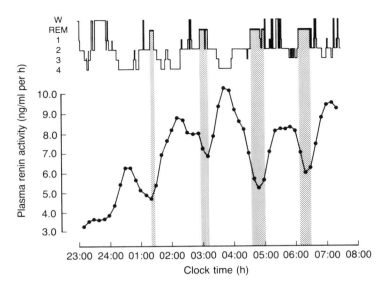

Fig. 11.3. Nocturnal plasma renin activity (ng/ml per h) versus time and REM–NREM sleep cycles. REM phases are within the shaded areas; declining plasma renin activity coincided with REM and increasing levels with NREM. *W*, wakefulness. (Reproduced, with permission, from Brandenberger et al. 1988.)

sampled every 15 min during 4 h in nine human volunteers, undergoing complete bed rest, showed fluctuations of this hormone with periods of 75–188 min. Brandenberger et al. (1985) showed that, in humans, during night sleep, plasma renin displayed sustained oscillations of periods ranging between 82 and 125 min which are closely related to the REM–NREM cycles (Fig. 11.3). This could be an important breakthrough in determining the oscillations of these sleep phases. Simon et al. (1987) found oscillations of plasma glucose, insulin and C-peptide with periods of 53–113 min in men under continuous enteral nutrition.

Iranmanesh et al. (1989) reported ultradian rhythms with periods of 100 min of β-endorphin in seven normal men; these rhythms showed a close temporal coupling to those of cortisol.

Metabolism

As stated in the Introduction, metabolic rate integrates all the energy variations of the organism and therefore reflects its global thermodynamic variations resulting from movement, absorption or excretion. Continuous recordings of oxygen consumption ($\dot{V}O_2$) performed by Kayser and Hildwein (1974) in rats showed 50-min period ultradian variations. Rubsamen and Hörnicke (1981) reported several ultradian rhythms of $\dot{V}O_2$ accompanied by synchronous variations of heart rate and of activity in rabbits during a light:dark alternation of 12 h:12 h (Fig. 11.4); the periods of these oscillations were between 5 and 40 min. Roussel and Bittel (1979) obtained, in sleeping rats, simultaneous

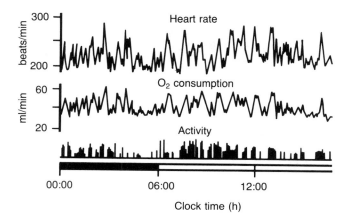

Fig. 11.4. *Heart rate* (beats/min), *oxygen consumption* (ml/min), *activity* (arbitrary scale) in a rabbit, in light from 06:00 till about 18:00 h. (Reproduced, with permission, from Rubsamen and Hörnicke 1981.)

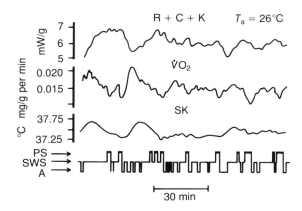

Fig. 11.5. Time course of sensible heat loss (*R*, radiation; *C*, convection; *K*, conduction; oxygen consumption ($\dot{V}O_2$), subcutaneous back temperature (*SK*) and awakeness–sleep (*PS*, paradoxical sleep (REM); *SWS*, slow wave sleep; *A*, awakeness) in a Sprague-Dawley rat. The bar (30 min) gives the scale of time. Measurements are made during light when the rat is mostly at rest. The oxygen consumption decreases at each phase of paradoxical sleep, while detectable heat loss diminishes at each phase of awakeness. T_a, ambient temperature. (Reproduced, with permission, from Roussel and Bittel 1979.)

recordings of $\dot{V}O_2$, electroencephalographs, subcutaneous temperature, and heat loss; the results showed synchronous variations of these four parameters during wakefulness, SWS and during each *paradoxical* REM sleep phase (Fig. 11.5).

In 1973, Bailey et al. measured, every 15 min over 6–12 h, the oxygen consumption of six men and four women at rest and after feeding. The recordings showed regular $\dot{V}O_2$ oscillations with intervals of 1–2 h, depending on the subject, and with amplitudes of 7%–20% of the mean $\dot{V}O_2$. In 1976,

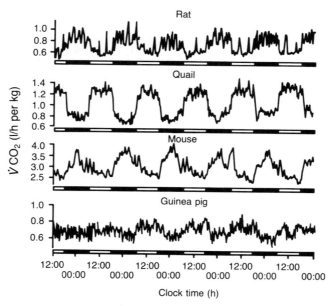

Fig. 11.6. Carbon dioxide emission ($\dot{V}CO_2$) of one Sprague-Dawley *rat*, one OF$_1$ *mouse*, one Hartley *guinea pig* and one Japanese *quail* in alternating light–dark cycle (light from 06:00 to 18:00 h). Circadian rhythms are obvious in the rat and in the mouse where $\dot{V}CO_2$ is greater during darkness than during light, and in the quail where conversely $\dot{V}CO_2$ is smaller during darkness than during light. No circadian differences between darkness and light are seen for the guinea pig. The tracings for the four animals show ultradian oscillations. (Reproduced, with permission, from Stupfel et al. 1990.)

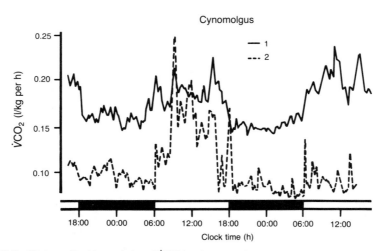

Fig. 11.7. Carbon dioxide emission ($\dot{V}CO_2$) of two male *Cynomolgus* monkeys, light (100 lux) from 06:00 to 18:00 h and dark from 18:00 to 06:00 h. Each tracing shows ultradian oscillations and circadian variations ($\dot{V}CO_2$ being greater during light than during darkness). (Reproduced, with permission, from Stupfel et al. 1984.)

Horne and Whitehead measured the respiratory rate of 12 people, men and women, at 3-min intervals over 6h. Although they noted substantial inter-individual differences, they concluded that the more frequently observed rhythm had a 90-min period, although they also found 60- and 30-min rhythms.

Stupfel and colleagues measured, under laboratory standardized environmental conditions (temperature 20–21 °C, relative humidity 60%–90%, food and water ad libitum), the respiratory exchanges of mice, rats, guinea pigs, monkeys, quail and chicks, and compared them with data from premature infants in a series of publication (Stupfel et al. 1979, 1980, 1981a,b, 1984, 1986a,b, 1987, 1989a,b, 1990, 1991; Stupfel and Pavely 1990). Continuous recordings of carbon dioxide emission ($\dot{V}CO_2$) and of oxygen consumption ($\dot{V}O_2$) showed ultradian oscillations, the amplitudes of which corresponded to 10% of the mean gaseous exchanges (Figs. 11.6, 11.7), with the exception of those of premature infants that were 30% (Fig. 11.8). To compare intervals between oscillating peaks, graphs were performed with samples of the CO_2 and O_2 concentrations every 20 min. In these graphs the time intervals, computed between two consecutive peaks of oscillations, proved to be nearly 1 h (between 0.80 and 1.55 h) with statistically significant differences between the seven endothermic species. The amplitudes of these peaks of oscillation were also significantly different and were strongly correlated to the logarithms of the body weights of the seven species investigated.

Origins

The origin of the biological ultradian rhythms is unknown, and speculations are welcome. Evidently the origin of circadian rhythms is the 24-h mean earth rotation which conditions the nycthemeral exogenous variations of light, temperature, relative humidity and other parameters and also life of an ecosystem at a fixed place in the terrestrial globe. It has been hypothesized that a biological circadian clock or circadian oscillator has been developed in order to keep a 20–28-h time independently of the cosmic daily signals. However, the animal or human endogenous clock varies from one individual to another one, and is not accurate enough to remain at the right time; it therefore needs regular external cues of cosmic synchronizers and/or of societal ones (e.g. for humans). Hence, it may be supposed that in the same way, ultradian rhythms also originate by synchronization with other cosmic, earthly or biological ultradian rhythms. But it is more simple to suppose that these long period ultradian biological rhythms are simply fractions (i.e. harmonics) of circadian rhythms. This hypothesis, supported by several authors and particularly for human rhythms by Hildebrandt (1988), is difficult to ignore. For instance there are many examples of bimodal circadian rhythms (also called bicircadian) which are repeated twice a day. Insects such as mosquitoes have two peaks of activity per day, one at dawn and the other one at sunset. Birds too have such two preferential times during the day for catching insects. Bimodal rhythms of activity, synchronized by tides, have been well investigated in marine animals, e.g. for crabs and shrimps living on beaches and seaside areas. Their ultradian sequences of activity may contain both lunar- and solar-day rhythmic components. Reviewing this subject, Palmer (1976) reported that the phase relationships of tidal rhythms are species specific and that they will often persist under constant laboratory conditions.

The origin of the high frequency (i.e. of small period) rhythms (nerve impulses, EEG or electromyograph) remains mysterious; Berridge and Rapp (1979) have proposed the concept of an electrochemical oscillator.

It is for such reasons that the possibility of entrainment of ultradian rhythms by cosmic earth and geophysical synchronizers of rapid ultradian frequencies has been considered.

Cosmic and Geophysical Synchronizers

Brown et al. (1970) in *The biological clock: two views* confronted two theories about the origin of biological rhythms: endogenous and exogenous. Supporting the second hypothesis, they suggested that "subtle environmental synchronizers" may be the cause of many kinds of biological rhythms. The difficulty in assessing this hypothesis comes from the lack of substantiated biological experimental data. In fact, they are impossible to realize. Moreover, the number of investigations in this domain appears discouraging because of the wide number of radiations and particles (electromagnetic and radiowaves, cosmic waves, ionized particles, photons, neutrinos, etc.) that have been identified on the terrestrial surface. Furthermore, it must be remembered that solar electromagnetic radiations are the source of all biological activities on our globe.

In connection with geophysical synchronization, Brown et al. (1970) asserted that the alpha rhythm of the human brain (10 Hz) has the same frequency as earth's crust vibrations. The greatest part of the earth's magnetic field comes from the ground; a small part, less than 1%, which comes from the ionosphere, is subject to circadian variations. These variations are quite small: 0.3 milligauss compared to the intensity of the earth's magnetic field (0.30–0.75 gauss). Moreover, magnetic oscillations of the earth, with periods of 10–40 s and also of about 10 min, both of very small amplitudes (1×10^{-4} gauss), have been recorded. However it must be stressed that geomagnetism is very difficult to measure in the field because of lack of sensitive methods and also because of interference from many environmental factors. Nevertheless biological actions of magnetism must not be neglected for it has been demonstrated that magnetite crystals embedded in certain bacteria (e.g. *Aquaspirillum magnetotacticum*), honeybees or carrier pigeons show that they are influenced by magnetic fields no stronger than that of the earth's electromagnetism. Wever (1979) demonstrated that, in humans, electromagnetic alternative fields of 10 Hz, and of relatively high intensity, may change circadian rhythms of activity.

The acceptance of the theory of subtle environmental synchronizers would necessitate the existence of a very great diversity of ultradian synchronizers of various periods, but it could be conceived that application of electromagnetic alternative fields of relatively high intensity may in humans change circadian rhythms of activity. Therefore, ultradian rhythms of longer periods would be the multiples of ultradian rhythms of shorter periods. It looks difficult, however, to go from ultradian rhythms of high frequencies (for example from 1 oscillation/s) to ultradian rhythms of low frequencies (e.g. 1 oscillation/h). In fact the spectral range of biological oscillations is wide: from neural rhythms (0.5– 30 Hz) to alternations of rest and activity (e.g. 12 cycles/day). The neural discharge corresponds to frequency modulation, the rest–activity corresponds to amplitude modulation (Fig. 11.1), a distinction that was well established by Hildebrandt (1988) in his classification of ultradian rhythms in humans.

Indeed the theory of partition of the 24-h rhythms and the one of accretion of ultradian rhythms of high frequencies could be better conceived and to some degree united if one considers biological temporal evolution.

Earth and Cosmic Evolution

The evolution of cosmic and geophysical synchronizers during palaeontology provides another approach to the possible origins of ultradian rhythms. The succession of life on earth has always been disrupted (by catastrophes?) and the appearance of new species at shorter and shorter intervals of time, which have been named periods, eras or cycles. The calendar actually admitted is: bacteria at 3500 million years before present (MYBP); cyanobacteria at 2500 MYBP; protozoans, accompanied by enrichment of the atmosphere with oxygen, first eukaryotes, stromatoliths, first metazoans, diversification of metazoans at 570 MYBP; proconsul monkey at 30 MYBP; *Australopithecus* at 3 MYBP; and *Homo sapiens* at 0.05 MYBP.

During the eons of time, numerous cosmic and geological events have occurred, many of them considered as cyclic or periodic: glaciations (1900, 600, 450, 280, 5 MYBP), epeirogeneses (700, 450, 360, 300, 145, 64, 6 MYBP), sea transgressions, etc. It has been possible to evaluate the lengths of these macroscopic periods, but it is actually impossible to have, for the moment, calendar dates for the small variations which have accompanied them. Shorter episodes have been dated by the varieties of fossils found in the earth layers (Termier and Termier 1979). For example, fossil corals several hundreds of millions of years old show more than 365 yearly growth rings. Therefore, given that the length of a year has not varied much, it can be supposed that the length of a day was, at this time, shorter than it is now. As mentioned by Klevecz (1984), it has been calculated that 350 MYBP, i.e. in the Precambrian, the year was 400 days long and therefore daylength was 22 h. It also seems likely that over such a long time there were variations in the lunar day which could be interpreted as being due to the moon being nearer to the earth than it is nowadays. These changes in lunar periodicity, as well as variations in coastal configuration, determine tidal rhythms and act as modifiers of synchronization of sea-dwelling organisms. As it is in seawater that life is generally thought to have evolved 350 MY ago, there is no doubt that the changes in tidal rhythms, accompanied latterly by modifications of tides in gulfs and river estuaries, could have resulted in a great diversity of circatidal ultradian rhythms.

During geological times, changes in the atmosphere, the appearance of oxygen, decreases in nitrogen and carbon dioxide, volcanic emissions (e.g. nitrous oxides, sulphur compounds, carbon oxides), ultraviolet radiations and radioactivity produced lethal and/or mutagenic (acting on nucleic acids) agents which locally either impeded or stimulated biological evolution.

Biological Evolution

The progressive increase in oxygen in the earth's atmosphere has finally produced a division between aerobic and anaerobic organisms. But the time was sufficient to allow the formation of symbiotic associations of organisms able to develop both anaerobic and aerobic metabolic pathways. This happened in organisms that lived in the superficial layers of the sea where they could receive radiative solar energy and carry out photosynthesis. In these sites were created associations, for example between molluscs or cnidarians and small flagellated algae (the zooxanthellae or the chloroxanthellae) that dwell inside their cells. The algae brought the benefit, and it can be supposed the photo-dependency, of the rhythms of chlorophyll, and their hosts provided metabolic products, especially carbon dioxide.

The ultrastructure and the biochemistry of mitochondria, in which are located most of the important energetic processes (e.g. the electron transport chain and Krebs cycle) of cells, are very similar to those of aerobic bacteria. The chloroplasts that house the electron transport chains and photosynthetic apparatus of higher plants show a great similarity to photosynthetic bacteria. It is now generally admitted that eukaryotic cells, which at the times of their origins were devoid of mitochondria and chloroplasts, have been invaded by or have phagocytosed bacteria during the development of endosymbiosis. The symbiont preserves a partial autonomy in the host cell relative to the cell nucleus whose role is mostly that of controlling coordination. At cell division, mitochondria or chloroplasts reproduce themselves independently of the other cell constituents. It may be assumed that these organelles would preserve the rhythmic autonomy of their genome; and this is true at least for the photoperiodicity of chloroplasts, which remain influenced by light–dark synchronization.

If we try to bring these multidisciplinary observations and experimental data together, it emerges that:

Evolution has modified life during millions of years with cyclic, but not strictly periodic, episodes in cosmic, solar, lunar and earthly parameters that might have acted as synchronizers.

These parameters should have imposed exogenous rhythms that could have become endogenous after such long spans of time.

Punctuated biological evolution proceeds in complexity through symbioses, mutations and genetic selections in ecological niches.

Ultradian rhythms as currently observed are therefore a combination of endogenous primitive rhythms (that might have originated from still unknown geophysical or cosmic synchronizers), and interactions between synchronized rhythms and those genetically transmitted.

Mechanisms

The mechanism of these long period ultradian rhythms is much disputed and, like their origin, is really unknown. Let us consider three hypotheses: neural, thermodynamic and chemical.

Neural Mechanism

The neural mechanism of biological rhythms seems to be the oldest hypothesis. Neural fibres show discharges coming from sensory organs when they respond to stimuli. However, when isolated, neural fibres may also discharge rhythmically at rapid frequencies (between 5 and 200 Hz, i.e. with periods of between 0.2 and 0.005 s). These discharges result from unstable equilibria arising from the differences in concentrations of cations, especially Na^+ and K^+, inside and outside the nerve fibre. This constitutes a chemical oscillator such as the one proposed by Berridge and Rapp (1979). The rhythms of neural action show frequency modulation that results in the transmission and communication of information to the brain. The electrical oscillations of nervous centres are particularly well marked in the brain cortex. It has been suggested that the electroencephalographic waves (1–60 Hz) arise from closed loops between the neurons of the brain cortex and those of the thalamus.

In 1980, Delcomyn reviewed experimental arguments showing that behavioural repetitive movements of animals (e.g. walking, swimming, scratching and breathing) need proper timing of muscle activation. He concluded that this is provided by the central nervous system without requiring sensory feedbacks. The only explanation that he gives is the existence of one or several neural oscillators.

Ultradian rhythms are connected to circadian rhythms through hypothalamic control. It has been demonstrated, mostly in the rat, that suprachiasmatic nuclei (situated in the hypothalamus) control circadian rhythms of displacement activity. In this rodent, ablation or lesions of these nuclei suppress or at least decrease the amplitude of the circadian rhythms, whereas the ultradian activity rhythms become more frequent and increase in amplitude.

Thermodynamic Mechanism

Thermodynamically, an endothermic animal is a system which, at a moderate ambient environmental temperature, maintains its internal temperature T_i between strict limits. It acts as an open system receiving and giving up energy. Therefore a heat flow Q/t (where Q is a quantity of energy and t time) runs through it; this heat flow may be expressed by the following equation:

$$\frac{Q}{t} = \lambda \frac{S}{l}(T_i - T_s)$$

where λ is the coefficient of heat conductivity; S, the body surface area of the animal; l, the thickness of its isolation surface layer (usually skin or feathers and underlying tissues); T_i, its internal and T_s its surface temperature, respectively.

This corresponds to a dissipative structure. The equilibrium expressed by this equation will be fulfilled at every moment, if the heat flow (i.e. the energy flow) remains constant. This will be assumed by physiological changes which could be rapid in S, l and T_s (movements, circulatory adjustments) but would be slower in T_i (metabolic adjustments). This creates a state of instability far from equilibrium, which according to Prigogine gives rise to oscillations

(Prigogine and Balescu 1956; Prigogine 1961). Furthermore, in endotherms such a flow of energy results from the differences between the energy intake and the energy expenditure (Lloyd and Stupfel 1991). The intake of energy comes from environmental radiative (principally infrared) absorption, gaseous exchanges and often discontinuous food and water uptake. The energy losses are the sum of heat dissipation, movements, activities of muscles and organs, excretion of water and faeces.

Discontinuous food intake is partly a consequence of limited stomach volume, and similarly, for respiratory rhythms, breathing rates are in some degree regulated by lung volumes. Regarding the oscillatory activity–rest episodes, Aschoff and Gerkema (1985) pointed out that ultradian rhythms of long periods may be an economic strategy to avoid continuous expense of energy, and to alternate energetic expenditure and restoration.

This biological energetic discontinuity has much in common with the physical quantum theory. In 1900, Planck formulated the principle that energy is not continuously radiated, but is discontinuously emitted by quanta of energy hv (h being the Planck constant and v the radiation frequency). Biologically speaking this would correspond to the intermittent, more or less periodic, exchanges of energy, heat, food intake, and rest–activity alternations between endotherms and their environment. Of course the overall thermodynamic exchanges of animals are achieved as infrared radiations (v) corresponding to their emissive, conductive and radiative heat losses. Furthermore, adoption of the principles of undulatory mechanics leads to an indeterminacy which necessitates the introduction of statistical probabilities. This is consistent with the situation and behaviour of animals surrounded by environmental stimuli; the two are often difficult both to predict and to measure. As aforementioned, ultradian as well as circadian biological variations are less periodic in animals living in the wild than in those in a laboratory controlled environment.

The thermodynamic exchanges of the whole organism constitute a generalized kinetics that integrates the sum of its temperature-regulatory peripheral reactions, its movements to control its heat loss or acquisition, its feeding, its self-defence, its excretory losses, in other words all that constitutes its behaviour. This kinetics, resulting partly from innate and partly from acquired conditioned reactions, is characteristic of each species and of each individual. The "thermodynamic behaviour" is, during the lifetime of an organism, modified by foetal development, and subsequent growth, maturity and ageing. Therefore, the observed amplitudes, periods and phases of ultradian metabolic "rhythms" can be very different and depend on the variety and the hierarchy of biochemical, cellular and physiological processes.

The Molecular Mechanism

Among the numerous investigations instigated by Kleitman's (1963) BRAC hypothesis, the real chronobiological breakthrough may be the ultradian endocrine research performed in the last 15 years. As discussed above, rhythms of 1–3 h have been found in "pulses" of secretion of hormones produced by the hypophysis, hypothalamus, suprarenal glands, pancreas and kidneys.

Although for the moment there are not enough data for a molecular blue-print of ultradian rhythms of behaviour and metabolism, there are a few indications that this will be attainable.

1. The long-period ultradian rhythms of several hormones have been related to sleep phases (Weitzman 1975), ultradian activity–rest alternations (e.g. for luteinizing hormone, renin, prolactin, thyroid-stimulating hormone) or to digestion (for neuropeptides).
2. It has been known for many years that hypophysial hormones control behaviour (e.g. hunger and aggression), and take part in temperature regulation; furthermore the suprachiasmatic nuclei that are proposed as the location of the circadian oscillators are parts of the hypothalamus, which is related to the hypophysis through a neurosecretory axis.
3. Various classes of compound are used in human therapy, e.g. neuroleptics and antidepressants to control behaviour, amphetamines to increase vigilance, hypnotics to induce sleep.

It is conceivable that chemical messengers, neurotransmitters (acetylcholine, adrenaline, noradrenaline, peptides, endorphins etc.) acting on specific targets could entrain or mask, synchronize or inhibit the periodicities of behavioural reactions and their metabolic consequences.

Conclusion

Genetics

The effects of genetic factors on long period ultradian rhythms arise from interspecific and intraspecific investigations. Stupfel et al. (1990, 1991) have published the results of 15 years of measurements. These were made continuously over 4–30 consecutive days of carbon dioxide emission by premature human infants, monkeys, rats, mice, guinea pigs, quail and chicks. By computing ultradian periods of between 40 min and 24 h, they concluded that ultradian periods and amplitudes were different in these seven species. Furthermore, they showed interspecific discrepancies for these respiratory parameters between several strains of mouse, rat and quail. With regard to the sleep phases, it has been reported that discrepancies in periods have been shown between 53 mammalian species (from the elephant to the brown bat) by Zepelin and Rechschaffen (1974) and between 39 mammalian species by Allison and Cichetti (1976). Several investigaters reported ultradian rhythmic differences between strains of the same species. In 1981, Lemner et al. showed ultradian discrepancies in the motor activity of six strains of grouped rats. In 1984, Büttner and Wollnik found distinctive ultradian rhythms with periods of 12 h, 6 h, 4.8 h, and 4 h of locomotor activity in rats characteristic of five inbred strains.

With respect to sleep phases, Valatx (1984) demonstrated, in an experiment on selection and hybridization, the occurrence of hereditary sleep traits in mice. This extended the results of Oliverio and Malorni (1979), who have reported differences of activity and of EEG-defined sleep between two strains of mouse. Zung and Wilson (1967) and Chouvet et al. (1983), investigating sleep phases in human twins, reported data that suggest a genetic influence on sleep parameters. However, the fact that there are many more genotypes than phenotypes involved suggests that some genes could be carriers of temporal information.

Edmunds (1976) has reviewed several genetic models for endogenous time keeping. Several of them use DNA transcription as means to measure biological time. The rate at which an RNA polymerase molecule moves along the DNA in the chromosome is, at $37\,^\circ$C, 30 to 40 nucleotides/s. It has been calculated that, at this rate, the total transcription time for the genome of *Escherichia coli* would be 33 h (assuming that several operons are not simultaneously transcribed). Therefore, only 1% of the genome would be needed to separate periodic events occurring once every 20 min. One 3-h period could thus be directly timed on the DNA. In 1974, Ehret proposed a genetic theory of this kind for circadian rhythms in eukaryotes. According to him, in the eukaryotic chromosome there are intense activities at certain times and over certain regions. This suggests a multireplicon structure with a period of about 1 day. In every cell there may be hundreds of replicons of a special type that Ehret called chronons which are the rate-limiting components of the basic transcriptional cycle. Similar entities may be imagined as regulators of various ultradian rhythms.

Ontogeny

Perhaps the best way to understand the mechanism of the temporal rhythmicity of complex organisms is to take an ontogenetic approach: as expressed by Haeckel (1834–1919), "Ontogeny recapitulates phylogeny". Temporal discontinuity appears at every moment of evolution of all kinds of species. Organisms have from their beginning a rich variety of endogenous, i.e. innate, rhythms that, despite several quantitative and qualitative changes, are retained throughout their lives.

According to Corner (1984), in early mammalian, as well as in chick embryo, behavioural development, motility begins by endogenous (i.e. non-reflexive) electric discharges of spinal interneurons which tend to spread in bursts from one part of the spinal cord to the other, recruiting neurons in unpredictable sequences and at variable intervals. This establishes a pattern of continuous but irregular twitches and in addition, intermittent stereotyped bursts. In vitro cultures of embryonic spinal cord or hindbrain neurons show similar spontaneous more or less rhythmic patterns. Corner (1990) compared these findings to similar activities that have been observed in coelenterate nerve nets. Furthermore, he considered, in human embryos and foetuses, the origin of active sleep (AS) which is the equivalent of REM sleep in adults. According to Horne (1988), extended episodes of REM sleep occur around birth in most mammals. Parmelee and Stern (1972) reported that human newborns spend nearly half their sleep time in REM phase; in 1-month and two-month premature infants this rises to 67% and up to 80%, respectively. It should be noted that precociously born mammals (e.g. sheep and guinea pigs) have very little REM sleep, whereas their brain and physical ability are in a fairly advanced state, particularly to allow walking within minutes of birth. Corner

(1990) stated that foetal motility is composed of several fast ultradian bursts of 5–10 s of AS, two motility cycles in the 1–5-min range, plus one of about 10 min. He reported that rhythmic motility has been demonstrated in premature infants as early as 24 weeks after conception and in utero as early as 15 weeks. Moreover, McGinty and Drucker-Colin (1982) observed that, in premature infants, the alternation between rest and activity phases is extremely variable and furthermore unrelated between different groups of muscles. Almost continuous sequences of jerky, irregular and largely uncoordinated movements are found in the 27-week human foetus, in the 13–14-day chick embryo and in the neonatal rat. The level of spontaneous motor activity gradually declines thereafter, probably under the influence of both local and descending inhibitory neural systems. As regards the arousal mechanisms, Corner (1990) indicated: (a) in the human foetus, behavioural states corresponding to quiet and active wakefulness in the last month of gestation; (b) in the sheep foetus, intermittent generalized rotary movements that last up to 5–10 s or less; (c) in the chick embryo, in the last month prior to birth, rather similar bouts of movement which appear at 17 days in ovo. According to Corner in chicks as in rats early postnatal rhythmic patterns of behaviour appear to be largely of spinal origin.

As mentioned above, Sterman and Hoppenbrouwers (1971) showed that a BRAC of 40–60 min is found in the human foetus and that it is distinguishable from the activity cycle of the mother, but similar in duration to that of the newborn infant (Figs. 11.8 and 11.9). In 1986, Hoppenbrouwers stated that this foetal BRAC is probably wired into the central nervous system at an early gestational age and appears to be uninfluenced by the mother, although foetal circadian manifestations are prenatally wholly dependent on maternal hormones. The endocrine secretions in the foetus are not yet sufficiently well known to estimate whether they could influence or instigate foetal ultradian motility.

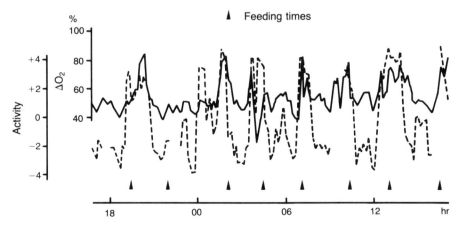

Fig. 11.8. Percentage variations in oxygen consumption (ΔO_2, continuous line) and in activity (dashed line) in a premature male infant (gestational age 32 weeks, postnatal age 37 days) over 24 h. Milk was provided eight times (black triangles) through a tube left in the stomach throughout the 24-h duration of the measurements. Constant light was maintained during all times. (Reproduced, with permission, from Putet et al. 1990.)

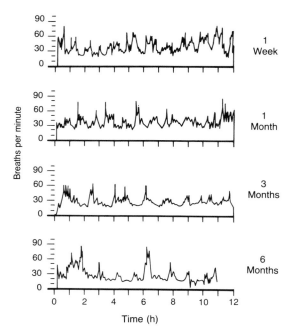

Fig. 11.9. Minute-by-minute respiratory rate (measured from a continuous recording of expired CO_2) in the same infant at the age of 1 week, 1, 3 and 6 months. (Reproduced, with permission, from Hoppenbrouwers et al. 1978.)

After birth, neonates are exposed to circadian and ultradian exogenous environmental and societal stimuli and rhythms to which they respond according to their degree of postnatal nervous system maturation. For example, ultradian rhythms develop according to feeding and nursing time intervals as it has been shown in premature human babies by Putet et al. (1990) and in fullterm babies by, for example, Sander et al. (1970).

After birth a provoked nervous activity superimposes itself on the spontaneous endogenous activity of the newborn. As stated by Changeux (1983), this is accompanied by an enormous increase in the number of glial brain cells and synapses, with changes in the configuration of their acetylcholine receptors. At this time the immature sensory organs of the foetus develop progressively. Then comes a critical time when the ultradian (endogenous) and circadian (mostly mother's and nursing) rhythms of the neonate are confronted by new environmental stimuli and circadian ecological, as well as social, rhythms. The foetal endogenous cycles which have been genetically determined are either going to remain as the inherited part of the temporal genome (already predisposed to an ancestral environment) or to be modulated by its new ecosystem: food, predators, hierarchy, commensalism, physicochemical and societal parameters for animals, and also societal factors for humans. Behaviour of precocious and altricious individuals will become different during this critical episode.

Interferences in rhythms will cause transiency, beat phenomena, suppression and enslavement (or phase locking) of periods. Entrainment or masking of rhythms will be, according to the species, under the circadian constraint at the arousal of polycyclic, bicircadian, or preponderantly diurnal or nocturnal activity. Several rhythms will remain in a subtle form and emerge during the waking behaviour. Such is the "archisleep" suggested by Jouvet (1961) to occur in humans. However, a somatic and primitive temporal basic pattern of behaviour will persist. Examples of this are provided by episodes of quiescence alternating with episodes with a high incidence of spontaneous movements, which, according to Corner (1977) almost certainly have their origins in the brainstem.

This rhythmic instability lasts during the whole of neural formation. It is a time of great vulnerability for the newborns. Moreover, because of the great energy demands related to their high metabolism due to their rapid growth and small size, and with their poor temperature regulation, young vertebrates are at risk in an ecosystem well organized in its temporal structure. Even for humans, protected by a sophisticated way of living, this time is dangerous. As stressed by Hoppenbrouwers (1989 and see Chap. 9) the peak age of sudden infant death syndrome (SIDS) is between 1 and 3 months of life, at an age characterized by many maturational changes: (a) a number of central and autonomic nervous functions undergo reintegration into new patterns; (b) at least three biological behaviour rhythms (circadian, prandial and the BRAC) undergo phase alternations which involve changes in their synchronization; (c) several components of this complex rhythmic system appear to be transiently more rigid than either before or after this critical age.

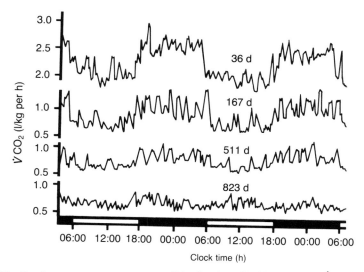

Fig. 11.10. Continuous measurements over 48 h of carbon dioxide emission ($\dot{V}CO_2$) of four rats of different ages: 36, 167, 511 and 823 days. Lighting (100 lux) from 06:00 till 18:00 h; ultradian rhythms are superimposed on circadian rhythms, $\dot{V}CO_2$ being higher during darkness than during light; the circadian difference between dark and light decreases progressively with age and is almost suppressed in the 823 day (*d*) old rat. (Reproduced, with permission, from Stupfel et al. 1986b.)

These ontogenetic considerations are incomplete because they are based only on movements, motility and EEG data. They neglect, inter alia, differences in thermodynamic exchange in avian or mammalian embryos (bathing in egg, or amniotic fluid) and neonates (in an aerial surrounding) that should result in transient temporal energy modifications that are still ignored.

Adulthood and maturity are discussed here only briefly because all the data for long period ultradian rhythms reviewed above concerned animals and humans at these two stages of life. At these ages it may be conceived that the temporal organization of an individual is stable, though always capable of responding to surrounding stimuli, which means an equilibrium between endogenous innate and acquired exogenous rhythms. Aschoff (1984) and Stupfel et al. (1990) have demonstrated in adult animals that ultradian rhythms are either synchronized or not synchronized by circadian periods, according to the species and the considered environmental parameter. For instance, the respiratory exchanges and motility of Japanese quail, of chickens and of Sprague-Dawley rats are synchronized by light–dark alternations, whereas those of Hartley guinea pigs and of OF_1 mice are not.

To a first approximation, the levels and amplitudes of the variations of all physiological functions decrease with age, but not in a similar manner and not at the same age. Only a few experimental investigations have been related to the influence of ageing on ultradian rhythms. Stupfel et al. (1986b) found in old (i.e. >500 days) rats a marked diminution of the ultradian amplitudes of their respiratory gaseous exchanges with no significant modifications in their frequencies (Fig. 11.10). Moreover, Stupfel et al. (1990) observed persistence of the frequency but progressive decrease in the amplitudes of respiratory ultradian rhythms in a few very old rats and guinea pigs, at the time of their spontaneous death; this indicates the fundamental nature of these long period ultradian rhythms.

In humans, the mean levels and the amplitudes of the respiratory exchanges, as well as those of most other physiological functions decrease with age, although at any particular age there are great interindividual variations in these diminutions. This variability is probably due in part to inherited genetic factors. Due to a decreased metabolism, ageing people become physically tired more and more quickly and they rest more often when accomplishing strenuous physical efforts. Many also eat smaller amounts but more often. However, these are anecdotal observations and precise data are lacking.

On the other hand more reliable data are available on sleep, where aged subjects have been monitored during closely controlled routines and conditions (for reviews, see Minors and Waterhouse 1984; Horne 1988). Thus, after the age of 60 years, the total amount of time spent asleep decreases and a fall in the amount of SWS (i.e. deep sleep) and an increase in lighter sleep can be detected in the EEG. Furthermore, old people wake often at night and take an increased number of naps in the day, particularly after lunch and at late afternoon. In many elderly people naps can add up to a total of 2 h of sleep per day. It has been supposed that this "sleeping" diurnal time, let us say 2 h, may be subtracted from the 7 h per day of a "usual" night sleep, which leaves only 5 h of sleep to be taken at night.

Wise, et al. (1990) have reported that LH endocrine pulses are less frequent in middle-aged than in young female rats.

Thus, the rather limited data we have for rats and those observations, mostly clinical, that we have for humans tend to show that ageing provokes a decrease in circadian synchronization accompanied by an increased occurrence of ultradian oscillations. If this is so, it is tempting to speculate.

Diminution of circadian rhythmicity due to ageing may arise from a diminution of the sensory perceptions of exogenous circadian synchronizers, light, odours, ecological constraints in animals, and a decrease in societal ones, loneliness in humans. Decreased circadian rhythmicity may also result from an increase in organic diseases associated with ageing (especially brain, vascular and kidney defects). Thus endothermic organisms return to the isolation from external stimuli, reminiscent of the time when they were, as foetus or embryo, protected inside their maternal uterus or egg. The primordial temporal ultradian structures of their oceanic phylogeny and behaviour can now again predominate.

Summary

Endothermic animals (mammals and birds) maintain their internal body temperature within the range 32–42 °C. Most of them show daily rhythms in many of their physiological functions with circadian periods (near 24 h) which, however, are only well marked when they are adults. In addition, all their functions show a great diversity of oscillations, with intervals of less than 24 h. These intervals are different and most often they are not periodic. Therefore they do not fit the usual definitions of "rhythms" or "periods". Furthermore, it is generally recognized that, for the moment, there are no mathematical procedures suitable for their time analysis. Ultradian "rhythms" of short periods (0.5–30 Hz) are of neural origin and related to information: they bring, by frequency modulation, messages from the environment to the brain and/or neural centres and conversely they give orders from these centres to the periphery of the organism. Ultradian "rhythms" of middle periods (seconds–minutes) are those of systems transporting and distributing air (respiratory rate), blood (heart rate), and food (gastrointestinal movements) in the organism; these have been known for a long time. However, they are not really "periodic", for they may be influenced by variations of environmental parameters (ambient temperature and humidity, barometric pressure, food availability) and also by endogenous factors (interfunctional connections, e.g. cardiac and respiratory rates, hormone secretions, psychological influences). Moreover they are most often entrained by external circadian ones.

Ultradian oscillations with intervals of between 10 min and a few hours have been found recently in most physiological functions. Thermodynamic and metabolic processes particularly show such ultradian oscillations at all times in the life of endotherms, from before birth and until death. Their intervals remain relatively constant with means around 1 or 2 h, so that they may be called circahoral. Their amplitudes are roughly related to the metabolic activity of the endotherms, so that they are of the amplitude-modulation type. Continuous recordings of respiratory exchanges that give a continuous index of the thermodynamic equilibrium of several species of small laboratory animals (mice, rats, guinea pigs, monkeys, quails, chicks) and of premature infants show that these ultradian oscillations are: (a) characteristic of each species, (b) independent of the circadian rhythmicity (as they are observed before birth and also in the absence of overt circadian rhythms), (c) entrained by exogenous parameters such as light, food, societal factors and environmental challenges, (d) related to ultradian behavioural acts, such as activity–rest, food intake, sleep phases and reactions to environmental challenges.

Endotherms, being open systems and performing far from thermodynamic equilibrium, develop these circahoral thermodynamic oscillations. The latter may be considered as energetic quanta which allow discontinuous restoration after energy dissipation provoked by random environmental stimuli. The inter-species and interstrain differences in these ultradian thermodynamic oscillations suggest that they are endogenous and have a genetic origin. It is hypothesized that ultradian oscillations have originated in response to discontinuous environmental changes and that the advantages that they confer have ensured their survival during the course of evolution from protozoa to human.

Acknowledgement

I would like to commemorate here Michel Davergne, who, in 1978, participated with youthful enthusiasm at the beginning of this ultradian research.

References

Allison T, Cichetti DV (1976) Sleep in mammals: ecological and constitutional correlates. Science 194:732–734

Angeli A, Carandente F (1988) An update on clinical chronoendocrinology. In: Hekkens WJMH, Kerkhoff GA, Rietveld WJ (eds) Trends in chronobiology. Pergamon Press, Oxford, pp 319–333

Aschoff J (1957) Aktivitätsmuster der Tagesperiodik. Naturwissenschaften, 13:361–367

Aschoff J (1981) Handbook of behavioral neurology, vol 4, Biological rhythms. Plenum Press, New York London

Aschoff J (1984) Circadian timing. Ann NY Acad Sci 423:442–468

Aschoff J, Gerkema M (1985) On diversity and uniformity of ultradian rhythms. In: Schulz H, Lavie P (eds) Ultradian rhythms in physiology and behavior. Springer, Berlin Heidelberg New York, pp 321–334

Aserinsky E, Kleitman N (1955) Regularly occurring periods of eye motility and concomitant phenomena during sleep. Science 118:273–274

Bailey D, Harry D, Johnson RE, Kupprat L (1973) Oscillations in oxygen consumption of man at rest. J Appl Physiol 34:467–470

Barnett SA (1963) A study in behaviour. Camelot Press, London Southampton

Benton LA, Berry SJ, Yates EF (1990) Ultradian rhythmic models of blood pressure variations in normal human daily life. Chronobiologia 17:95–116

Berridge M, Rapp P (1979) A comparative survey of the function, mechanism and control of cellular oscillations. J Exp Biol 81:217–286

Blinowska K, Marsh DJ (1985) Ultra- and circadian fluctuations in arterial pressure and electro-myogram in conscious dogs. Am J Physiol 18:R720–R725

Bowden DM, Kripke DF, Wyborney G (1978) Ultradian rhythms in waking behavior of Rhesus monkeys. Physiol Behav 21:929–933

Brandenberger G, Follenius M, Muzet A, Ehrhart J, Schieber JP (1985) Ultradian oscillations in plasma renin activity: their relationships to meals and sleep stages. J Clin Endocrinol Metab 61:280–284

Brandenberger G, Simon C, Follenius M (1987) Ultradian endocrine rhythms: a multi-oscillatory system. J Interdiscipl Cycle Res 18:307–315

Brandenberger G, Follenius M, Simon C, Ehrhart J, Libert JP (1988) Nocturnal oscillations in plasma renin activity and REM–NREM sleep cycles in humans: a common regulatory mechanism? Sleep 2:242–250

Broten TP, Zehr JE (1989) Baroreflex modulation of ultradian oscillations of blood pressure and heart rate in unanesthetized dogs. Chronobiologia 16:241–255

Brown FA Jr, Hastings JW, Palmer JD (1970) The biological clock: two views. Academic Press, New York

Bueno L (1986) Brain neuropeptides and ultradian motor activity of the gut. J Interdiscipl Cycle Res 17:125–162

Büttner D, Wollnik F (1984) Strain differentiated circadian and ultradian rhythms in locomotor activity of the laboratory rat. Behav Genet 14:138–152

Changeux JP (1983) L'homme neuronal. Fayard, Paris

Chouvet G, Blois R, Debilly G, Jouvet M (1983) La structure d'occurrence des mouvements oculaires rapides du sommeil paradoxal est similaire chez les jumeaux homozygotes. CR Acad Sci Paris 296:1063–1068

Corner MA (1977) Sleep and the beginnings of behavior in the animal kingdom. Studies of ultradian motility cycles in early life. Prog Neurobiol 8:279–285

Corner MA (1984) Maturation of sleep mechanisms in the central nervous system. Exp Brain Res 8 [Suppl]:50–65

Corner MA (1990) Brainstem control of behavior: ontogenic aspects. In: Klemm WR, Vertes RP (eds) Brainstem mechanisms of behavior. John Wiley, Chichester, pp 239–269

Cozzi B, Ravault JP, Ferrandi B, Reiter RJ (1988) Melatonin concentration in cerebral vascular sinuses of sheep and evidence for its episodic release. J Pineal Res 5:535–543

Crowcroft P (1954) The daily cycle of activity in British shrews. Proc Zool Soc Lond 123:715–729

Daan S, Aschoff J (1975) Circadian rhythms of locomotor activity in captive birds and mammals: their variations with season and latitude. Oecologia 18:269–316

Del Pozo F, De Feudis FV, Jimenez JM (1978) Motilities of isolated and aggregated mice. A difference in ultradian rhythmicity. Experientia 34:1302–1304

Delcomyn F (1980) Neural basis of rhythmic behavior in animals. Science 210:492–498

Delgado-Garcia JMR, Del Pozo F, Montero P, Monteagudo VL, O'Keeffe JI, Kline N (1978) Behavorial rhythms of gibbons on Hall's Island. J Interdiscipl Cycle Res 9:147–168

Dierschke DJ, Bhattacharya AN, Atkinson LE, Knobil E (1970) Circhoral oscillations of plasma LH levels in the ovariectomized rhesus monkey. Endocrinology 87:850–853

Eccles R, Maynard RC (1975) Studies on the nasal cycle in the immobilized pig. Proc Physiol Soc, Middlesex Hospital Meeting, 1–2 and 17–18

Edmunds LN Jr (1976) Models and mechanisms for endogenous time keeping. In: Palmer JD (ed) An introduction to biological rhythms. Academic Press, New York San Francisco London, pp 280–361

Ehret CF (1974) The sense of time: evidence for its molecular basis in the eukaryotic gene-action system. Adv Biol Med Phys 15:47–77

Ellis GB, Desjardins C, Fraser HM (1983) Control of pulsatile LH release in male rats. Neuroendocrinology 37:117–183

Gardner R, Grossman WJ (1976) Normal patterns in sleep in man. In: Weitzman ED (ed) Advances in sleep research vol 2. Spectrum, New York, pp 66–107

Globus GG, Phoebus EC, Humphries J, Boy R, Sharp R (1973) Ultradian rhythms in human telemetered gross motor activity. Aerosp Med 44:882–887

Gordon CR, Lavie P (1985) Urinary ultradian rhythms in dogs. In: Schulz H, Lavie P (eds) Ultradian rhythms in physiology and behavior. Springer, Berlin Heidelberg New York, pp 110–124

Hiatt JF, Kripke DF (1975) Ultradian rhythms in waking gastric acidity. Psychosom Med 34:320–325

Hildebrandt G (1988) Temporal order of ultradian rhythms in man. In: Hekkens WTJM, Kerkhof GA, Rietveld WJ (eds) Trends in chronobiology. Pergamon Press, Oxford, pp 107–122

Hoogenboom I, Daan S, Dallinga JH, Schoenmakers M (1984) Seasonal change in the daily timing of behavior of the common vole Microtus arvalis. Oecologia 61:18–21

Hoppenbrouwers T (1986) Ontogenesis of ultradian respiratory rhythms. J Interdiscipl Cycle Res 17:140–141

Hoppenbrouwers T (1989) Sudden infant death syndrome (SIDS) and sleep. Proc IEEE Conf Eng Med Biol 11:310–312

Hoppenbrouwers T, Harper RM, Hodgman JE, Sterman MB, McGinty DJ (1978) Polygraphic studies of normal infants during the first six months of life. II. Respiratory rate and variability as a function of state. Pediatr Res 12:120–125

Horne J (1988) Why we sleep. The function of sleep in humans and other mammals. Oxford University Press, Oxford

Horne J, Whitehead M (1976) Ultradian and other rhythms in human respiration. Experientia 32:1165–1167

Hughes GP, Reid D (1951) Studies on the behavior of cattle and sheep in relation to the utilization of grass. J Agric Sci 41:360–366

Iranmanesh A, Lizarralde G, Johnson ML, Veldhuis JD (1989) Circadian, ultradian and episodic release of β-endomorphin in man and its temporal coupling with cortisol. J Clin Endocrinol Metab 68:1019–1026

Jalife J (1990) Mathematical approaches to cardiac arrhythmias. Ann NY Acad Sci Vol 591

Jilge B, Friess L, Stähle H (1986) Internal coupling of five functions of rabbits exhibiting a bimodal circadian rhythm. J Interdiscipl Cycle Res 17:7–28

Johnson LM, Gay LV (1981) Luteinizing hormone in the cat. I. Tonic secretion. Endocrinology 109:240–246

Jouvet M (1961) Telencephalic and rhombencephalic sleep in the cat. In: Wolstenholme GEW, O'Connor MJA (eds) The nature of sleep. J.A. Churchill, London, pp 188–206

Katz RJ (1980) The temporal structure of motivation. III. Identification and ecological significance of ultradian rhythms of intracranial reinforcement. Behav Neurol Biol 30:148–159

Kayser R (1895) Die exacte Messung der Luftdurch-gängigkeit der Nase. Arch Laryngol 3:101–120

Kayser C, Hildwein G (1974) Le rythme circadien de la consommation d'oxygène et de l'activité locomotrice du rat; ses relations avec les deux formes de sommeil: le sommeil à ondes lentes et le sommeil paradoxal. Dev Psychobiol 2:216–239

Kleitman N (1963) Sleep and wakefulness. University of Chicago Press, Chicago, IL

Klevecz RR (1984) Cellular oscillators as vestiges of a primitive circadian clock. In: Edmunds LN Jr (ed) The cycle clocks. Marcel Dekker, New York, pp 47–61

Kripke D (1982) Ultradian rhythms in behavior and physiology. In: Brown FM, Graeber RC (eds) Rhythmic aspects of behavior. Lawrence Erlbaum Associates, Hillsdale, NJ, pp 313–343

Lavie P, Kripke DF (1977) Ultradian rhythms in urine flow in waking humans. Nature 269:142–144

Lavie P, Kripke DF (1981) Ultradian circa $1\frac{1}{2}$ hour rhythms; a multioscillatory system. Life Sci 29:2446–2450

Lefcourt AM (1990) Circadian and ultradian rhythms in ruminants: relevance to farming and science. In: Hayes DK, Pauly JE, Reiter RJ (eds) Chronobiology: its role in clinical medicine, general biology and agriculture, pt B. Wiley–Liss, New York, pp 729–742

Lemner BG, Caspari-Irving G, Weimer R (1981) Strain dependency in motor activity and in concentration and turnover of catecholamines in synchronized rats. Pharmacol Biochem Behav 15:173–178

Levin BE, Rappaport M, Natelson BH (1979) Ultradian variations of plasma noradrenaline in humans. Life Sci 25:621–627

Livnat A, Zehr JE, Broten TP (1984) Ultradian oscillations in blood pressure and heart rate in free-running dogs. Am J Physiol 246:R817–R824

Lloyd D, Stupfel M (1991) The occurrence and functions of ultradian rhythms. Biol Rev 66: 275–299

Lovett Doust JW (1979) An ultradian periodic servosystem of thermoregulation in man. J Interdiscipl Cycle Res 10:95–103

Lovett Doust JW, Payne WD, Podnieks I (1978) An ultradian rhythm of reaction time measurements in man. Neuropsychobiology 4:93–98

Maxim PE, Storrie M (1979) Ultradian barpressing for rewarding brain stimulation in rhesus monkeys. Physiol Behav 22:683–687

McGinty DJ, Drucker-Colin RR (1982) Sleep mechanisms: biology and control of REM sleep. Int Rev Neurobiol 23:391–436

Minors DS, Waterhouse JM (1981) Circadian rhythms and the human. Wright, Bristol London Boston

Minors DS, Waterhouse JM (1984) The sleep–wakefulness rhythm, exogenous and endogenous factors (in man). Experientia 40:410–416

Moore-Ede MC, Sulzman FM, Fuller CA (1982) The clocks that time us. Physiology of the circadian timing system. Harvard University Press, Cambridge, MA

Mori T, Inoue S, Minami S, Egawa M, Wakabayasi I (1986) Episodic growth hormone secretion following exogenous growth hormone-releasing factor in rats. Biomed Res 7:371–377

Morin LP (1986) A concept of physiological time: rhythms in behavior and reproductive physiology. Ann NY Acad Sci 474:331–351

Odell WD, Griffin J (1987) Pulsatile secretion of human gonadotropin in normal adults. N Engl J Med 317:1688–1691

Okudaira N, Kripke DF, Webster JB (1984) No basic rest–activity cycle in head, wrist or ankle. Physiol Behav 32:843–845

Oliverio A, Malorni W (1979) Wheel running and sleep in two strains of mice: plasticity and rigidity in the expression of circadian rhythmicity. Brain Res 163:121–133

Ortega SM, Cabrera MC (1990) Ultradian rhythms in EEG and task performance. Chronobiologia 17:183–194

Palmer JD (1976) An introduction to biological rhythms. Academic Press, New York San Francisco London

Pang SF, Yip PC (1988) Secretory patterns of pineal melatonin in the rat. J Pineal Res 5:279–292

Parmelee AH, Stern E (1972) Development of states in infants. In: Clemente CD, Purpura DP, Mayer FE (eds) Sleep and the developing nervous system. Academic Press, New York, pp 199–215

Pelletier J, Thiery JC (1986) The LH pulsatility: a reflection of a centrally inhibited rhythm? J Interdiscipl Cycle Res 17:148–149

Pöllman L, Pöllman B (1988) Ultradian rhythms (about 1.5 hours) in pain thresholds. Pflugers Arch 412 [Suppl]:R49

Prigogine I (1961) Introduction to the thermodynamics of irreversible progress. Interscience, New York

Prigogine I, Balescu R (1956) Phénomènes cycliques dans la thermodynamique des processus irréversibles. Bull Acad R Belg Clin Sci 42:256–263

Putet G, Stupfel M, Gourlet V, Salle B, Court L (1990) Respiratory and metabolic ultradian (40 min < period < 6h) variations in normal premature infants periodically fed through a gastric tube. Chronobiologia 17:1–13

Rasmussen DD (1986) Physiological interaction of the basic rest–activity cycle of the brain: pulsatile luteinizing hormone secretion as a model. Psychoneuroendocrinology 11:389–405

Rasmussen DD, Jacobs W, Kissinger PT, Malven PV (1981) Plasma luteinizing hormone in ovariectomized rats following pharmacologic manipulation of endogenous brain serotonin. Brain Res 229:233–235

Richter CP (1922) A behavioristic study of the activity of the rat. Comp Psychol Monogr 1:1–15

Richter CP (1927) Animal behavior and internal drives. Q Rev Biol 2:307–343

Roberts W (1860) Observations on some of the daily changes of the urine. Edin Med J 5:817–825

Rossi EL (1982) Hypnosis and ultradian cycles: a new state(s) theory of hypnosis. Am J Clin Hypn 25:21–32

Rossi EL (1986) Altered states of consciousness in every day life: the ultradian rhythms. In: Wolman B, Ullman M (eds) Handbook of altered states of consciousness. Van Nostrand, New York, pp 97–132

Roussel B, Bittel J (1979) Thermogenesis and thermolysis during sleeping and waking in the rat. Pflugers Arch 382:225–231

Rubsamen K, Hörnicke H (1981) Herzschlag-frequenz und O_2-Verbrauch als Belastungs-indikatoren bei Kaninchen. In: Aktuelle Arbeiten zur artgemässen Tierhaltung. Kuratorium für Technik und Bauwesen in der Landwirtschaft, 5th edn, pp 84–93

Sander LW, Stechler G, Burns P et al. (1970) Early mother–infant interaction and 24-hour patterns of activity and sleep. J Am Acad Child Psychiatry 9:103–123

Schulz H (1988) Schlafforschung. In: Kisker K, Lauter H, Meyer JE, Müller C, Stromgrens S (eds) Psychiatrie der Gegenwart, Band 6, Organische Psychosen. Springer, Berlin Heidelberg New York, pp 401–442

Schulz H, Lavie P (1985) Ultradian rhythms in physiology and behavior. Springer, Berlin Heidelberg New York

Shimada SG, Marsh DJ (1979) Oscillations in mean arterial blood pressure in conscious dogs. Circulat Res 44:692–700

Simon C, Brandenberger G, Follenius M (1987) Ultradian oscillations of plasma glucose, insulin and C-peptide in man under continuous enteral nutrition. J Clin Endocrinol Metab 64:669–674

Stanbury SW, Thompson AE (1951) Diurnal variations in electrolyte excretion. Clin Sci 10:267–293

Sterman MB, Hoppenbrouwers T (1971) The development of sleep–waking and rest–activity patterns from fetus to adult in man. In: Sterman MB, McGinty DJ, Adinolfi AM (ed) Brain development and behavior. Academic Press, New York, pp 203–227

Sterman M, Lucas E, MacDonald L (1972) Periodicity within sleep and operant performance in the cat. Brain Res 38:327–341

Stupfel M, Pavely A (1990) Review. Ultradian, circahoral and circadian structures in endothermic vertebrates and humans. Comp Biochem Physiol 96A:1–11

Stupfel M, Davergne M, Perramon A, Lemercerre C, Gourlet V (1979) Rythmes ultradiens (5 < τ < 10 minutes) respiratoires ($\dot{V}O_2$, $\dot{V}CO_2$) de quatre petits vertebrés utilisés pour la recherche biomédicale. CR Acad Sci Paris 289D:675–678

Stupfel M, Molin D, Thierry H, Busnel MC (1980) Respiratory activity variations induced in groups of LD 12:12 synchronized Sprague-Dawley rats by a 100 dB white noise emitted at 12-hour intervals. Chronobiologia 7:337–342

Stupfel M, Perramon A, Mérat P, Demaria Pesce VH, Massé H, Gourlet V (1981a) Grouping and respiratory behavior induced in rats and quails by LD 12:12 illumination. Physiol Behav 25:439–447

Stupfel M, Perramon A, Gourlet V et al. (1981b) Light–dark and societal synchronization of respiratory and motor activities in laboratory mice, rats, guinea-pigs and quails. Comp Biochem Physiol 70A:265–274

Stupfel M, Perramon A, Gourlet V, Thierry H, Ali M, Lemercerre C (1983) Harmonic analysis of ultradian respiratory rhythms in four small laboratory vertebrates lit in LD 12:12. Comp Biochem Physiol 75A:293–297

Stupfel M, Gourlet V, Zeitoun G, Maral R, Bourut C, Chenu E (1984) Effects of B16 melanoma transplantation on the respiration of grouped C57 B1 female mice. Biomed Pharmacother 38:389–397

Stupfel M, Gourlet V, Court L, Demaria Pesce VH (1986a) Starvation and respiratory rhythmic behavior in groups of light–dark synchronized Sprague-Dawley rats. Physiol Behav 58:265–274

Stupfel M, Gourlet V, Court L (1986b) Effects of aging on circadian and ultradian respiratory rhythms of rats synchronized by a LD 12:12 lighting (L = 100 lux). Gerontology 32:81–90

Stupfel M, Gourlet V, Court L, Mestries J, Perramon A, Mérat P (1987) Periodic analysis of ultradian (40 min < τ < 24 h) respiratory variations in laboratory vertebrates of various circadian activities. Chronobiologia 14:365–375

Stupfel M, Gourlet V, Perramon A, Monvoisin JL, Court L (1989a) Societal synchronization in groups of rats or quail submitted to various lighting regimens. Chronobiolgia 16:215–228

Stupfel M, Gourlet V, Court L, Perramon A, Mérat P, Lemercerre C (1989b) There are basic rest–activity ultradian rhythms of carbon dioxide emission in small laboratory vertebrates characteristic of each species. Prog Clin Biol Res 341A:179–184

Stupfel M, Gourlet V, Perramon A, Mérat P, Court L (1990) Ultradian and circadian compartmentalization of respiratory and metabolic exchanges in small laboratory vertebrates. Chronobiologia 17:275–304

Stupfel M, Gourlet V, Demaria Pesce VH, Plétan Y (1991) Are there behavioral sequelae following an acute carbon monoxide intoxication? An animal model. Int J Environ Health Res 1:87–102

Sulzman FM, Fuller CE, Moore-Ede MC (1978) Comparison of synchronization of primate circadian rhythms by light and food. Am J Physiol 234:R130–R135

Szymanski JS (1918) Versuche über Aktivität und Ruhe bei Saüglingen. Pflugers Arch 172: 424–429

Tannenbaum GS, Martin JB (1976) Evidence for an endogenous ultradian rhythm governing growth hormone secretion in the rat. Endocrinology 98:562–570

Termier H, Termier G (1979) Histoire de la terre. PUF, Paris

Tobler I (1984) Evolution of sleep process: a phylogenetic approach. In: Borbély A, Valatx JL (eds) Sleep mechanisms. Springer, Berlin Heidelberg New York, pp 207–238

Valatx JL (1984) Genetics as a model for studying the sleep–waking cycle. Exp Brain Res 8 [Suppl]:135–143

Vaughan GM, Bell R, De la Pena A (1979) Nocturnal plasma melatonin in humans: episodic pattern and influence of light. Neurosci Lett 14:81–84

Weigelin J (1868) Versuche über die Harnstoffauscheidung wahrend und nach der Muskelhetigkeit. Arch Anat Physiol Wiss Med, pp 207–223

Weitzman ED (1975) Neuroendocrine pattern of secretion during the sleep–wake cycle of man. Prog Brain Res 42:93–102

Weitzman ED, Hellman L (1974) Temporal organization of the 24-hour pattern of the hypothalamic pituitary axis. In: Ferin M, Halberg F, Richart RM, Vandewiele RL (eds) Biorhythms and human reproduction. Wiley, New York, pp 371–395

Weitzman ED, Fukushima D, Nogeire C (1974) Studies in ultradian rhythms in human sleep and associated neuroendocrine rhythms. In: Scheving LE, Halberg F, Pauly JE (eds) Chronobiology. Igaku Shoin Ltd, Tokyo, pp 503–505

Werntz DA, Bickford RG, Bloom FE, Shannahoff-Khalsa DS (1983) Alternating cerebral hemispheric activity and the lateralization of autonomic nervous function. Human Neurobiol 2:39–43

Wever RA (1979) The circadian system of man. Springer, Berlin Heidelberg New York

Wilson DM, Kripke DF, McClure DK, Greenburg GA (1977) Ultradian cardiac rhythms in surgical intensive care unit patients. Psychosom Med 39:432–435

Wise PM, Weiland NG, Scarbrough K, Larson GH, Lloyd JM (1990) Contribution of changing rhythmicity of hypothalamic neurotransmitter function to female reproductive aging. Ann NY Acad Sci 592:31–43

Wollnik F, Döhler KD (1986) Effects of adult or perinatal hormonal environment on ultradian rhythms in locomotor activity of laboratory LEW/Ztm rats. Physiol Behav 38:229–240

Yates EF (1982) Outline of a physical theory of physiological systems. Can J Physiol Pharmacol 60:217–248

Yen SSC, Tsai CC, Naftolin F, Vandenberg G, Ajabor L (1972) Pulsatile patterns of gonadotropin release in subjects with and without ovarian function. J Clin Endocrinol Metab 34:671–675

Zepelin H, Rechschaffen A (1974) Mammalian sleep, longevity, and energy metabolism. Brain Behav Evol 10:425–470

Zung WWK, Wilson WP (1967) Sleep and dream patterns in twins. In: Wartis J (ed) Recent advances in biological psychiatry, vol IX. Plenum Press, New York, pp 119–130

Part III

The Behavioural and Psychosocial Level

What, then, scientifically speaking is a "feeling-toned complex?". . . Certain experimental investigations seem to indicate that its intensity or activity curve has a wavelike character, with a wavelength of hours, days, or weeks. This very complicated question remains as yet unclarified.

Carl Gustav Jung, The structure and dynamics of the psyche

From the firm foundation of ultradian rhythm research on the molecular, genetic, cellular and neuroendocrinal levels in Parts I and II, we now turn to studies on the behavioural and psychosocial levels for a broad appreciation of their role in human health, performance, stress and illness.

Alfred Meier-Koll (Chap. 12) begins this section with a historical survey of the developmental, social and cultural aspects of ultradian behaviour in humans that sets an appropriate attitude of enquiry for all the chapters that follow. Nathaniel Kleitman's concept of the basic rest–activity cycle (BRAC) is seen as playing a pivotal role in an integrative view of 90–120 min ultradian rhythms, which range from the molecular–genetic to the psychosocial, that is a major theme of this volume. The interview with 96-year-old Kleitman (Chap. 14) highlights the fact that for 32 years his still controversial BRAC concept has stimulated a fertile series of studies of the implications of ultradian rhythms for human experience in everyday life.

The chapters by Lavie, Sing, Rossi and Lippincott (Chaps. 13, 17 and 18) in this section all describe new methodological approaches to assessing Kleitman's BRAC. These new approaches have important implications for our understanding of the multi-faceted dynamics of ultradian rhythms on the behavioural and psychosocial levels in humans. What is most provocative about these contributions is their relevance for formulating a new foundation for the psychobiology of human health, performance and illness. Rütger Wever's carefully documented analysis (Chaps. 15 and 16) of the sleep–wake threshold in "disorders of the ultradian sleep rhythm, independent of the actual circadian rhythmicities" has potentially profound implications for the next generation of clinicians studying and treating sleep disorders.

Many practical applications of these new methodological approaches to ultradian rhythms in performance are pioneered by Helen Sing in her profusely illustrated fractal analysis of ultradian rhythms in human movement behaviour (Chap. 17). Sing's beautifully generated computer figures were also instrumental in illustrating Rossi and Lippincott's discussion of "The wave nature of being" (Chap. 18). The potentially seminal "Unification hypothesis of chronobiology" proposed by Rossi and Lippincott outlines a deep theoretical structure for a new science of psychobiology that integrates information, transduction and communication on all levels from molecule to mind. The surprising associations that they document between ultradian rhythms at the cellular–genetic, neuroendocrinal and psychosocial levels suggest a challenging paradigm of research for elucidating the cybernetic relationships between ultradian rhythms, stress and psychosomatic problems. Such research may lay the foundation for a new generation of integration between traditional medicine and the leading edges of a truly psychobiological psychiatry and clinical psychology. What is required is a grand unification of the phenomenological and molecular approaches that are now made possible by their sister science of chronobiology.

Chapter 12

Ultradian Behaviour Cycles in Humans: Developmental and Social Aspects

A. Meier-Koll

There is no doubt that circadian rhythms common to plants, animals and humans serve to adjust all organisms to the geophysical day–night regimen. But why ultradian cycles? What benefit is there in the short-term rhythms which subdivide the 24-h interval into several periods of a few hours each? For humans, particularly, ultradian cycles, well documented under laboratory conditions, are generally disguised in daily life by modern socioeconomic schedules which are regulated according to clock time. For instance, most civilized people are accustomed to fixed meal times even though endogenously regulated appetite, demonstrated by ad libitum feeding as well as by monitoring of gastric contractions in adult subjects, follows a cycle of about $1\frac{1}{2}$ hours (Friedman and Fisher 1967; Hiatt and Kripke 1975). Consequently it can be argued that ultradian cycles represent a topic of more academic than practical value, since they seem to play a less important part in organizing human behaviour, at least in the daily life of modern humans. There are, however, several conditions in human life under which ultradian cycles become a prominent time structure in both physiological and behavioural functions. One of them is early infancy. The human newborn lives predominantly according to its intrinsically determined demands. Its pattern of sleep and waking states shows manifestations of endogenous ultradian cycles just before circadian rhythmicities appear and become fully established during the first months of life. A second way to obtain insight into quasi free-running ultradian behaviour cycles is through the study of mentally impaired children and adults. In cases of severe mental retardation, deficiencies in cognitive functions are frequently accompanied by a loss of social skills, so that mentally impaired subjects behave in an autistic manner and can be considered as isolated from their social environment. Additionally, such persons are predominantly engaged in meaningless stereotyped acts which can easily be defined as specific items of behaviour. A third approach is the study of temporal dynamics obvious in the social play of preschool children from different cultures. Finally, individual and common daily activities in small communities of so-called primitive races living without artificial time cues can be investigated with respect to ultradian time structures. Several examples for each of these approaches are described in this chapter.

Biological Rhythms in Early Infancy

The human infant, well protected during its time in utero, is suddenly delivered to a world of innumerable stimuli which then begin to affect the immature nervous system. The flow of environmental influences, however, is limited and timed according to the sequence of waking and sleep epochs. The pattern of sleep–waking behaviour thus constitutes a natural time frame within which the newborn infant forms the earliest impressions of its world. This pattern, however, is determined by several endogenous rhythms. In the literature on this chronobiological aspect of the human postnatal development three main types of rhythm have been described: a cycle of rapid eye movement (REM) phases, with periods of about 50 min, a ca 4-h rhythm of feeding demand; and a circadian rhythm of sleep and waking pattern.

Kleitman (1963) was the first to describe the 50-min cycle of alternating sequences of REM and NREM (non-rapid eye movement) stages during the sleep of newborn infants and called it the "basic rest–activity cycle" (BRAC). An early precursor of this cycle is manifested in periodic variations of the foetal motility which can be observed during the second half of pregnancy (Sterman and Hoppenbrouwers 1971). As emphasized by Gesell and Amatruda (1945) human newborns wake and sleep according to an approximate 4-h period, when nursed under self-demand conditions. Morath (1974) has demonstrated that the occurrence of feeding demands of the young infant is related to this 4-h cycle, which seems to be already established at birth and does not depend on the degree of hunger (Emde et al. 1975). Besides these ultradian cycles, circadian periodicities can be observed in the infant's sleep–waking pattern as it develops during the first months of life (Kleitman and Engelmann 1953; Parmelee 1961; Hellbrügge et al. 1964; Hellbrügge 1974).

There is reason to believe that different endogenous rhythms in the same organism do not run independently of each other, but form a system of hierarchically organized and interacting oscillators. In particular, with respect to circadian rhythms of different body functions, this has been emphasized by Aschoff (1965). Despite the fact that many observations of individual behavioural rhythms of the human newborn have been reported, little is known about their interactions. A longitudinal study of one normal newborn during the first months of life seemed to be a suitable way of analysing these endogenous rhythms and some possible mechanisms of their interaction.

A Single-Case Follow-up Study

One male infant, "Korbi", who was normal with respect to pregnancy and delivery, was studied. During the first 12 days in the maternity hospital the newborn was not submitted to any clinical nursing routine but was cared for by his mother and members of a study group in a separate room. During the waking spans the newborn had sufficient contact with his mother and was fed by bottle according to his spontaneous feeding demand (Morath 1974). Behaviour observations including polygraphic recordings were initiated at the day of birth. In particular the newborn's sleep and waking spans were recorded. At home, behaviour observations were continued around the clock by the

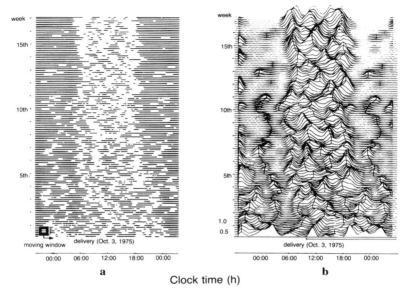

Fig. 12.1. **a** Observed pattern of sleep (bars) and waking spans (blanks) of the infant "Korbi" from birth until the fourth month of age. **b** Distribution of wake density corresponding to the pattern of sleep and waking spans. (Reproduced, with permission, from Meier-Koll et al. 1978.)

parents and staff members. This day-by-day study lasted throughout the first year of life. However, here only the development of sleep and waking behaviour during the first 4 months are described. Polygraphic recordings of electroencephalographic (EEG) activity, eye movements and heart rate were conducted several weeks later in order to study the cycle of sleep stages as it develops.

Sleep-Waking Pattern

Sleep and waking intervals were plotted on a day-chart (Fig. 12.1a) adopted from that used by Kleitman and Engelmann (1953). Each line of this chart represents 1 day. The first line at the bottom corresponds to the first day of life. Sleep intervals are marked by black lines and waking states are represented by blanks. The day-chart reflects the development of the sleep–waking behaviour during the first 4 months of life. If this pattern is visualized through a small window moving horizontally with time of day, the waking epochs will be found to be more or less accumulated in different regions of the behavioural day-chart. A measure of this accumulation of waking states can be defined by the number of 10-min samples of waking state found within the window. The window was chosen to cover 5 days vertically and 2 h horizontally. Employing this method for averaging the data of 5 consecutive days, smooth functions could be plotted which represented the "density" of waking states (called wake density from now on, Fig. 12.1b). The wake density alternates maximally between values of 1 and 0. Values approximating to 1 were scored when the infant was mainly in the waking state. By contrast, if the values approximated to 0, the infant was predominantly asleep. In order to emphasize prominent peaks of wake density, this function was plotted by using a squared ordinate. Finally, corresponding maxima of wake density of consecutive days were connected by crest lines.

The map of wake density can be subdivided roughly into three parts with different structures:

1. During the first 4 weeks, wake density increases and decreases periodically with respect to the time of day, suggesting that the sleep and waking epochs were organized mainly according to a mean period of about 4 h. Correspondingly, six periods from peak to peak, three during the daytime (06:00 to 18:00 h) and three during the night (18:00 to 06:00 h) appear within a 24-h interval.

2. During the fifth to the eleventh week, a clear difference between day and night becomes obvious. During the daytime (06:00 to 18:00 h) three periods appear, as can be seen during the first 4 weeks. By contrast, only two periods can be discerned during the night (18:00 to 06:00 h).

3. Beyond the eleventh week the nocturnal peak of wake density breaks off, indicating that the infant has now slept throughout the whole night. However, the wake density still oscillates two or three times during the daytime.

The map of wake density makes obvious that the ultradian 4-h rhythm established at birth did not disappear during the first month of life. However, with respect to day and night this ultradian rhythm of sleep–waking behaviour was altered by a circadian component. The variations of the 4-h rhythm seem to result from a modulation of period which accelerates and decelerates the ultradian cycle with respect to the half-phases of a circadian rhythm. Specifically, the ultradian period becomes increasingly longer during the night as the circadian component of the sleep–waking behaviour develops during the first month of life. Additionally, a related development of the winding structure of crest lines can be emphasized. Following one of the crest lines which connect the corresponding peaks of the wake density, for instance the crest line starting at 06:00 h on the first day, a winding structure can be seen. Windings to the right or left are found at an approximately constant interval of about 10 days. This structure is prominent till the eleventh week, when the circadian rhythm becomes fully established and obviously synchronized with the external period of day and night.

The Spectrum of Wake Density

In order to emphasize ultradian and circadian periodicities inherent in the infant's sleep–waking pattern, sequential power spectra were determined (Blackman and Tukey 1958). Functions of autocovariance of wake density were computed by time lag analysis for consecutive samples of 5 days. Corresponding power spectra resulted from Fourier transformation and were plotted consecutively against age using a hidden line plot routine (Fig. 12.2a). The resulting spectral map reflects roughly the three parts of the sleep–waking

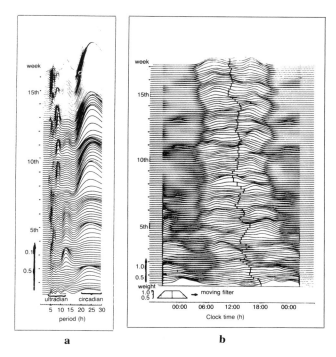

a   **b**

Fig. 12.2. a Sequential power spectra of the wake density during the first 4 months of age. Note that circadian amplitude rises continuously after the fourth week (linear calibration of power and period). **b** Circadian component of the wake density. A median line indicates the phase drift of the circadian periodicity. (Reproduced, with permission, from Meier-Koll et al. 1978.)

pattern mentioned above. Spectra of wake density corresponding to the first 4 weeks show that power amplitudes are minimal or absent within the circadian range of periods. After the fifth week the power within the circadian band rises continuously till the end of the eleventh week and builds up to its final amplitude beyond the eleventh week. Besides the circadian component, which increases from birth to the third month of life, several small peaks of power were found at ultradian periods of about 4, 8 and 12 h.

The distribution of wake density as well as the results of spectral analysis suggest that the circadian component of the sleep–waking behaviour developed together with its ultradian periodicities. When the data of sleep and waking epochs were filtered by a moving window of larger width, the ultradian components could be partly suppressed and a further map resulted, yielding the circadian component of the wake density exclusively (Fig. 12.2b). Each segment of this distribution is divided into halves by a median line. The medians of consecutive days constitute a line which drifts slightly from night to daytime until the eleventh week, when it reaches a relatively stable position with respect to the 24-h time scale. This drift of the median line describes the development of a free-running circadian rhythm of the infant's sleep–waking behaviour which was synchronized by an external zeitgeber not earlier than the eleventh week of life.

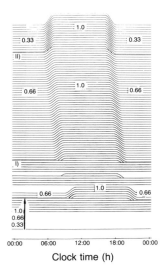

00:00   06:00   12:00   18:00   00:00

Clock time (h)

Fig. 12.3. Computerized simulation of the development of the circadian rhythmicity during the first 3 months of age. (Reproduced, with permission, from Meier-Koll et al. 1978.)

A simple model is proposed to describe the main results of the behaviour analysis. The infant's sleep–waking behaviour seems to have been governed mainly by two endogenous rhythms, ultradian and circadian, constituting a system of interconnected oscillators. As the circadian rhythm has not yet developed during the 4 weeks following birth, the ultradian cycle performs within a mean period of about 4 h. However, after the fourth week, the frequency of the ultradian cycle seemed to be modulated according to a circadian variation: specifically, the less the circadian variation, the longer the ultradian period. Since its period is modulated by the circadian rhythm, the latter becomes phase-locked to the former. Finally, as soon as the circadian rhythm is synchronized to the external day and night regimen, the ultradian rhythm also shows a stable relationship to the time of day. Should modulation of period, as hypothesized, be the actual mechanism connecting the ultradian cycle to the circadian rhythm of the infant in this study, it should be possible to generate a distribution of wake density similar to the one empirically observed by a simulation model of two coupled oscillators.

Computerized Model

The development of the circadian rhythm during the first 3 months was represented by a simple simulation (Fig. 12.3) corresponding to the map of the observed circadian rhythm (Fig. 12.2b). The circadian rhythm is represented by distinct levels of amplitude of its diurnal and nocturnal half phase. These amplitudes are arbitrarily calibrated using 1.0 as a reference value. During the first week, spectral analysis provided no evidence of circadian periodicity (Fig. 12.2a). This lack of circadian periodicity is represented in the simulation by straight lines at a level of 1.0, indicating a lack of difference in amplitude between day and night (Fig. 12.3). By contrast, during the second week, a

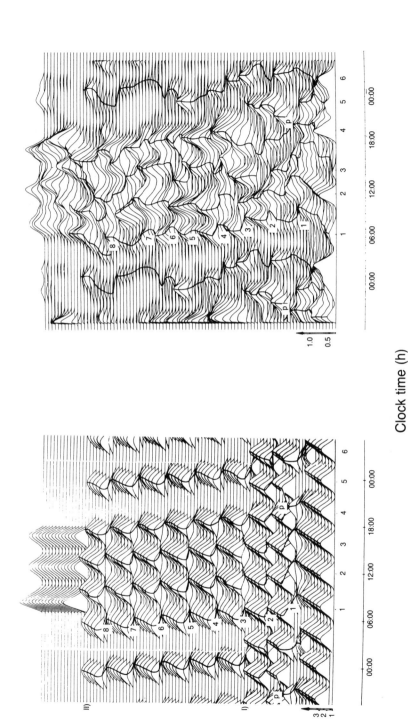

Clock time (h)

Fig. 12.4. a Theoretical wake density determined by a computerized model of two coupled rhythms. The frequency of a sine wave representing the ultradian cycle is modulated according to the simulated circadian variation (see Fig. 12.3). **b** Observed wake density (as shown in Fig. 12.1b). Note the striking similarity between the theoretical and the observed distribution of wake density. Compare windings of crest lines and corresponding bifurcation points P. (Reproduced, with permission, from Meier-Koll et al. 1978.)

circadian periodicity was obvious in the power spectra. Accordingly, this is reflected in the simulation by a lower amplitude of 0.66 during the nocturnal phase. After the second week, the initial, weak, circadian periodicity disappeared and then reappeared after the fourth week (compare Figs. 12.2a and 12.3). Consequently the simulation indicates, by straight lines at a level of 1.0, this transitory lack of circadian periodicity after the second week. Subsequent to the fourth week the circadian amplitude in the power spectra is pronounced. This is simulated by different levels of amplitude, specifically at a level of 1.0 during the diurnal and 0.66 during the nocturnal phase. The entire simulated pattern shows a mild drift from right to left reflecting the observed drift (compare Figs. 12.2b and 12.3). Finally, after the eleventh week the theoretical pattern stabilizes with respect to the time of day, thus simulating the observed synchronization of the circadian rhythm with an external zeitgeber. Beyond the eleventh week there is a considerable increase in the power of the circadian periodicity (Fig. 12.2a). This is depicted in the simulation by a decrease in the nocturnal amplitude to a level of 0.33, indicating a more prominent change of the circadian variation between day and night (Fig. 12.3). The ultradian cycle was represented by a simple sinusoidal oscillation with a period of 4 h and 20 min. This period was modulated according to the levels of the simulated developing circadian rhythm. Additionally the sinusoidal oscillations were superimposed on the simulated circadian variations. The net result is a wave pattern representing the theoretical distribution of wake density (Fig. 12.4a) and can be compared with the observed wake density of the infant investigated in this study (Fig. 12.4b).

The striking similarity between the theoretical and observed patterns of wake density becomes obvious when corresponding crest lines are compared. Note that the crest lines of both patterns correspond well in showing eight windings at corresponding points. Furthermore the bifurcation of crest line (4) at the point P as well as the breaking off in crest line (5) is similar in both maps (Fig. 12.4a and b). Note also the similar drift of crest lines from left to right during the first week. In the theoretical pattern this drift results from the free-running ultradian oscillation.

The circadian modulation of an ultradian cycle causes a phase-locking of the ultradian with the circadian rhythm. However, during the fifth and the eleventh weeks the winding structure of the crest lines indicates that the circadian rhythm is still unable to phase-lock the ultradian cycle completely. Phase-locking is most effective when a maximum of the ultradian cycle intersects with the left slope of the circadian variation (Fig. 12.3). By contrast, phase-locking is least effective when a minimum of the ultradian cycle intersects with the left slope of the circadian variation. In this condition the ultradian cycle is not altered with respect to its phase and proceeds free-running. Consequently, during the following days the ultradian cycle drifts slightly from left to right. Finally, after a few days, the ultradian cycle has reached a position in which a maximum coincides with the left slope of the circadian variation. In this position the effective phase-locking traps the phase of the ultradian cycle and shifts it towards the left such that a minimum of the ultradian cycle once more intersects with the slope of the circadian variation. Thus another winding can start. It should be emphasized here that the proposed model describes what was called the winding structure of crest lines. From this point of view, the 10-day periodicity of winding crest lines is not caused by a distinct endogenous

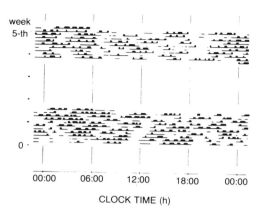

week
5-th

0

00:00   06:00   12:00   18:00   00:00

CLOCK TIME (h)

Fig. 12.5. Distribution of polygraphically determined REM stages (thick black bars) within the sleep intervals during different weeks of age. (Reproduced, with permission, from Meier-Koll 1979.)

rhythm, but results from the hypothesized interaction of the coupled ultradian and circadian rhythms. After the eleventh week the circadian rhythm was fully developed and synchronized to the day–night regimen. Accordingly the simulated circadian variations show a low level (0.33) during the nocturnal phase. This prominent circadian variation phase-locks the ultradian cycle completely. The nocturnal period of the theoretical ultradian cycle now extends over the whole night. Note that the winding crest lines are now substituted by three crests which show a fixed relationship to the time of day (compare Fig. 12.4a and b). The striking similarity of the model and the actually observed wake density of the infant "Korbi" demonstrates that the proposed period modulation of the ultradian cycle by a circadian rhythm could be a mechanism underlying the observed pattern of sleep and wakefulness. Furthermore the study implies a high degree of order in the development of the infant's sleep–waking behaviour.

REM Cycle and Sleep–Waking Behaviour

As mentioned, in addition to the observation of the sleep–waking behaviour polygraphic sleep recordings were conducted during several weeks of life. Thus the intervals of sleep could be subdivided into REM and NREM stages (Fig. 12.5). Small black bars within the sleep intervals (black lines) indicate the observed REM stages. The amount of data collected enabled the analysis of statistical relationships between the occurence of both REM phases and waking states. The data collected during the first 11 days of life and data from the fifth week were analysed separately with respect to day and night time.

Fig. 12.6a summarizes the analysis of data collected during the daytime (06:00 to 18:00h) of the first 11 days of life. The upper diagram indicates the frequency distribution of waking states as a function of time passed since the last awakening. One hour after awakening, the infant was frequently found to be still in the wakeful state. Some time later, the frequency of the waking state

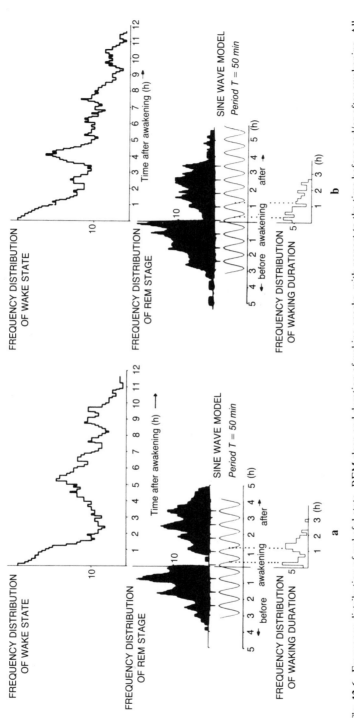

Fig. 12.6. Frequency distribution of wakeful state, REM sleep and duration of waking epochs, with respect to the time before and/or after awakening. All data are of polygraphic measurements during the first 11 days. **a** Daytime (06:00 to 18:00 h), **b** night-time (18:00 to 06:00 h). *T* is the period of the REM cycle. (Reproduced, with permission, from Meier-Koll 1979.)

decreases continuously for 2–3 h and increases for about 4–5 h after awakening. This distribution of waking states as a function of time passed since the last awakening is another expression of the ultradian 4-h cycle of the sleep–waking pattern.

The middle diagram of Fig. 12.6a represents the distribution of frequency in the occurrence of REM sleep before and after awakening. The diagram is characterized by periodically occurring peaks. Using a cross-correlation method it was demonstrated that the periodic structure of this distribution can be interpreted as a manifestation of a 50-min cycle continued through the wakeful state. This 50-min periodicity is represented by a sine wave. Its maxima correspond to individual peaks of frequency in the occurrence of REM sleep. This means that REM stage tends to occur most frequently at 150, 100 and 50 min before awakening. This periodicity continues throughout the wakeful state and becomes manifest again by the occurrence of REM stage at a periodic interval of 50-min, since the infant has started a new sleep epoch.

A further aspect of this analysis is the distinct phase relationship of the time of awakening to the 50-min cycle. It seems that awakening occurred predominantly at times during which an REM stage would have probably occurred, if the infant had kept on sleeping. Similarly the end of a waking epoch seemed also to be correlated with the 50-min cycle. The bottom diagram of Fig. 12.6a represents the frequency histogram of the durations of waking states. It can be seen that the duration of waking epochs is distributed bimodally, in that there is one interval of 20 min and one of 70 min. This means that re-entrance into sleep occurred predominantly 20 or 70 min after awakening. These intervals correspond to the transitional phases, during which the 50-min cycle of REM occurrence sweeps down. It seems, that the 50-min cycle manifested by the occurrence of REM phases triggers a gate function for both transition from sleep to wakeful state and the re-entrance into sleep. Corresponding relationships of the 50-min cycle of REM stage and time of wakeful state were found at night-time (18:00 to 06:00 h) during the first 11 days of life (Fig. 12.6b).

The same statistical analysis was performed with respect to waking states and REM phases observed during the fifth week (see Fig. 12.5). The results are shown in the diagrams of Fig. 12.7a and b, which correspond to those of Fig. 12.6a and b. The distributions of waking states were considerably different with respect to day and night during the fifth week of life (upper diagrams of Fig. 12.7). However, the beginning and the end of waking states seemed again to be triggered by upward and downward transitional phases, respectively, of a 50-min cycle as already described for the first 11 days of life. As a result of this trigger mechanism a trimodal distribution of waking duration may be seen (see bottom diagrams of Fig. 12.7a and b). The third maximum in trimodal frequency distributions of waking duration is correlated with the third downward transitional phase of the 50-min cycle.

Conclusions

The main results of the present one-case study can be summarized schematically by the graphs of Fig. 12.8 depicting the hypothesized interactions of three endogenous rhythms as they develop during the first 3 months of life.

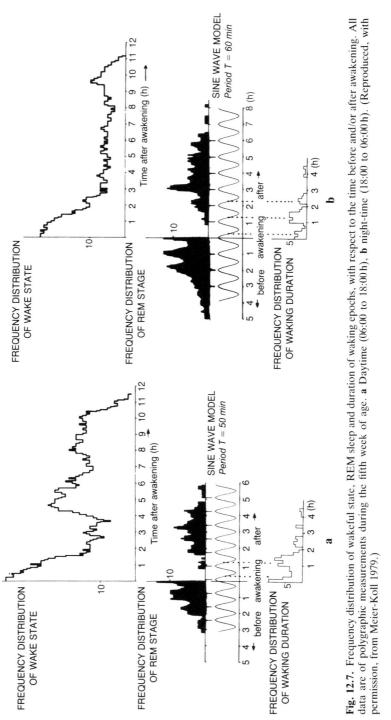

Fig. 12.7. Frequency distribution of wakeful state, REM sleep and duration of waking epochs, with respect to the time before and/or after awakening. All data are of polygraphic measurements during the fifth week of age. **a** Daytime (06:00 to 18:00h), **b** night-time (18:00 to 06:00 h). (Reproduced, with permission, from Meier-Koll 1979.)

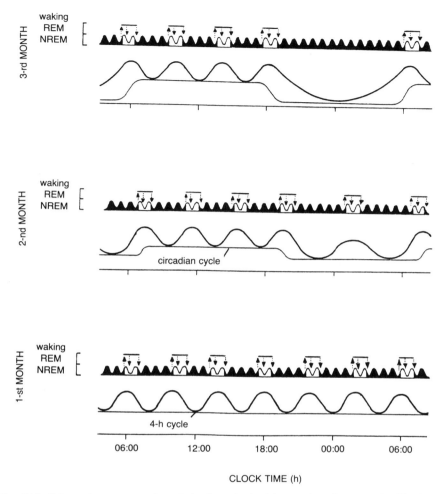

Fig. 12.8. Schematic representation of the hypothesized interactions between the 50-min cycle of REM stages, the 4-h cycle and the gradually developing circadian rhythm of sleep–waking behaviour during the first 3 months of age. (↑) Awakenings, (↓) re-entrance into sleep. (Reproduced, with permission, from Meier-Koll 1979.)

During the first month after birth a 4-h cycle modulates the level of arousal as indicated by the approximately periodic occurrence of wakeful states. In addition, the distinct times of awakenings and of re-entrance into sleep depend on the upward and downward transitional phases, respectively, of a 50-min cycle, as indicated by arrows in Fig. 12.8. The 50-min cycle, as manifested by periodic occurrence of REM phases during sleep, is represented by a black sinusoidal wave. It continues throughout the short waking states, as demonstrated by statistical analysis. Thus, the 50-min cycle corresponds exactly to Kleitman's concept of a BRAC operating continuously throughout the wakeful state and sleep. Correspondingly, one to three latent REM phases occur during one waking state (interrupted black waves, Fig. 12.8). Since the end of each latent REM phase coincides with the downward transitional phase of the 50-min cycle which seems to open a gate for the re-entrance into sleep, the infant can immediately fall into a short REM phase when starting a new sleep epoch.

This interaction of both ultradian rhythms, the 50-min cycle and the 4-h rhythm seem not to be altered during the first 3 months of life. During the second and third month, however, the 4-h rhythm is modulated with respect to its period by a developing circadian one, thus showing longer periods during the night than during the day (Fig. 12.8). Finally the whole pattern of sleep and waking states is completely adapted to, and synchronized with, the external day–night regimen. It can be argued, that the sleep–waking behaviour becomes gradually adjusted to the day–night regime as the infant's brain develops postnatally (Meier-Koll et al. 1978). During the first weeks after birth, however, the sleep–waking behaviour is timed mainly by ultradian cycles. Particularly the 4-h cycle of waking states constitutes a framework within which the newborn experiences its world and forms its earliest impressions. In addition it can be thought that consolidation of these first impressions takes place during sleep. This view is supported by a study by Paul and Dittrichova (1975), who demonstrated that the process of learning can influence sleep patterns, at least in 6-month-old infants. Head turns of the infants in their study were reinforced by blinking coloured bulbs. After an infant had learned to turn his or her head to the appropriate side in one session, the conditions for reinforcement were changed in the next. Immediately after the conditioning procedure the infant was put to bed and the sleep stages were registered polygraphically. REM phases during the first REM–NREM cycle following successful learning were significantly longer than in control sessions, whereas the mean period of the REM–NREM cycle remained unchanged. However, when the tasks were so difficult that the infant did not acquire the correct response, there were no significant differences from the control session. From these learning experiments in human infants, it can be concluded that sleep is involved in consolidation of basic experiences in early infancy.

One can speculate that the 4-h rhythm of the sleep–waking pattern and the 50-min cycle of REM phases represent periodically operating gate mechanisms mediating both the acquisition of new experiences and their long-term consolidation.

Ultradian Behaviour Cycles in the Mentally Retarded

Higher cognitive functions, which normally enable humans to organize their lives and to communicate appropriately with others, are impaired in mentally retarded patients. The resulting cognitive deficiencies are frequently accompanied by a loss of social skills, so that mentally defective subjects refuse social contact and behave predominantly in an autistic manner. Additionally, in cases of severe mental retardation a variety of deviant behaviours (including multiple stereotyped acts and self-mutilations) can be observed. Mentally impaired patients are engaged predominantly in repetitions of meaningless stereotyped

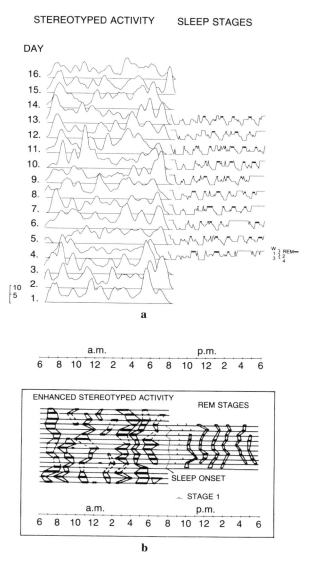

Fig. 12.9. a Stereotyped activity (smoothed data) of the mentally impaired child "Réné" observed for 16 days, and sleep stages for 10 nights. Scale of stereotyped activity: burst of hand waving/ 10 min. W, wake; 1,2,3,4, stages of N-REM sleep. **b** Zones of enhanced stereotyped activity and REM phases (for details, see the text). (Reproduced, with permission, from Meier-Koll et al. 1977.)

movements, such as body rocking, hand waving and clapping. Frequently such stereotypies are accompanied by vocalizations and oral automatisms.

Since severe mental impairment implies both loss of cognitive skills and autistic self-isolation from the social environment, the stereotyped behaviour of the mentally handicapped can be considered as an appropriate clinical model to obtain further insight into human free-running ultradian oscillator systems.

Observation Technique and Data Handling

A mentally impaired girl "Réné", $6\frac{1}{2}$ years old, was observed in the clinical ward of a nursing home with which she was familiar. She refused eye and body contact and exhibited no speech or language comprehension. Since the girl performed stereotyped hand waving in intermittent bursts, each lasting about 5 to 10 s, the stereotyped activity could be determined by the number of bursts observed during consecutive sample intervals of 10 min. The stereotyped activity of both hands was observed continuously over 16 consecutive days. The observations were begun at 06:00h, when the girl was still asleep, and finished at about 20:00h, when she slept again. The resulting time series were smoothed by means of a 5-point filter (weights: 0.5, 1, 1, 1, 0.5) in order to minimize the statistical noise and were finally plotted against the time of day (Fig. 12.9a). In addition to the observations of stereotyped activity during wakefulness, the child's sleep stages were scored by means of polygraphic recordings during 10 consecutive nights each following 1 day of observation. Despite the girl having suffered from an infantile brain damage all EEG patterns specifying the sleep stages appeared. Consequently the sleep stages could be scored according to the criteria of Rechtschaffen and Kales (1968). The sleep stages were plotted against time of day using consecutive sample intervals of 5 min (Fig. 12.9a). The REM phases are emphasized by black bars.

During all the days of observation the activity of stereotyped hand waving showed considerable variations. The peak-to-peak intervals are distributed around a mean value of 87 min (s.D. = 33 min). This indicates the existence of an ultradian periodicity. A periodic time structure can be emphasized, when for each day of observation only such time spans are marked during which the stereotyped activity exceeds a distinct level. For each time series plotted in Fig. 12.9a the zero line of the subsequent day can be arbitrarily chosen as the cut-off. Employing this procedure on the data of all 16 days, a new pattern results within which time spans of enhanced stereotyped activity appear to be arranged according to periodic zones (Fig. 12.9b). Since with the exception of only occasional nursing, nothing was done to prevent the child from spontaneous stereotyped behaviour, the periodic pattern cannot be due to a corresponding schedule of nursing routine. In addition to the zones of enhanced stereotyped activity the REM phases observed during 10 consecutive nights are plotted in Fig. 12.9b. As with the periodic pattern of stereotyped activity the REM phases of consecutive nights appear within distinct periodically arranged zones.

As already mentioned above, the relative frequencies determined for peak-to-peak intervals of stereotyped activity observed during all 16 days are found to be distributed around a mean interval of 87 ($\pm$33) min (Fig. 12.10a). The range of peak-to-peak intervals corresponds to that of ultradian periods found in a variety of physiological and behavioural functions of the normal adult man, such as urine flow and electrolyte concentration (Lavie and Kripke 1977), gastric motility (Hiatt and Kripke 1975) and spontaneous oral intake (Friedman and Fisher 1967), several perceptive performances (Globus et al. 1971; Lavie 1976) and day-dreaming (Kripke and Sonnenschein 1978). Similar periods could also be detected in stereotyped activities of other severely handicapped children by means of autocorrelation and spectral analysis (Meier-Koll and Pohl 1979). A corresponding histogram of the relative frequencies of

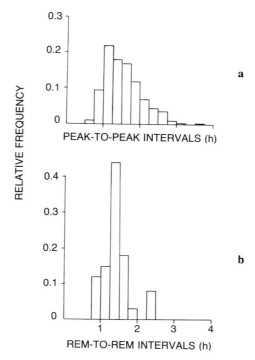

Fig. 12.10. Histograms of peak-to-peak intervals of **a** stereotyped activity and **b** REM-to-REM intervals.

REM-to-REM intervals observed in the subject "Réné" (Fig. 12.10b) yields a distribution around a mean value of 77 ($\pm$22) min, which is not significantly different from that of the stereotyped activity (χ^2-test: d.f. $= 11$, $\chi^2 = 18.9$, $p = 0.063$).

This provides evidence that two cyclic phenomena, ultradian variations in stereotyped behaviour during wakefulness and the cycle of REM sleep, could be related to each other (Meier-Koll et al. 1977). More specifically, they could be interpreted as manifestations of a common BRAC operating throughout wakefulness and sleep, as previously emphasized by Kleitman (1963). However, Kleitman's concept of a BRAC has been repeatedly challenged by studies demonstrating that the REM cycle is a sleep-dependent process triggered by sleep onset (Moses et al. 1977, 1978; Campbell 1987). Consequently, ultradian behaviour cycles of the waking organism and the REM cycle have to be considered as separated components of a multioscillator system. However, since in addition to several behavioural functions the degree of vigilance can be modulated with ultradian periods (Broughton 1975), it might be argued, that the onset of sleep is linked to a distinct phase of an ultradian behaviour cycle. This implies that the ultradian behaviour rhythm operating during wakefulness can trigger the onset of sleep and consequently the REM cycle too. As a net result both cycles can appear as if they were one rhythm running continuously throughout the waking state and sleep.

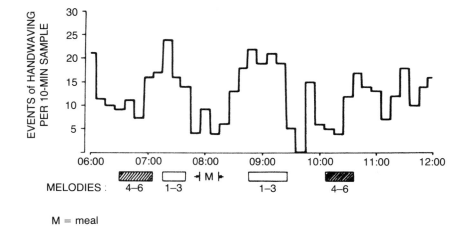

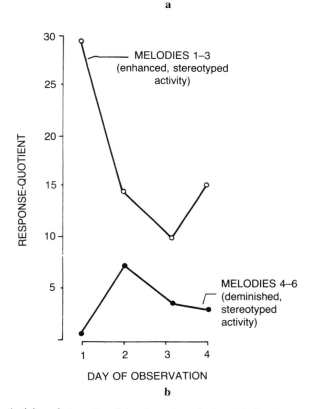

Fig. 12.11. a Activity of stereotyped hand waving of the girl "Réné" during a 6-h interval. Exposition periods of different melodies are represented below the time axis: white bars (melodies *1* to *3*) indicate presentation during maxima, hatched bars (melodies *4* to *6*) indicate presentation during minima of stereotyped hand waving. **b** Differences in response quotient as a function of phase position (maxima versus minima) of stereotyped activity during 4 days of consecutive observation. (Reproduced, with permission, from Meier-Koll and Pohl 1979.)

A Clinical Implication

The ultradian variations in stereotyped motor behaviour can be taken into account with respect to therapeutic measures. An observation concerning vocal imitation in subject "Réné" pertains to possible clinical implications. Like other mentally impaired individuals, "Réné" showed a particular gift for music. As a result she was able to listen and respond to musical stimuli by correct vocal imitation. During 4 consecutive days several weeks subsequent to the study described above, the child's hand waving was observed and scored again. Additionally a set of six simple melodies recorded on magnetic tape was presented repeatedly to the child to be imitated correctly. Melodies 1–3 were presented during phases of an increasing or enhanced stereotyped activity, whereas melodies 4–6 were played back during phases of minimal hand waving (Fig. 12.11a). During each exposition interval, the individual melodies were presented in 15-s intervals.

A response quotient was determined as a measure of vocal imitation. The quotient consists of the frequency of correct imitations of melodies per 2-h exposition interval surrounding peaks or troughs of the stereotyped activity. The response quotients determined for the 4 days of observation are plotted in Fig. 12.11b. The upper curve depicts the response quotients during phases of enhanced stereotyped activity, and the lower curve represents the response quotients for phases of diminished stereotyped activity. The child's ability for correct vocal imitation of simple melodies was obviously enhanced during phases of increased stereotyped hand waving, whereas vocal imitation was considerably less pronounced during phases of diminished stereotyped activity. These findings imply that certain skills can be facilitated during specific phases of ultradian cycles of stereotyped motility.

Ultradian Multioscillations

A second single-case study yields an example of a more complex ultradian time-structure. The boy "Ulf", 20 months old, was preterm, delivered at the thirty-sixth week of gestation. The infant had obviously suffered cerebral damage, since he developed epileptic seizures during the third month of life. Additionally his psychomotor development was severely retarded. Except at times of nursing and feeding, the infant lay on his bed and was occupied with stereotyped movements of his fingers. The single stereotyped act was characterized by a movement bringing one of his hands in front of the face and spreading three fingers.

The number of these stereotyped acts was determined for consecutive 10-min sample intervals from 07:00 to 20:00 h during 17 days of observation. The raw data were smoothed again by means of a 5-point filter. As with the procedure already described above, the resulting time series were cut off at an arbitrarily chosen level (10 stereotyped acts/10 min). This level corresponded roughly to the mean value of the daily stereotyped activity. Finally, the intervals within which the stereotyped activity exceeded the level of cut-off were plotted against the time of day; epochs of sleep were also plotted (Fig. 12.12). Both behavioural states, phases of enhanced stereotyped activity and sleep, seem to be arranged according to distinct periodic patterns. Corresponding histograms

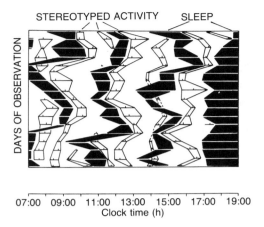

Fig. 12.12. Day-chart of sleep–waking behaviour and stereotyped activity of the mentally impaired boy "Ulf". Seventeen consecutive days of observation; first day at the bottom. Phases of enhanced stereotyped activity ($\perp$). Small vertical marks indicate the position of maxima in between. Note the periodic arrangement of both sleep epochs and zones of enhanced stereotyped activity.

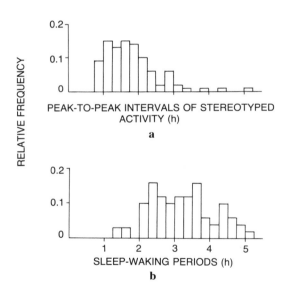

Fig. 12.13. Histograms of relative frequency for peak-to-peak intervals of **a** stereotyped activity and **b** sleep–waking periods.

of the relative frequencies of peak-to-peak intervals in stereotyped activity and sleep–waking periods yield different distributions (Fig. 12.13a and b). The mean peak-to-peak interval of stereotyped activity is 104 ($\pm$49) min and corresponds to that found in the infant "Réné". There was no significant difference between the distributions of the corresponding histograms shown in Figs. 10a and 13a (χ^2-test: d.f. = 15, χ^2 = 14.14, p = 0.515).

The distribution of sleep–waking periods was clearly different from that of the peak-to-peak intervals of the stereotyped activity (χ^2-test: d.f. = 17, χ^2 = 69.4, p = 0.0001). The mean interval of 184 ($\pm$55) min determined for the sleep–waking periods of the infant "Ulf" corresponds more to the ca 4-h rhythm of sleep–waking behaviour found in newborns and very young infants. This ultradian rhythm seems to be phase-locked to a circadian one and/or to the day–night regimen, since the sleep and waking epochs form distinct periodic zones. This would not be the case if the 4-h rhythm were not roughly synchronized with the day–night regimen. A corresponding phase-locking with the day–night regimen becomes evident when the periodic zones of enhanced stereotyped activity are considered.

In conclusion, the behaviour of the handicapped infant "Ulf" seems to be influenced by two distinct ultradian rhythms, one of them underlying the sequences of sleep and waking epochs and another which determines how frequently a distinct stereotyped act occurs.

Ultradian Rhythms and Cerebral Dominance

A third single-case study provides evidence that stereotyped activities of the right and left hand can reflect ultradian cyclic variations in the relative predominance of one cerebral hemisphere compared with the other.

A 23-year-old woman, "Doris", had been born with anoxia due to complications during delivery. "Doris" had obviously suffered from brain damage, since she developed as a mentally impaired child. For some years she has been living in a special institution for mentally handicapped patients. She has no speech and language comprehension and refuses social contact. Predominantly she is engaged in multiple stereotypies.

"Doris" was observed in the ward in her familiar environment. Her behaviour was recorded on videotapes continuously from waking in the morning to sleep onset at night. During sleep EEG and eye movements were recorded polygraphically in order to score her sleep stages (Rechtschaffen and Kales 1968). The whole study included 10 consecutive days and nights. The number of intermittent bursts of hand waving was determined separately for each hand during consecutive time samples of 5 min and plotted against the time of day (Fig. 12.14). Time series were plotted of stereotyped activity of the left hand (upward) and that of the right hand (downward). In order to emphasize the main structure of stereotyped activities, highly frequent variations of the raw data were eliminated by a moving averaging procedure (5-point filter). Employing this method of filtering smoothed diagrams of hand waving activities resulted. The band between the smoothed curves represents the stereotyped activity of both hands and shifts around the midline, indicating temporal variations in the relative dominance of one hand compared with the other. The shifting in laterality is emphasized by a dotted line representing the balance between right- and left-hand stereotyped activity. Since each hand is controlled by its contralateral cerebral hemisphere, shifts of activities from one hand to the other can be interpreted as shifts of motor activation between the related hemispheres.

As determined during several days of observation stereotyped activities of both hands can be modulated periodically with intervals of about 2 h (Fig.

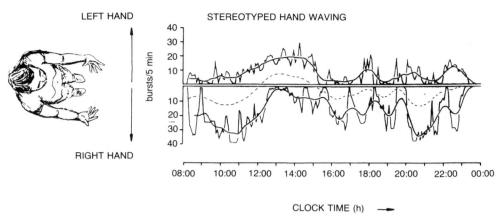

Fig. 12.14. One-day time series of left- and right-hand stereotyped activities (raw and smoothed data) of the mentally impaired woman "Doris". Dotted line, balance between left- and right-hand activities. (Reproduced, with permission, from Meier-Koll 1989.)

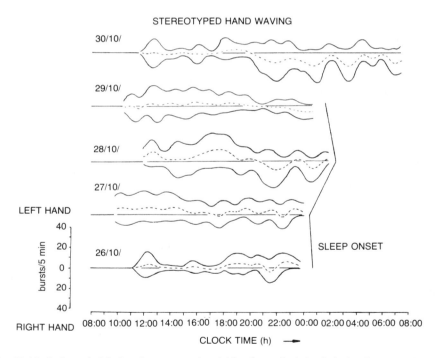

Fig. 12.15. Left- and right-hand stereotyped activities (smoothed data) during 5 consecutive days of observation. (Reproduced, with permission, from Meier-Koll 1989.)

12.15). Furthermore, time series of right-hand activities can tend to oscillate in phase or in counterphase to those of the left hand. Predominantly, during the morning hours, right- and left-hand activities showed a tendency of bilateral synchrony which continued during the day. In the late afternoon, however, ultradian cycles of both hand activities developed a more counterphase course. The development from a bilateral synchrony to a counterphase organization during time of day is obvious in the data of 30 October (Fig. 12.15). On this day the subject did not sleep, but engaged in stereotyped hand waving throughout the whole night.

Since EEG and eye movements were recorded during sleep, REM and NREM stages could be scored for each night following the day of observation. Like normal subjects, the mentally impaired woman showed a cyclic organization of sleep with intervals of about 1.5–2 h. A period interval of 108 (±14) min was found in sleep cycles of the first five consecutive nights (Fig. 12.16). This period corresponds to the mean interval (114 (±17) min) between adjacent maxima of oscillating left-hand activities observed during the late afternoon and evening of preceding days. It seems that an oscillating process which modulates the stereotyped activities of both hands during the day-

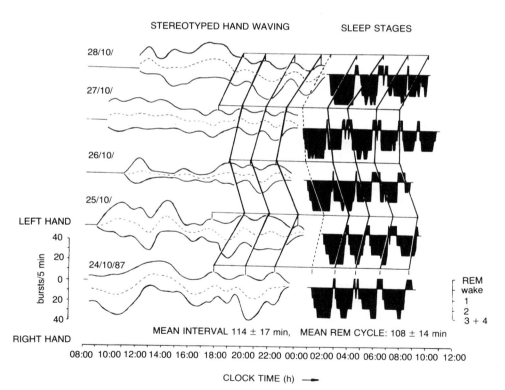

Fig. 12.16. Time series of left- and right-hand stereotyped activities (smoothed data) and sleep stages of 5 consecutive days. Extrema of the left-hand stereotyped activities observed during the evenings and REM phases of sleep cycles are connected by grid lines in order to illustrate equivalent periodicities. (Reproduced, with permission, from Meier-Koll 1989.)

time continues in sleep at night or interacts with a separate neuronal system involved in the generation of the REM sleep cycle (McCarley and Massaquoi 1986).

Neuroanatomical and electrophysiological studies provide evidence that neuronal systems involved in the generation of self-sustained ultradian oscillations are situated in the pontine and mesencephalic brainstem (McCarley and Hobson 1975). Such a brainstem oscillator system can influence the electrocortical arousal of both hemispheres. Since the brainstem contains bilateral neuronal nuclei, the oscillator system can be thought of as being subdivided into two functionally homologous parts which are bilaterally arranged on the right and left side. Consequently, the cortical arousal of each hemisphere is exclusively influenced by its ipsilateral brainstem oscillator. When the two parts of the bilateral oscillator system show functional asymmetry, like the hemispheres, they might operate with slightly different ultradian periods. As a result different and changing phases would be found between both oscillators. From this point of view the ultradian variations in the stereotyped activities of the right and left hand of the subject "Doris" might be interpreted as manifestations of an bilateral ultradian oscillator system (Meier-Koll 1989; Bohl and Meier-Koll 1991).

Ultradian Cycles in Normal Children's Play

Since endogenous ultradian rhythms are able to influence the daily behaviour pattern considerably, at least as demonstrated in several handicapped individuals, one may expect that ultradian time structures could also be detected in the behaviour of normal, healthy children. However, in contrast to the limited and predominantly stereotyped behaviour of the mentally impaired the behaviour of a healthy child shows a great variety of motor acts. Most of them are goal-directed by cognition or by the focus of interest and influenced by innumerable external events. Consequently it could be argued that ultradian behaviour cycles are unlikely to be detected when behaviour scoring is used as a method. However, there is some evidence that ultradian cycles can be detected in undisturbed play activities of preschool children.

Laboratory Studies of Free-play in Normal Children

A quiet and spacious room of my laboratory was appropriately furnished and equipped abundantly with different toys. A pair of preschool children, one boy of 6 years and his sister, 3 years old, were allowed to play, eat and drink when and how they pleased. Their father was present continuously in an adjacent room and available if necessary. However, the father and all staff members avoided interventions as far as possible. The children's spontaneous and self-determined behaviour was continuously recorded on videotapes for about 12 h (07:30 to 19:00 h). The records were analysed with respect to the frequency of individual locomotion, social contacts and object changes observed during

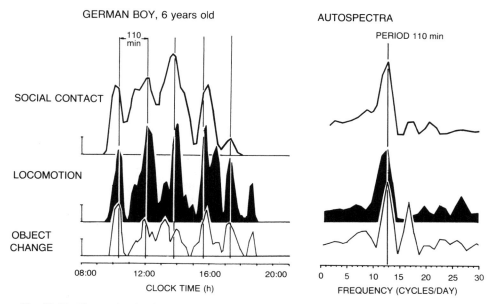

Fig. 12.17. Time series (left) of social contact, locomotion and object changes in the behaviour of a healthy German boy, 6 years old. Spectral analysis shows (right) that the different time series were influenced by the same rhythm, with a period of 110 min. (Reproduced, with permission, from Gast and Meier-Koll 1987.)

consecutive sample intervals of 10 min. In order to score the locomotion the laboratory room was subdivided into quarters by cross lines drawn with chalk on the floor. Whenever a child crossed one of these lines, it was scored as a unit of locomotion. Events of social contact were counted, each time a child addressed his or her playmate, father or any member of the staff. Finally, in order to define a degree of concentration or focussing on special objects, we counted the frequency with which the object of play was changed during the sample interval.

Typical examples of time series (smoothed data) determined for the boy's locomotion, social contact and object change are shown in Fig. 12.17. Corresponding spectra of variance (autospectra) revealed an ultradian periodicity of 110 min common to all these time series. This common periodicity is represented by vertical lines intersecting the time series at distances of 110 min. Obviously the child's locomotor activity increased and decreased in phase with a tendency of social contact and the frequency of object changes. As could be demonstrated by video replay, the boy alternated between two phases of behaviour: an exploratory one with frequent locomotion, social contacts and object changes, and a phase which can be characterized by a withdrawal to more solitary activities focussed on one or very few objects.

By contrast, the boy's younger sister showed somewhat lower levels of locomotion, social contacts and object changes and considerable variations of these activities could not be found. Therefore, the corresponding time series are not presented here.

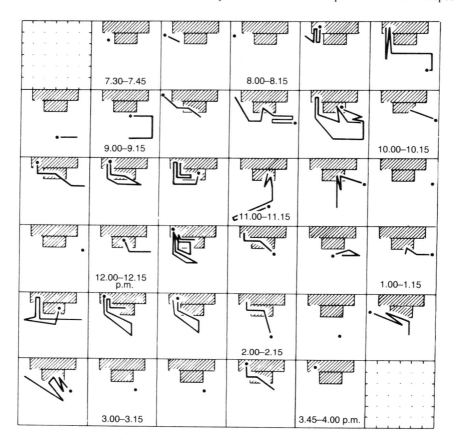

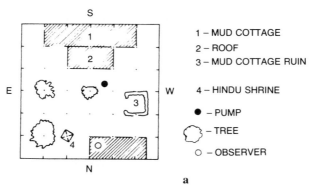

1 – MUD COTTAGE
2 – ROOF
3 – MUD COTTAGE RUIN
4 – HINDU SHRINE
● – PUMP
– TREE
○ – OBSERVER

a

Fig. 12.18. Recorded locomotion of a 2-year-old Indian boy, "Ram", 07:30 to 16:00. **a** Matrix of consecutive samples showing traces of locomotion. In each sample the final position of the boy is indicated by a dot. The number of position changes between adjacent square-units during the sample interval was used as a measure of locomotion. Below: sketch of observation area showing features used as reference points. The whole area was subdivided into a grid of 6 × 6 square-units of 10 m × 10 m each. The boy's movements were mapped on the grid during consecutive time

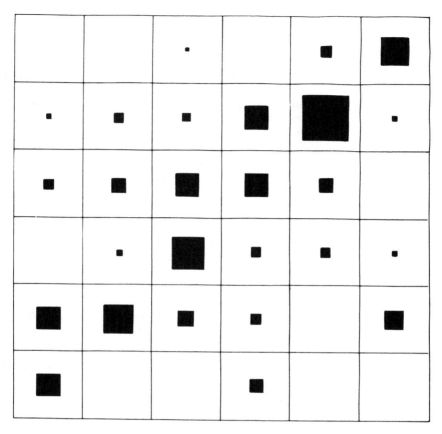

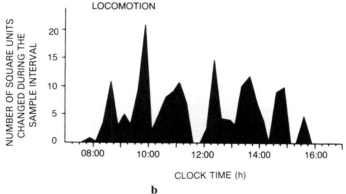

b

samples of 15 min. **b** Periodic variations of locomotion during the day. The matrix above corresponds to that of **a**. The amount of locomotion observed during the consecutive 15-min samples
is represented by black squares. Lengths correspond to the number of position changes between
adjacent square-units in the area of observation. The locomotion of the boy was enhanced with a
period of five samples, resulting in a clustering of phases of enhanced locomotion in the diagonal of
the matrix. Below: time series of locomotion obtained when the measure of locomotion is plotted
against time of day. (Reproduced, with permission, from Gast and Meier-Koll 1987.)

A Field Study in East India

In addition to laboratory studies, I had the opportunity to investigate the daily activities of children living in a small village 24 km from Lucknow in Uttar Pradesh, northwest India. I occupied quarters on the flat roof of a brick-built house belonging to the family of the owner of a small-holding. From this position I had an excellent view over an area of about 60 m × 60 m (Fig. 12.18a). Cottages, a pump, trees etc. were used as markers to localize different activities of observed subjects. All the activities of a boy, "Ram", 2 years old, were timed and recorded in a notebook continually from 07:30 a.m. to 16:00 h. Although the child and other subjects were aware of the observer, they had no idea of the purpose of my presence. With some modifications the observed activities were classified according to the same categories of behavioural change already used in the laboratory studies described above. For example, the observation area was divided arbitrarily into a grid of 6 units × 6 units. The changes of position of the small boy were plotted on a map for consecutive sample times of 15-min duration (Fig. 12.18a). The number of movements between adjacent squares in a time sample was used as a measure of locomotor activity. The variations of locomotion of the boy "Ram" are illustrated in Fig. 12.18b, where the upper matrix corresponds to the maps in Fig. 12.18a. The number of movements is represented by the size of the black squares, the length of edges representing the number of position changes in a given time sample. The maxima of locomotion are clustered on the diagonal of the matrix, indicating a peak of movements about every 75 min. The same periodicity is additionally illustrated by the time series in Fig. 12.18b, in which the frequency of locomotions is plotted against the time of day.

The raw data corresponding to this time series were analysed by means of an autocorrelogram and autospectrum (spectrum of variance). Moving averages were calculated for the daily locomotion and their plot represents a mean level around which the values of locomotion fluctuate (Fig. 12.19a). The autocorrelogram of these fluctuations shows an inherent periodicity. Finally a Fourier transformation of the autocorrelogram yields a spectrum with a clear peak at the frequency corresponding to a period of 80 min (shaded spectrum in Fig. 12.19b).

Besides locomotion, the occurrences of object change and social contacts were recorded simultaneously. The number of object changes was determined in consecutive 15-min intervals (Fig. 12.19a centre) and so was the frequency of social contacts (Fig. 12.19a lower). As with locomotion, autocorrelograms and spectra yield values for dominant periodicities (Fig. 12.19b). The fact that different periodicities were found in the time series of the Indian boy "Ram" seems to contradict the view of a common rhythm modulating locomotor activity as well as the frequency of social contacts and object changes, as was demonstrated in the laboratory study of the older German boy. However, the

---

Fig. 12.19. a Time series of locomotion, object changes and social contacts for "Ram". Light lines, mean trends. **b** Corresponding autocorrelograms (left) demonstrating periodic structures of the time series. Dotted lines, level of confidence: p = 0.05. Corresponding autospectra (right) show prominent ultradian periodicities of 80, 65 and 103 min. (Reproduced, with permission, from Gast and Meier-Koll 1987.)

2 year-old BOY "RAM"

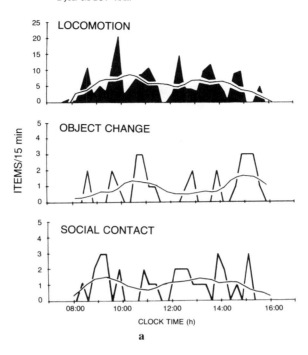

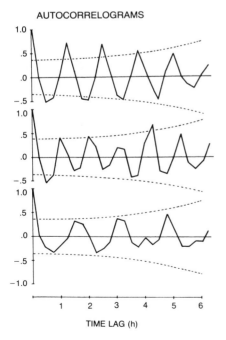

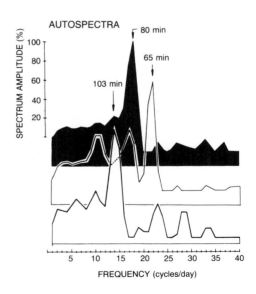

b

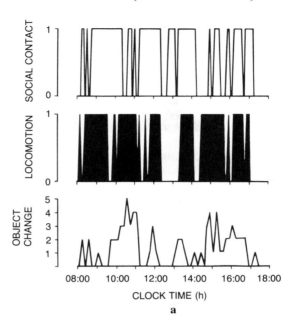

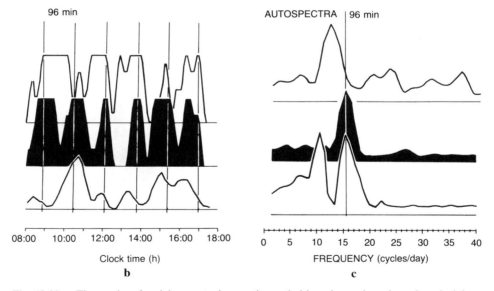

Fig. 12.20. a Time series of social contacts, locomotion and object change (raw data of scoring) for the 4-year-old girl "Omadana" (Kalahari !Ko bushmen). The child's behaviour was observed and documented by Sbrzesny (1976). The protocol was scored by B. Schardl, a co-worker of the author, using a sample interval of 10 min. **b** Smoothed data (5-point filter) and **c** corresponding autospectra. Note an ultradian periodicity of 96 min common for locomotion and object change.

results of autocorrelogram and spectral analysis carry error terms arising from random variation from one period to the other in the different time series. For "Ram", the 65-min period of object change and the 103-min period of social contact could be interpreted as statistical deviations about the significant period of 80 min in locomotion (for more details, see Gast and Meier-Koll 1987).

"Omadana", a Child of the !Ko Bushmen

In her book *Die Spiele der !Ko-Buschleute*, Heide Sbrzesny (from the Max Planck Institute of Human Ethology, Erling-Andechs, Germany) published protocols of the daily behaviour of individual children of the Kalahari bushmen. One of these protocols lists all the activities that a girl, "Omadana", 4 years old, performed in 1 day from 08:10 to 17:15 h (Sbrzesny 1976). Since the events were registered in detail with an accuracy of 1 min, the child's behaviour could be scored and coded according to the categories of locomotion, social contact and object changes. However, since detailed information about spatial distances in locomotor activity were not available, locomotion had to be determined by means of a binary code. This means that locomotion was scored as 1, since the child had moved from one site to another during a 10-min interval; otherwise it was coded as 0. A corresponding binary code had to be used in order to quantify the child's social contact. Time series (raw and smoothed data) of locomotion, social contact and object change were plotted (Fig. 12.20a and b). Finally, after eliminating linear trends of the raw data, the spectra of variance could be computed for each of these behaviour categories (Fig. 12.20c).

The spectra of locomotion and object change yield a common periodicity of 96 min, whereas the dominant spectral peak of the variations in social contact appears to be shifted slightly to a slower periodicity of about 120 min. However, taking the limitation of spectral resolution into account the variations in all three categories of behaviour can be interpreted as manifestations of a common ultradian cycle. This view is supported by the fact that the time series of locomotion, social contact and object changes oscillate nearly synchronously, as already observed in the German boy.

In summary, the few examples of time series analysis of children's play and spontaneous behaviour revealed ultradian behaviour cycles with periods of 1.5–2 h. One may postulate that in distinct states roughly described in terms of exploratory or solitary behaviour confusion will be avoided if they are organized as other partly contradictory states according to several phases of a common cycle. Thus ultradian cycles seem to provide a time frame within which the young child collects manifold experiences and acquires the basic skills necessary for its life.

Social Synchronization of Ultradian Behaviour Cycles in a Village Community of Columbian Indians

Humans, as a biological species, have developed on the sociocultural level of small hunter–gatherer communities over a period of about 2 million years.

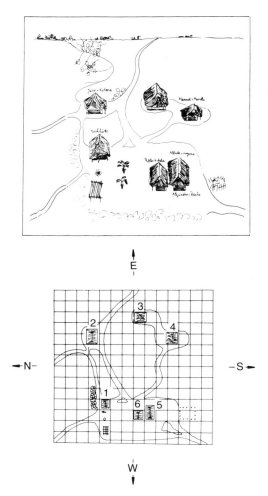

Fig. 12.21. Sketch of the Indian village "Corocito" (above), and its map with an overall projected grid (below). Grid unit: 8 m × 8 m. 1, cooking hut; 2–6, huts of families.

Recent tribes of so-called primitive races still living at the cultural level of the stone age are thought to represent a model of that ancient state of human evolution. Many anthropologists have therefore addressed their efforts to investigating the distinct forms of social organization, economic and cultural patterns which characterize hunter–gatherer populations all over the world (for a review, see Lee and DeVore 1972). All hunter–gatherer communities in different parts of the world live in small groups of about 30 individuals. They exploit natural resources, but do not keep the collected food in stock. Since machines or technical equipment are not available, their survival depends crucially on their mutual cooperation and economy of physical effort. This implies a particular organization of their daily life which enables them to synchronize their individual activities and provide a means for optimal coordination.

Since, as demonstrated in numerous laboratory studies, different individuals carry endogenously regulated ultradian rhythms, modulating their behaviour according to comparable periods, it is possible that different individuals of small communities, such as those of hunters and gatherers, are able to synchronize their activities within a common ultradian time frame. This suggests that ultradian behaviour cycles of individuals living under natural conditions without artificial time schedules may serve to organize daily behaviour patterns for optimal coordination and satisfaction of individual and common demands.

As a first step in the search for ultradian behaviour cycles, in a group of recent hunter–gatherers Barbara Schardl and I studied daily activity patterns in a community of Guahibo-Sicuani, living in the Llaños Orientales of Columbia. Nowadays Guahibos are slightly acculturated and have developed settlements in small villages, but they still exploit nature by hunting and gathering. Beside this ancient form of subsistence economy, traditionally they practise a primitive slash and burn horticulture by clearing the forest with fire and planting yucca palms.

The observed community of about 40 people was settled in a village, Corosito, about 20 km from Puerto Gaitan at Rio Meta. The observations undertaken in March 1986 were focussed on six huts situated in an area of about 100 m × 100 m (Fig. 12.21). Schardl and I cooperated in focussing on predetermined subjects and in recording their activities over 7–9 h each day. In order to define a measure of individual locomotion, a grid of arbitrary 14 × 14 square-units was projected over a map of the village (Fig. 12.21). The area of a square-unit, 8 m × 8 m, corresponds to the mean size of the huts. The traces of individuals observed during consecutive sample times of 10 min were transcribed on to the map. The number of square-units visited during a sample interval could be used as a measure of locomotion. Thus, individual time series of locomotion were obtained (Fig. 12.22). In order to define a group locomotion the time series of several persons observed simultaneously were averaged. After eliminating a trend line from the raw data of group locomotion, corresponding autocorrelograms and autospectra were calculated (Fig. 12.22, right side).

In addition to the individual locomotions, the number of people assembled at special sites, such as family huts or the common cooking hut, was counted for consecutive sample intervals of 5 min. Finally the number of assembled persons was plotted against time of day. Corresponding time series are referred to as "social aggregation", below.

The time series of group locomotion and social aggregation determined for several days were analysed with respect to periodicities using autocorrelograms and spectra of variance. Time series analysis of the raw data of both group locomotion and social aggregation yields common ultradian periodicities in the range 1.5–2 h or subharmonic periods of 3–4 h (Fig. 12.23a and b). As demonstrated by the time series analysis, group locomotion and social aggregation show periodic variations approximately in counter-phase. This means that, for instance, the members of one family spread out periodically and moved separately or in small subgroups in and around the village in order to gather or prepare food. Afterwards they congregated again in their hut for short rests, social activities and meals.

The observed ultradian periodicities were obviously not generated by external time-cues, since the Columbian Indians were neither familiarized to

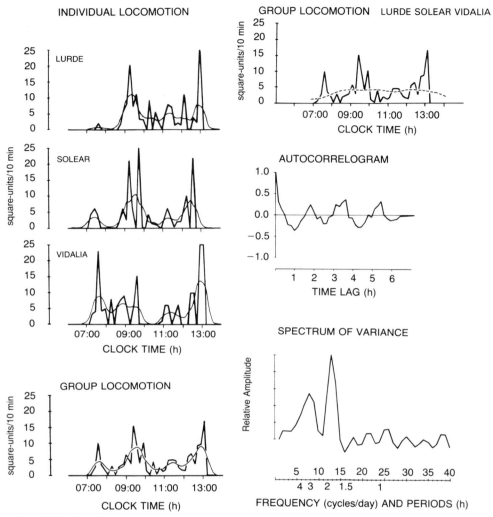

Fig. 12.22. Individual time series of locomotion for three women (Lurde, Solear and Vidalia) cooperating during the day of observation. Group locomotion resulting from the averaged individual time series. Raw data (thick), smoothed data (thin lines). Right side: group locomotion with mean trend (dotted line), autocorrelogram and autospectrum. Note prominent ultradian periodicity of about 2 h.

Fig. 12.23. a and **b** Time series of group locomotion and social aggregation observed during 4 days. Corresponding autocorrelograms and spectra of variance show prominent ultradian periodicities. Note counter-phase between group locomotion and social aggregation. *n*, number of individuals contributing to the observed group locomotion. Calibration of social aggregation: number of persons assembled at special sites (huts 1, 3, 6). Spectral amplitude is in arbitrary units.

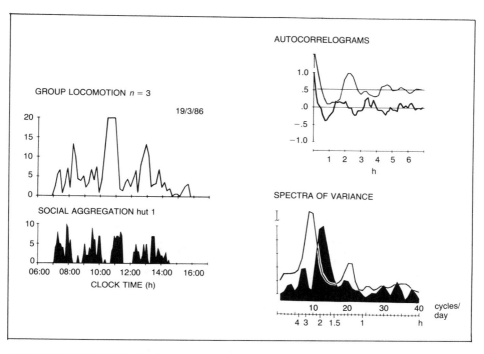

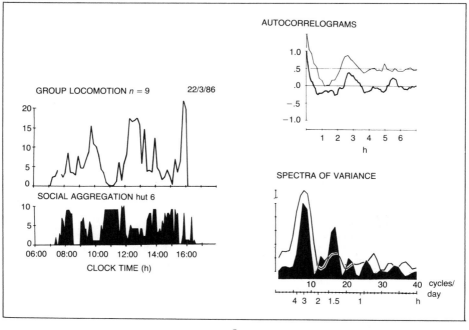

a

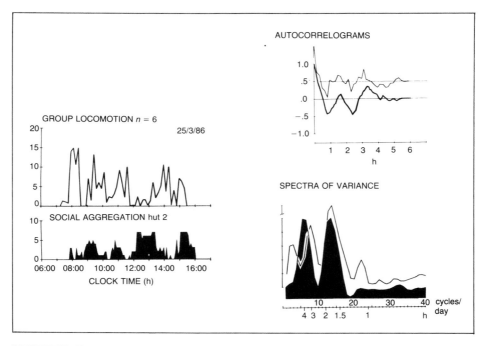

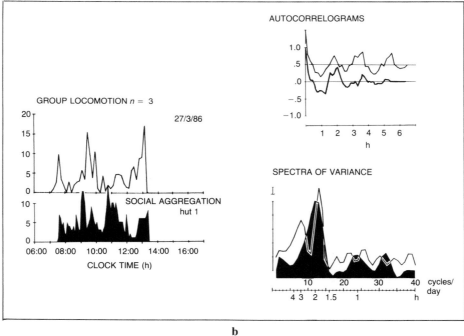

b

Fig. 12.23. (*Continued*)

time measuring by means of watches nor subjected to artificial schedules. The observed periodicities in both group locomotion and social aggregation could result from synchronization of endogenously regulated behaviour cycles of several individuals. From this point of view, the social synchronization of individual ultradian behaviour cycles may be regarded as a mechanism influencing the time structure of daily activities in small communities over and above the cognitive conceptualizations and plannings of their members.

Animal studies have previously provided evidence that social synchronization of ultradian behaviour cycles operates. For instance, synchronization of ultradian rhythms of locomotion, foraging and social activities was described for groups of voles (Daan and Slopsema 1978) and monkeys (Delgado-Garcia et al. 1976; Delgado et al. 1978) even in animals living under natural conditions or with experimentally controlled zeitgebers. As shown in my Indian field study, described above (Gast and Meier-Koll 1987), social synchronization of ultradian behaviour cycles in humans seems to operate in couples and families.

Animal studies suggest that socially synchronized behaviour cycles might serve for survival, as synchrony of foraging or exploratory behaviour of social-living animals provides a greater security for the individual (see Daan and Slopsema 1978). Since humans as a biological species have developed as hunters and gatherers, it might be hypothesized that the social synchronization of individual ultradian behaviour cycles has served a survival strategy for early hominid groups.

Summary

Ultradian cycles of about 4 h underlying the spontaneous sleep–waking behaviour of the human neonate provide a basic time frame within which the newborn infant forms the first impressions of its world. This early pattern of sleep and waking changes as the postnatal development proceeds, and ultradian periods become synchronized by circadian rhythms. In the preschool child ultradian cycles can be detected in time series of free playing and related activities. The child's spontaneous behaviour roughly described in terms of exploratory and solitary phases seems to be organized according to the alternating half-phases of an ultradian 2-h cycle. In both mentally handicapped children and adults, several stereotyped activities were found to be modulated according to ultradian periods. Additionally, the frequency of stereotyped hand waving determined separately for the right and left hand during day-time can reflect ultradian shifts in the relative predominance of one cerebral hemisphere compared with the other. Finally, social synchronization of individual, ultradian behaviour cycles could be observed in a community of Columbian Indians living as hunters and gatherers. This suggests, that social synchronization of ultradian behaviour cycles enables small groups of so-called primitive races to organize their daily life with respect to optimal cooperation and the economy of their physical efforts. As *Homo sapiens* and some of his ancestors have developed on the sociocultural level of small hunter–gatherer communities, one might speculate that socially synchronized behaviour cycles served for the survival of early hominid groups.

References

Aschoff J (1965) Circadian rhythms in man. Science 148:1427–1432

Blackman RB, Tukey JW (1958) The measurement of power spectra. Dover Publications, New York

Bohl E, Meier-Koll A (1991) A mathematical model of a bilateral ultradian oscillator system of human cerebral hemispheres. In: Mosekilde E (ed) Modelling and simulation. Proceedings of the 1991 European Simulation Multiconference, Copenhagen, pp 775–780

Broughton R (1975) Biorhythmic variations in consciousness and psychological functions. Can Psychol Rev 16:217–239

Campbell SS (1987) Evolution of sleep structure following brief intervals of wakefulness. Electroencephal Clin Neurophysiol 66:175–184

Daan S, Slopsema S (1978) Short term rhythms in foraging behaviour of the common vole, Microtus arvalis. J Comp Physiol 127:215–227

Delgado JMR, Del Pozo F, Montero P, Monteagudo JL, O'Keefe T, Kline NS (1978) Behavioral rhythms of gibbons on Hall's Island. J Interdiscipl Cycle Res 9:147–168

Delgado-Garcia JM, Grau C, DeFeudis P, del Pozo F, Jimenez JM, Delgado JMR (1976) Ultradian rhythms in the mobility and behavior of rhesus monkeys. Exp Brain Res 25:79–91

Emde R, Swedberg J, Suzuki B (1975) Human wakefulness and biological rhythms after birth. Arch Gen Psychiatry 32:780–783

Friedman S, Fisher C (1967) On the presence of a rhythmic, diurnal, oral instinctive drive cycle in man. J Am Psychoanal Assoc 15:317–343

Gast B, Meier-Koll A (1987) Biosocial behaviour cycles: a field study in an east indian village community. J Biosoc Sci 19:27–47

Gesell A, Amatruda CS (1945) The embryology of behavior. The beginnings of the human mind. Harper & Brothers, New York

Globus GG, Drury RL, Phoebus EC, Boyd R (1971) Ultradian rhythms in human performance. Perceptual and Motor Skills 33:1171–1174

Hellbrügge T (1974) The development of circadian and ultradian rhythms of premature and full-term infants. In: Scheving LE, Halberg F, Pauly J (eds) Chronobiology. Igaku Shoin Ltd, Tokyo, pp 339–341

Hellbrügge T, Ehrengut-Lange J, Rutenfranz J, Stehr K (1964) Circadian periodicity of physiological functions in different stages of infancy and childhood. Ann NY Acad Sci 117: 361–373

Hiatt JF, Kripke DF (1975) Ultradian rhythms in waking, gastric activity. Psychosom Med 37:320–325

Kleitman N (1963) Sleep and wakefulness. Chicago University Press, Chicago, IL

Kleitman N, Engelmann TG (1953) Sleep characteristics of infants. J Appl Physiol 6:269–282

Kripke DF, Sonnenschein D (1978) A biologic rhythm in waking fantasy. In: Pope KS, Singer IL (eds) The stream of consciousness. Plenum Press, New York, pp 321–332

Lavie P (1976) Ultradian rhythms in the perception of two apparent motions. Chronobiology 3:214–218

Lavie P, Kripke DF (1977) Ultradian rhythms in urine flow in waking humans. Nature 269:142–144

Lee RB, DeVore I (eds) (1972) Man the hunter. Aldine-Atherton, Chicago, IL

McCarley RW, Hobson IA (1975) Neuronal excitability over the sleep: a structural and mathematical model. Science 189:58–60

McCarley RW, Massaquoi SG (1986) A limit cycle mathematical model of the REM sleep oscillator system. Am J Physiol 251:R1011–1029

Meier-Koll A (1979) Interactions of endogenous rhythms during postnatal development: observations of behavior and polygraphic studies in one normal infant. Int J Chronobiol 6: 179–189

Meier-Koll A (1989) Ultradian laterality cycle in stereotyped behavior of a severely mentally retarded patient. Int J Neurosci 47:103–113

Meier-Koll A, Pohl P (1979) Chronobiological aspects of stereotyped motor behaviour in mentally retarded children. Int J Chronobiol 6:191–209

Meier-Koll A, Fels T, Kofler B, Schulz-Weber U, Thiessen M (1977) Basic rest activity cycle and stereotyped behavior of a mentally defective child. Neuropaediatrie 8:172–180

Meier-Koll A, Hall U, Hellwig U, Kott G, Meier-Koll V (1978) A biological oscillator system and the development of sleep–waking behavior during early infancy. Chronobiologia 5:425–440

Morath M (1974) The four-hour feeding rhythm of the baby as a free running endogenously regulated rhythm. Int J Chronobiol 2:39–45

Moses J, Johnson A, Johnson LC, Naitho P (1977) Rapid eye movement cycle is a sleep-dependent rhythm. Nature 265:360–361

Moses J, Naitho P, Johnson LC (1978) The REM cycle in altered sleep–wake schedules. Psychophysiology 15:569–575

Parmelee AH Jr (1961) Sleep pattern in infancy. A study of one infant from birth to eight months of age. Acta Paediatr (Uppsala) 50:160–170

Paul K, Dittrichova J (1975) Sleep patterns following learning in infants. In: Sleep. 2nd Eur Congr Sleep Res, Rome. Karger, Basel, pp 388–390

Rechtschaffen A, Kales A (1968) A manual of standardized terminology and scoring system for sleep stages of human subjects. Public Health Service US Government Printing Office, Washington, DC

Sbrzesny H (1976) Die Spiele der !Ko-Buschleute. Piper, Munich Zurich

Sterman MB, Hoppenbrouwers T (1971) The development of sleep–waking and rest–activity patterns from fetus to adult in man. In: Sterman MB, McGinty DJ, Adinolfi AM (eds) Brain development and behavior. Academic Press, New York, pp 203–227

Chapter 13

Ultradian Cycles in Sleep Propensity: Or, Kleitman's BRAC Revisited

P. Lavie

Kleitman's BRAC

In 1961, and then in 1963, Kleitman speculated that the cyclic variations in brain activity during sleep manifested as the rapid eye movement–non-rapid eye movement (REM–NREM) cycles are only a nocturnal fragment of an ongoing 24-h cycle which he termed the "basic rest–activity cycle" (BRAC). Although Kleitman was somewhat vague about the properties of the BRAC, he predicted that it would be manifested in cyclic variations in alertness. In his words: "This cycle is obscured during wakefulness by the great surge of cortical activity, but suggestions of its presence may be discerned in daytime oscillations in alertness, the often irresistible drowsiness after a big meal, and the relief that some persons get from brief catnaps" (Kleitman 1961, p. 361). No doubt Kleitman's BRAC concept captured the imagination of many. This is evident from the large number of studies searching for ca 1.5-h cycles, which is the dominant periodicity of the REM–NREM sleep cycle in a variety of physiological, behavioural and endocrine functions (for a review see Lavie 1982). A large number of these studies have shown ultradian cycles with periodicities centred around this privileged periodicity, in support of Kleitman's hypothesis. Such cycles were shown in various electroencephalographic (EEG) frequency bands such as α (8–9 Hz; Kripke and Sonnenschein 1978; Gertz and Lavie 1983) and δ (0.5–3 Hz; Kripke 1972), or the total power of the EEG (Manseau and Broughton 1984), in the tendency to fall asleep during the day (Lavie and Scherson 1981), and in autonomic indices of arousal such as pupillary diameter and reactivity to light (Lavie 1979), respiratory rate (Horne and Whitehead 1976) and heart rate (Orr et al. 1976).

These findings were corroborated by findings of ultradian cycles with similar periodicities in behavioural responses such as reaction time (Orr et al. 1976), the intensity of visual illusions (Lavie et al. 1974), and in the accuracy of motor coordination (Gopher and Lavie 1980).

However, there is as yet no firm evidence that the ultradian cycles observed during wakefulness are regulated by the same underlying neural mechanism that is responsible for the REM–NREM cycle. To further complicate the picture, it has become evident that physiological functions unrelated to arousal also show cyclicities of approximately 1.5 h. Such cycles were shown in urine flow in humans (Lavie and Kripke 1977), and in gastric motility in waking (Hiatt and Kripke 1975) and sleeping humans (Lavie et al. 1978), as well as in other species (Grivel and Ruckebusch 1972). Elsewhere, Lavie and Kripke (1981) and Lavie (1982) proposed that ultradian cycles of approximately 1.5 h/cycle represent a multioscillatory phenomenon rather than the activity of a single oscillator.

This chapter re-examines the sleep REM–NREM cycle and the ultradian cycles in alertness during wakefulness in view of the experimental data obtained in my laboratory in recent years. It will be argued that, in spite of the fact that REM–NREM cycles and the waking cycles in alertness most probably reflect independent processes, they nevertheless evolved along similar evolutionary lines subserving similar regulatory functions which complement each other.

REM–NREM Cycle: A Concerted Tide of Brain Activities

Although there is general agreement regarding the cyclic nature of the REM–NREM sequences, the nocturnal component of the BRAC, actually this cycle is far from being precise and stationary. Data accumulated over the last 30 years have shown that the REM–NREM cycle represents intricate and coordinated changes in multiple brain subsystems, all obeying the same basic cyclicity of approximately 1.5 h/cycle. Occulomotor activity, muscle tonus, EEG activities, autonomic dominance, brain energy utilization, and the subjective experience of dreaming, all oscillate nightly in a concerted cycle. Occasionally, however, disharmony occurs when some of the REM components are "spilled over" to neighbouring sleep stages. The opposite may also occur: occasionally REM sleep is fragmented by brain activities belonging to NREM sleep stages. But these irregularities cannot refute the fact that REM–NREM sequences alternate cyclically; moreover, in view of its complexity, the regularity of the cycle is rather impressive.

An important feature of the REM–NREM cycle is its phase control. In normals retiring to sleep around their habitual sleep time, the phase of the REM–NREM cycle is determined to a great extent by the time of falling asleep. Thus, the first REM period occurs approximately 90($\pm$20) min after falling asleep. Subsequent REM periods recur every 90–80 min thereafter. Although there are some exceptions where the phase control of the REM–NREM cycle is not functioning properly, such as in narcoleptics who have REM periods at sleep onset, or in patients with major depression who also have abnormally short REM latencies, ca 1.5 h REM latency is considered to be a robust marker of adult sleep.

In 1975, McCarley and Hobson proposed a model of REM sleep that was physiologically based on the activity of two neuronal populations: "REM-on" cells, which appear to promote REM sleep; and "REM-off" cells, which appear to inhibit REM sleep. They proposed that the periodic occurrence of REM sleep might be a function of a reciprocal interaction between the populations of these two cells. The Lotka–Volterra equations (see McCarley and Hobson 1975) were used to describe this interaction. The model was later refined by McCarley and Massaquoi (1986) as a limit cycle model which also included a consideration of the circadian effects on the REM generator. It should be pointed out, however, that this model relates only to the appearance of the traditional markers of REM sleep, i.e. rapid eye movements, low voltage fast EEG activity, loss of muscular tonus, and the appearance of high voltage, monophasic ponto-geniculate–occipital waves (in the cat). As is shown below, it is not at all clear that all REM components are generated by the same oscillator.

REM Sleep: A Single or Multioscillatory Phenomenon?

As described above, REM sleep is manifested by a variety of physiological phenomena all occurring in synchrony. Is the REM–NREM cycle run by a single "clock" having different "hands", or does it reflect the activity of multiple clocks all beating in synchrony? There is compelling evidence suggesting that at least some of the REM components represent the temporal confluence of multiple oscillations all sharing the same periodicity. The following case history exemplifies the complexity of the control mechanisms of the REM–NREM cyclicity and its heterogeneity.

Penile erection is one of the most prominent and marked changes associated with REM sleep (Fisher et al. 1965). Erections accompanied some 80%–90% of all REM periods and, when measured independently of other REM variables, they clearly displayed a 90-min periodicity. In fact, the 90-min periodicity in erections was described 9 years before the discovery of REM sleep (Ohlmeyer et al. 1944). In spite of the near-perfect synchronization between the REM–NREM cycle and the penile erection cycles, under certain circumstances they may be completely dissociated from each other. In recent years, in my laboratory, we have followed a unique patient who suffered a highly localized brainstem lesion which damaged his executive REM mechanism (Lavie et al. 1984b; Lavie 1990). During multiple laboratory sleep recordings, this patient was found to have astonishingly low amounts of REM sleep (2%–4%). Nevertheless, the patient showed an intact 90-min periodicity in penile erections (Fig. 13.1). Episodes of erections occurred at the expected times of the REM periods without any concomitant REM components. It is also interesting to note that this patient shows no behavioural deficits and conducts a normal working life – in spite of the minute amounts of REM sleep. This case clearly indicates that because two measures share the same periodicity and are synchronized with each other it does not necessarily mean that they are regulated by the same oscillatory mechanisms. Most probably several neural generators participated in the regulation of the REM–NREM cycles, all sharing the same dominant periodicity of ca 1.5 h.

Besides the problem of the singularity of the control mechanism of REM sleep, there is a question concerning the nature of the REM–NREM cycle which is of no lesser importance: is the REM–NREM cycle a sleep-dependent or a sleep-independent cycle? Any attempt to examine the BRAC model must deal first with this question.

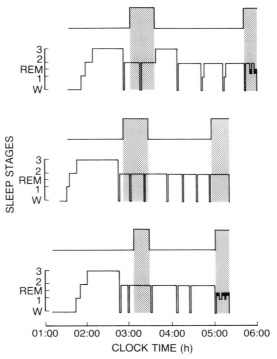

Fig. 13.1. Ultradian cyclicity in penile erections in a patient with almost total absence of *REM* sleep. Note that each night the first episode of erection (*stippled area*) occurred at approximately the same time that REM sleep would be expected i.e. 90 min after sleep onset. (Reproduced, with permission, from Lavie 1989b.)

REM–NREM Cycle: A Sleep-Dependent or a Sleep-Independent Phenomenon?

Whether the oscillator regulating REM sleep functions only during sleep and ceases to function during wakefulness is the subject of an on-going debate in the literature (Schulz et al. 1975; Moses et al. 1977; McPortland and Kupfer 1978; Ursin et al. 1983; Carmen et al. 1984; Lavie 1987). This question touches upon issues of the existence of a waking counterpart to the REM–NREM cycle and upon the nature of the sleep cycle itself. Most studies attempting to investigate this question did so by experimental interruptions of the sleep cycle at different phases with respect to the REM–NREM cycle, and observing the effects of the interruption on the appearance of subsequent REM periods. Although it was shown that sleep interruptions lengthened subsequent REM–NREM cycles (Brezinova 1974; Gaillard and Tuglular 1976), Schulz (1985) demonstrated that the lengthening of the cycle is proportional to the length of the NREM period before the interruption. Schulz interpreted these results as indicating that a REM-inducing process is taking place inbetween two successive REM episodes. This process may be halted temporarily but not destroyed by the sleep interruption.

Experimental results from our laboratory provided evidence that the oscillator regulating REM periodicity during sleep is not halted by short periods of wakefulness. In a series of studies we utilized ultrashort sleep–wake cycles in order to investigate circadian and ultradian variations in sleep propensity (Lavie and Scherson 1981; Lavie and Zomer 1984; Lavie 1986; Lavie and Segal 1989, Lavie and Weler 1989; Lavie and Zvuloni 1992). By this technique, subjects were instructed to attempt either to fall asleep or to resist sleep for 5 or 7 min every 20 min. The total amount of sleep obtained in each trial was used to construct the 24-h sleep propensity function (SPF). In recent experiments we preferred the 7/13 sleep–wake ratio (see p. 289) because it is close to what is considered to be the optimal sleep/wake ratio of 8/9, while at the same time it does not allow a significant accumulation of sleep. Such an accumulation could obscure the shape of the SPF.

Besides the information about the shape of the SPF and the occurrence of ultradian cycles in sleep propensity obtained by the 7/13 paradigm which will be described shortly, an unexpected experimental outcome of the studies utilizing the 7/13 paradigm was the occurrence of multiple REM periods within the 7-min trials. These occurred in about half of the subjects, particularly during the late night–early morning period. Although there is no immediate explanation of why sleep fragmentation induced REM periods resembling the "sleep onset REM periods" in narcolepsy, analysis of the temporal distribution of trials containing REM allowed examination of the properties of the REM generator (Lavie 1987). Fig. 13.2 shows the occurrence of REM sleep under such conditions in an "REM-prone" subject. As can be seen, the first REM episode appeared at approximately 02:00 h, which was followed by two subsequent REMs at 70–80-min intervals thereafter. Analysis of inter-REM intervals of more than 42 series of data, each constructed from 72 consecutive 7-min trials, revealed a dominant 60–80-min periodicity. Identical periodicity was also found in narcoleptic patients tested under a similar 7/13 paradigm after a night of normal sleep in the laboratory (Lavie 1991).

These results led to the conclusion that the oscillator regulating REM continues to function during short periods of wakefulness. This lends credence to the notion that under certain conditions there may be a continuation of ultradian cycles across the sleep/wakefulness boundary.

REM Sleep – A Gating Mechanism

REM sleep, no doubt, is an outstanding physiological phenomenon. There is no other comparable physiological phenomenon which has been described in such great detail, and whose function is still an enigma. The possible function of REM sleep has attracted the imagination of many. Memory consolidation, regulation of effect and motivation, and genetic programming have all been raised as possible candidates for REM function. However, none of these theories is sufficiently comprehensive to account for all the diverse manifestations of this sleep state. Like the blind people who examined the elephant and saw in it different animals depending on the part they were touching,

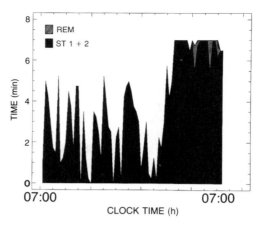

Fig. 13.2. Periodicity in the occurrence of REM sleep in a normal young adult tested under the 7/13 sleep–wake paradigm (raw data). ST, stages.

the proposed function of REM sleep is greatly dependent upon the specific REM aspects that were examined. There is nothing wrong with finding several functions of REM sleep which are seemingly independent of each other. In many cases, evolutionary processes utilize existing structures in widely disparate ways. I would like to propose that REM sleep has yet another function which utilizes its unique physiological characteristics. This function, I believe is linked with Kleitman's BRAC concept.

The periodic occurrence of REM provides the organism with privileged gates which facilitate a smooth transition from sleep to waking. Elsewhere I have elaborated some of the supportive evidence for this assertion (Lavie 1985, 1989a, 1992). Briefly, it was shown that subjects in isolation who rely only on internal cues to wake up from sleep, woke up significantly more often from REM sleep than from NREM sleep (Weitzman et al. 1980). Likewise, subjects requested to wake up from sleep at predetermined times without the aid of an alarm clock, significantly woke up more often from REM sleep (Lavie et al. 1979; Zepelin 1986). Furthermore, there is evidence that awakening from REM sleep has a distinct evolutionary advantage over awakening from NREM sleep, because it is associated with an activation of the right hemisphere's orienting mechanisms (Gordon et al. 1982; Bertini et al. 1984; Lavie et al. 1984a). Thus, upon awakening from REM the organism can immediately orient itself in space and avoid maladaptive behaviours. This view of REM sleep as a gating mechanism to ensure a smooth transition from sleep to wakefulness can be seen as an enlargement upon Snyder's (1966) sentinal theory of REM sleep. Snyder viewed REM sleep as a privileged state during which the organism can search the environment for possible predators. In the following sections, I argue that the REM gating mechanism is complemented by a similar mechanism during wakefulness which facilitates the transitions in the reverse direction, from wakefulness to sleep.

"Sleep Gates"

Although we tend to disregard introspection and personal experience as unscientific, personal experience can, nevertheless, provide valuable information about the nature of falling asleep. In spite of the fact that most people maintain a 24-h sleep–wake cycle which in the adult comprises a single sleep episode and a single waking episode, on many occasions sleepiness and readiness to sleep are experienced more than once a day. Most people feel drowsy during midafternoon and, if permitted, fall asleep much faster at that time of the day than at any other. However, 2–3 h later, readiness to sleep diminishes whether or not one has slept. Such transient feelings of "being ready to sleep", which rapidly disappear, may also come at other times of the day. Some people describe it metaphorically as: "Once my sleep gate is open I must take a nap, even a brief one. This charges me for hours." These brief "cat naps" – lasting for no more than a few minutes but offsetting sleepiness and fatigue for prolonged periods of time – were one of the clues that led Kleitman in 1961 to the BRAC hypothesis. In contrast to nocturnal sleep periods, "cat naps" occur abruptly without any preceding sensations of impending fatigue.

If sleep–wake cycles are regulated by a 24-h clock, why do these diurnal sleepiness crises occur? I return to this question at the end of this chapter.

Multiple Sleep Latency Tests and Sleep Propensity

Experimental data support personal experience by demonstrating that sleep propensity systematically varies across 24 h (for a review, see Broughton 1989). If prior wake-time is held constant, the 24-h course of sleep propensity (SPF) can be plotted. This, in fact, has been attempted by a number of investigators using a variety of experimental techniques. These included methods such as continuous or frequent sampling of EEG and frequent measurements of sleep latency during the day. The best known and most widely used method is the multiple sleep latency test (MSLT), first introduced by Carskadon and Dement in 1977. The MSLT measures the speed of falling asleep in a standard setting. The first test of the MSLT is usually conducted 2 h after the subject has woken up from nocturnal sleep, and then at 2-h intervals until 20:00 h. Although the MSLT provides reliable data on diurnal sleep latencies, and its clinical validity has been convincingly demonstrated (Carskadon and Dement 1979; Richardson et al. 1978; Mitler et al. 1982), a sampling frequency of once every 2 h is too slow to provide an accurate description of the diurnal variations in sleep propensity. The MSLT will miss any cycle in sleep propensity faster than six per day.

Continuous recording techniques, on the other hand, lack standardization, and their results are heavily dependent on subjects' activity and level of tonic arousal (for further discussion, see Lavie 1989b). Furthermore, since in normal subjects the act of falling asleep is under voluntary control, continuous recording techniques usually rely on indirect measurements of sleep propensity. To overcome these problems we have developed the ultrashort sleep–wake paradigm, described above, which allows measurements of diurnal variations in

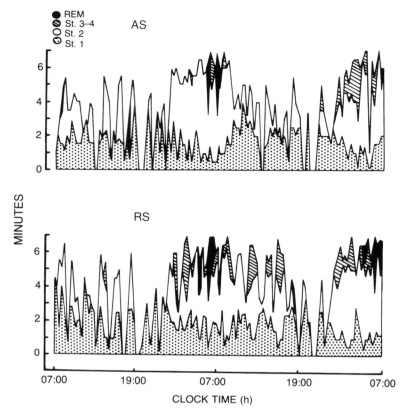

Fig. 13.3. Forty-eight hour sleep propensity functions of a normal young adult tested with the 7/13 sleep–wake paradigm under the attempting sleep (*AS*) and resisting sleep (*RS*) conditions. Note the prominent ultradian cyclicities during the periods 07:00 to 19:00 h on both days of the experimental periods and in both experimental conditions. *St.*, stage.

sleep propensity directly. Testing subjects under this paradigm indeed showed ultradian cycles in sleep propensity with a periodicity of approximately 90–120 min (Lavie and Scherson 1981; Lavie and Zvuloni 1992). Before these cycles are described, it should be made clear that, in comparison with the impressive changes in the state of brain activities associated with the circadian sleep–wake cycle, the ultradian cycles in sleep propensity are of a relatively low amplitude and on many occasions are unstable both with respect to frequency and/or with respect to phase. Fig. 13.3 depicts examples of ultradian cycles in sleep propensity obtained by the 7/13 paradigm.

The existence of sleep propensity cycles during the day, which is congruent with Kleitman's proposal, leads to the discussion of the relationships between the REM–NREM cycles and the sleep propensity cycles. Are they both part of an ongoing cycle which continues uninterrupted from sleep to waking as envisaged by Kleitman? Or, are perhaps the sleep and waking cycles regulated independently? If, for instance, it could be shown that the phase of the waking cycle is locked to the phase of the preceding REM–NREM cycle, then it would support the possibility of an ongoing cycle, presumably with a common origin. Preliminary experimental evidence indicated that the phase of the waking ultradian rhythms may have a consistent phase relationship with the preceding REM–NREM cycle (Lavie and Zomer 1984). Awakening from REM sleep was shown to be associated with a higher level of arousal than awakening from NREM sleep. These carry-over effects, however, were short-lasting and became progressively unstable. Other studies showing differential effects of awakening from REM and NREM sleep on alertness and performance were carried out for periods that were too short to allow any conclusion concerning the phase continuity of the sleep and waking cycles (Lavie 1974; Lavie and Sutter 1975). Furthermore, the fact that awakening from REM and NREM sleep was found to be associated with differential activation of the two cerebral hemispheres (Gordon et al. 1982; Bertini et al. 1984; Lavie et al. 1984a) placed these earlier observations within the context of REM sleep as a gating mechanism.

Recent results from our laboratory have provided evidence, however, that the phase of the ultradian cycles in sleep propensity may be determined by the starting time of the experimental paradigm itself. These results are now described in greater detail.

Living on a 20-Min "Day"

Evidence for the dependency of the phase of the ultradian cycles in sleep propensity on the starting time of the experimental paradigm was obtained in a study where subjects were tested with the 7/13 sleep–wake paradigm for 48 consecutive hours. Details about the design of this study can be found in the article by Lavie and Zvuloni (1992), therefore they are described only briefly here. Eight healthy young male adults who regularly slept 7–8 h per day participated in two 48-h experimental periods. In each of them, subjects arrived at the sleep laboratory at 21:00 h on the evening before the start of the 7/13 sleep–wake paradigm. They were fitted with electrodes to measure EEG, electrooculographic and electromyographic activity, and slept in the laboratory from approximately 23:00 until 06:40 h. At 06:40 h, they were wakened up from sleep, and at 07:00 h they started a 7/13 sleep–wake paradigm that lasted for 48 consecutive hours. As described before, under this paradigm, every 20 min subjects were instructed to lie in bed with eyes closed and to attempt to fall asleep. In total, the experimental period comprised 144 cycles of 7-min sleep attempts followed by 13-min intervening wake periods outside the bedroom.

The second experimental period was conducted at least 10 days after the first. Here, too, subjects spent the night before the 7/13 paradigm asleep in the laboratory. At 07:00 h they started a 48-h schedule of 7 min resisting sleep attempts, 13 min awake outside the bedroom. The specific instructions to the subjects were to lie quietly awake in bed for 7 min with eyes closed and to resist sleep. Each 7-min trial of attempting or resisting sleep was scored for sleep stages 1, 2, 3–4, REM and wake, according to conventional criteria.

Synchronized Cycles in Sleep Propensity

All subjects successfully completed the two 48-h experimental periods. In both the AS (attempting sleep) and RS (resisting sleep) conditions the temporal structure of sleep propensity revealed a pronounced 24-h cycle, on which faster variations were superimposed. A median smoothing procedure was used to emphasize the 24-h cycles. This procedure is aimed at removing the fast variations from the data (Figs. 13.4 and 13.5). It is evident from the Figures that there was a great similarity between the 24-h structure of sleep propensity in the AS and RS conditions. The nocturnal sleep "gates" i.e. the first nocturnal trial during which subjects were able to initiate sleep very easily, indicated in the Figures by a steep increase in sleep propensity, opened at exactly the same times on both days of the study and in both experimental conditions.

What about the ultradian variations? These are first exemplified in a single subject and then for the entire group. Fig. 13.3 (see above) presents the 48-h data of a single subject for the AS and RS conditions. In both conditions there were prominent ultradian variations in sleep propensity during the first day

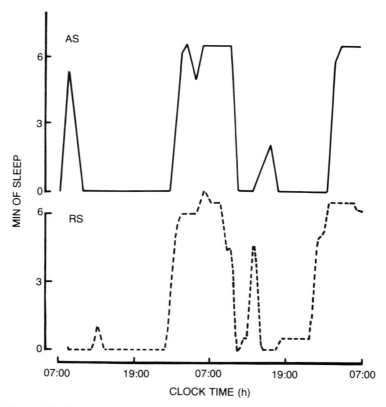

Fig. 13.4. Forty-eight hour sleep propensity functions for the *AS* and *RS* conditions smoothed by a median smoothing procedure.

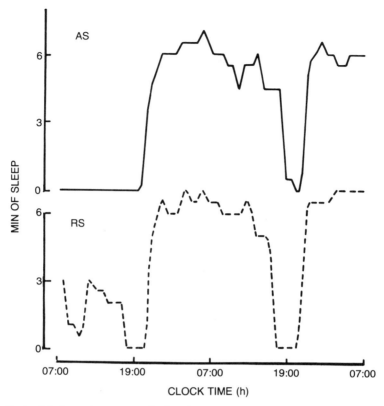

Fig. 13.5. Forty-eight hour sleep propensity functions for the *AS* and *RS* conditions smoothed by a median smoothing procedure.

(07:00 to 19:00 h). These were mostly manifested in prominent increases in sleep stage 2.

The ultradian variations were masked during the night period by the high levels of sleepiness, but they reappeared during the second day of the study, particularly during the period 09:00 to 19:00 h. The relationships between the ultradian variations in sleep propensity in the AS and RS conditions were investigated by a cross-correlation analysis. First, cross-correlations were calculated for the first 12 h of the experimental periods (36 data points), during which the ultradian variations were most pronounced (Fig. 13.6). This revealed a significant in-phase relationship between the two conditions, with a common periodicity of 120 min. Fig. 13.7 presents the 72-h lag cross-correlation function based on the entire 48-h time series of the same subject (144 data points). This revealed both the 24-h component and the 2-h component which were both in-phase across the two experimental conditions. A similar pattern emerged when data from all eight subjects were averaged before the cross-correlational analysis. Figs. 13.8 and 13.9 present the averaged SPFs for the two experimental conditions and the corresponding cross-correlation function. Although in comparison with the cross-correlation function of the single subject pre-

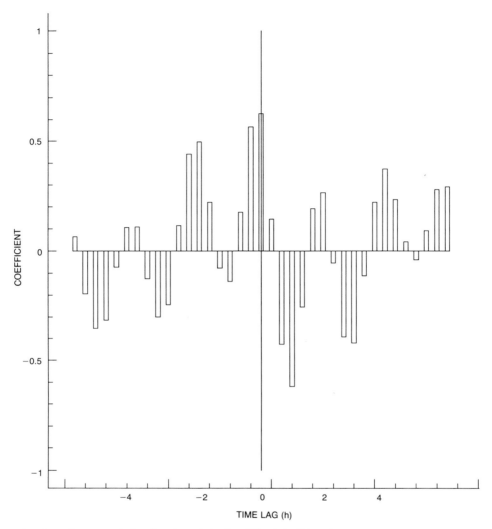

Fig. 13.6. Cross-correlation function of the first 12 h of the RS and AS sleep propensity functions presented in Fig. 13.3.

sented in Fig. 13.7 the magnitude of the ultradian component in the averaged data is relatively small, it is nevertheless clearly evident.

The results of the cross-correlational analysis yield two important conclusions. First, in agreement with the visual impression, it shows that sleep propensity under experimental conditions of AS and RS was modulated by both a 24-h and a 2-h component and, second, that subjects were synchronized with each other with respect to both components. The finding that subjects were synchronized with respect to the 24-h component is hardly a surprise in view of the fact that all were students having approximately the same daily

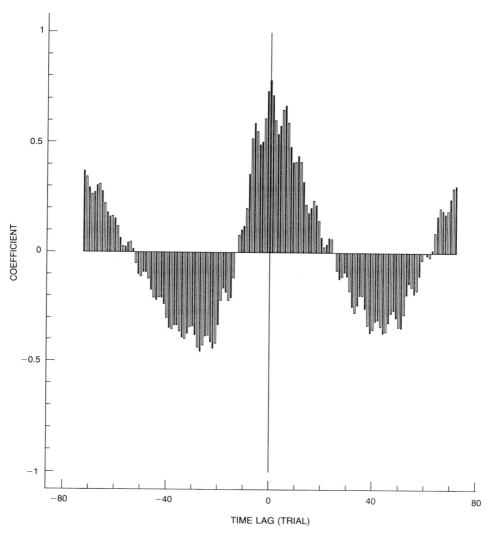

Fig. 13.7. Cross-correlation function of the 48-h RS and AS sleep propensity functions presented in Fig. 13.3.

routines. The synchronization among subjects with respect to the 2-h component, however, is unexpected. As mentioned before, previous results did not provide strong evidence for a consistent phase relationship between the REM–NREM cycle and the ultradian variations in arousal. If such relationships exist, then in order to explain the present findings it must be assumed that all subjects woke up from sleep at precisely the same phase of the REM–NREM cycle. But examination of the nocturnal sleep patterns revealed that this was not the case; different subjects woke up from sleep at different phases of the REM–NREM cycle.

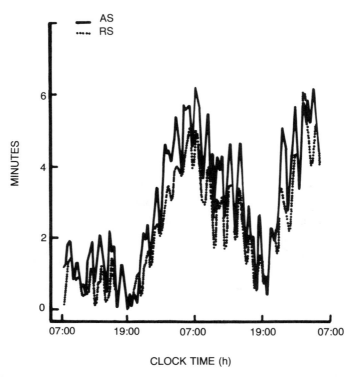

Fig. 13.8. Average 48-h sleep propensity function ($N = 8$) for the *AS* and *RS* conditions.

At least two other possible explanations can be invoked for this surprising finding. First, awakening from sleep itself, regardless of the specific sleep stage, acted as the synchronizing trigger. This can be seen as analogous to the synchronization of the REM–NREM cycle by the process of falling asleep. Since subjects were awakened at the same time, their ultradian rhythms which were triggered by the awakening process were synchronized. Second, the phase of the ultradian cycles was determined by the start of the testing procedure itself. In order to decide between these possibilities, data of subjects awakened from sleep at different times should be analysed. Such a study had been performed in our laboratory previously (Lavie and Scherson 1981). In that study, which was the first to utilize an ultrashort sleep–wake cycle, subjects were given 5-min sleep attempts every 20 min from 09:00 until 19:00 h. Since subjects spent the night before the start of the study asleep at their homes, wake-up time varied between subjects. Fig. 13.10 presents the mean of total sleep time calculated across all subjects, and the same data after removing the quadratic trend. The autocorrelation function of the detrended data revealed a 2-h component. The existence of a 2-h component in the average data indicates that subjects were synchronized with each other. Otherwise, the phase dispersion among subjects would have resulted in the disappearance of the cycle in the averaged data.

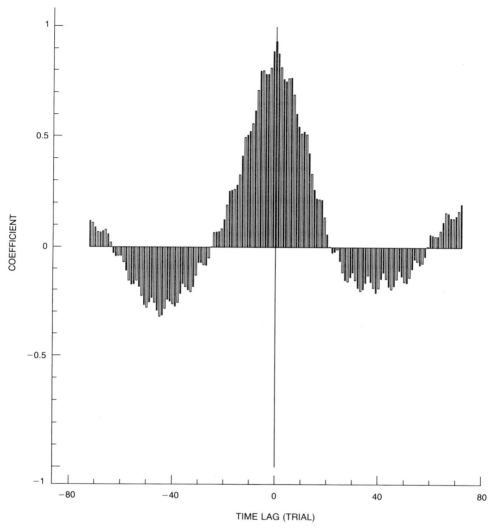

Fig. 13.9. Cross-correlation function of the average 48-h RS and AS sleep propensity functions presented in Fig. 13.8.

Increasing the Sleep Pressure

In both studies described above sleep pressure was low because subjects were not sleep deprived before the start of the study. Increasing the sleep pressure resulted in the disappearance of the 2-h cycle (Lavie 1986). This is exemplified in Fig. 13.11, which depicts the averaged data of 36 subjects tested with the 7/13 paradigm after a night of sleep deprivation. The important features of this Figure are the midafternoon increase in sleepiness, the decrease in

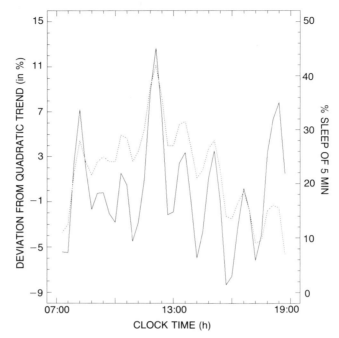

Fig. 13.10. Mean total sleep time of 11 subjects investigated with a 5/15 sleep–wake paradigm from 09:00 to 19:00 before (*broken line*) and after (*continuous line*) removing the quadratic trend.

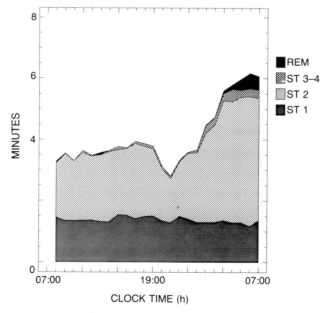

Fig. 13.11. Average sleep propensity function of 36 subjects tested with the 7/13 sleep–wake paradigm after an "overnight" sleep deprivation (smoothed by a three-point moving average). ST, stage(s).

sleepiness during the evening hours ("forbidden zone" for sleep) and the steep increase in sleepiness at approximately 23:00 h (sleep gate). This general pattern was found in more than 70 subjects tested with the 7/13 sleep–wake paradigm after a night of sleep deprivation. It should be noted, however, that in spite of the increasing sleep pressure, some individual subjects still showed prominent ultradian cycles, but these were the exception rather than the rule.

The Sleep and Waking Ultradian Cycles – A Unifying Concept

The existence of ultradian cyclicities in sleep propensity which appeared to be synchronized across individuals takes us back to Kleitman's BRAC. To recapitulate, Kleitman hypothesized that the ultradian REM–NREM sleep cycles are only a fragment of a 24-h cycle which, during wakefulness, is manifested as fluctuations in cortical alertness, and during sleep as the REM–NREM cycle. Although the accumulated findings cannot attest to a common origin for the sleep–wake cycles, they nevertheless support the existence of orderly fluctuations in cortical alertness during sleep and wakefulness subserving a similar function.

I propose that the ultradian cycles in sleep propensity can be best described as reflecting cyclic fluctuations in the balance between arousal-producing mechanisms and brainstem or diencephalic sleep-promoting structures. The balance between the two opposing systems determines cortical alertness in much the same way that the balance between "on" cells and "off" cells in the brainstem determines the balance between REM and NREM sleep. The function of the waking cycles is to provide multiple gates during which the transition from waking to sleep is greatly facilitated. These are complementary to the REM gating mechanism in sleep which facilitates transition in the opposite direction and is described above. Such a system ensures that, if the need should arise, sleep can be initiated at any one of many times across the day. Obviously, in order to be adaptive such a system should be highly flexible and highly responsive to environmental stimuli. This dual gating mechanism provides the sleep–wake system with great flexibility which is of obvious adaptive value.

Although Kleitman was not so explicit about what determines the phase of the BRAC, it appears from his writings that he envisaged, at least implicitly, a continuation between the sleep and waking components of the cycle. The present results indicate that the phase of the waking cycle was dependent upon the start of the experiment. This was evident from the fact that different subjects were synchronized with each other although testing occurred on different days. What could be the explanation for such a synchronization? It is possible that the ultradian cycles in sleep propensity reflect an inherent responsivity cycle of neural structures involving the regulation of sleep propensity. Thus, when subjects repeatedly attempt to fall asleep, the recruitment of sleep-promoting structures can be done only intermittently. Periods during which somnogenic structures are activated are followed by refractory periods during which the sleep-promoting structures are unresponsive. The fact that the same cyclicity was observed under both the attempting and the resisting sleep conditions may indicate that such intermittent recruitment of somnogenic structures can be done either actively or passively by sleep-inducing environmental conditions.

It can be concluded therefore that the temporal structure of both sleep and wakefulness has evolved evolutionarily in such a way that maximizes flexibility of the sleep–wake cycle by ensuring smooth transitions from one state to another. During both states of existence there is an ultradian gating mechanism which facilitates these state transitions. This cyclic process rightfully deserves to be termed the basic rest–activity cycle.

Summary

This chapter examines Kleitman's basic rest-activity cycle (BRAC) model in view of data accumulated in recent years on the sleep REM–NREM cycles and on the waking ultradian cycles in alertness. It is argued that the REM–NREM cycles represent a temporal confluence of changes in diverse brain systems which are potentially dissociable from each other. It is still unclear whether all REM components subserve the same function. Likewise, the 1.5-h ultradian periodicities during waking are a multioscillatory phenomenon which includes, besides cycles in alertness, ultradian cycles in kidney excretions and gastrointestinal motility which appear to be independent of each other. The waking ultradian cycles in alertness are shown in a variety of arousal-related indices and can be best described as reflecting periodic variations in the balance between the activity of brain structures concerned in sleep and waking. The cycles disappear when sleep pressure increases. Analysis of the phase relationships between the ultradian cycles in alertness of different individuals, or the relationships between cycle phases of the same individual tested at different times, led to the surprising conclusion that the phase of the ultradian cycles in alertness is locked to the start of the testing period. This finding is interpreted in the light of the proposal that the sleep and waking ultradian cycles function as a gating mechanism, providing the organism with smooth transitions from sleep to wakefulness and vice versa.

References

Bertini M, Violani C, Zoccolotti P, Antonelli A, Di Stefano L (1984) Right cerebral activation in REM sleep: evidence from unilateral tactile recognition test. Psychophysiology 21:418–423

Brezinova V (1974) Sleep cycle content and sleep cycle duration. Electroencephalogr Clin Neurophysiol 36:275–282

Broughton RJ (1989) Chronobiological aspects and models of sleep and napping. In: Dinges D and Broughton R (eds) Napping: biological, psychological and medical aspects. Raven Press, New York, pp 71–98

Carmen GJ, Mealey L, Thompson ST, Thompson MA (1984) Patterns in the distribution of REM sleep in normal human sleep. Sleep 7:347–355

Carskadon MA, Dement WC (1977) Sleep tendency: an objective measure of sleep loss. Sleep Res 6:200

Carskadon MA, Dement WC (1979) Effects of total sleep loss on sleep tendency. Percept Mot Skills 48:495–506

Fisher C, Gross J, Zuch J (1965) Cycle of penile erection synchronous with dreaming (REM) sleep. Arch Gen Psychiatry 12:29–45

Gaillard JM, Tuglular I (1976) The orthodox-paradoxical sleep cycle in the rat. Experientia 32:718–719

Gertz J, Lavie P (1983) Biological rhythms in arousal indices: a potential confounding effect in EEG biofeedback. Psychophysiology 20:690–695

Gopher D, Lavie P (1980) Short-term rhythms in the performance of a simple motor task. J Mot Behav 12:207–221

Gordon HW, Fruman B, Lavie P (1982) Shift in cognitive asymmetries between waking from REM and NREM sleep. Neuropsychologia 20:99–103

Grivel ML, Ruckebusch Y (1972) The propagation of segmental contractions along the small intestine. J Physiol 227:611–625

Hiatt JF, Kripke DF (1975) Ultradian rhythms in waking gastric activity. Psychosom Med 37: 320–325

Horne J, Whitehead M (1976) Ultradian and other rhythms in human respiration rate. Experientia 32:1165–1167

Kleitman N (1961) The nature of dreaming. In: Wolstenholme GEW, O'Connor M (eds) The nature of sleep. Churchill, London, pp 349–364

Kleitman N (1963) Sleep and wakefulness. University of Chicago Press, Chicago, IL

Kripke DF (1972) Ultradian biologic rhythm associated with perceptual deprivation and REM sleep. Psychosom Med 34:221–234

Kripke DF, Sonnenschein D (1978) A biologic rhythm in waking fantasy. In: Pope D, Singer JL (eds) The stream of consciousness. Plenum Press, New York, pp 321–332

Lavie P (1974) Differential effects of REM and non-REM awakenings on the spiral after effect. Physiol Psychol 21:107–108

Lavie P (1979) Ultradian rhythms in alertness – a pupillometric study. Biol Psychol 9:49–62

Lavie P (1982) Ultradian rhythms in human sleep and wakefulness. In: Webb WB (ed) Biological rhythms, sleep and performance. Wiley, Chichester London, pp 239–271

Lavie P (1985) Ultradian rhythms: gates of sleep and wakefulness. In: Schulz H, Lavie P (eds) Ultradian rhythms in physiology and behavior. Springer, Berlin Heidelberg New York, pp 110–124

Lavie P (1986) Ultrashort sleep–waking schedule. III. "Gates" and "forbidden zones" for sleep. Electroencephalogr Clin Neurophysiol 63:414–425

Lavie P (1987) Ultrashort sleep–wake cycle: timing of REM sleep. Evidence for sleep-dependent and sleep-independent components. Sleep 10:62–68

Lavie P (1989a) To nap, perchance to sleep – ultradian aspects of napping. In: Dinges D, Broughton R (eds) Napping: biological, psychological, and medical aspects. Raven Press, New York, pp 99–120

Lavie P (1989b) Ultradian rhythms in arousal – the problem of masking. Chronobiol Int 6:21–28

Lavie P (1990) Penile erections in a patient with nearly total absence of REM: a follow-up study. Sleep 13:276–279

Lavie P (1991) REM periodicity under ultrashort sleep–wake cycle in narcoleptic patients. Can J Psychol 45:185–193

Lavie P (1992) The 24-h sleep propensity function (SPF): theoretical and practical implications. In: Monk T (ed) Sleep, sleepiness and performance, pp 65–96

Lavie P, Kripke DF (1977) Ultradian rhythms in urine flow in humans. Nature 269:142

Lavie P, Kripke DF (1981) Ultradian circa 1.5 hour rhythms: a multioscillatory system. Life Sci 29:2445–2450

Lavie P, Scherson A (1981) Ultrashort sleep–waking schedule. I. Evidence of ultradian rhythmicity in "sleepability". Electroencephalogr Clin Neurophysiol 52:163–174

Lavie P, Segal S (1989) 24-h structure of sleepiness in morning and evening persons investigated by the ultrashort sleep–wake cycle. Sleep 12:522–528

Lavie P, Sutter D (1975) Differential responding to the beta movement following awakenings from REM and non-REM sleep. Am J Psychol 88:595–603

Lavie P, Weler B (1989) The timing of naps: relationship with post-nap sleepiness. Electroencephalogr Clin Neurophysiol 72:218–224

Lavie P, Zomer J (1984) Ultrashort sleep–waking schedule. II. Relationship between ultradian rhythms in sleepability and the REM–NONREM cycles and effects of circadian phase. Electroencephalogr Clin Neurophysiol 57:35–42

Lavie P, Zvuloni A (1992) The 24-h sleep propensity function (SPF): experimental bases for somnotypology. Psychophysiology (in press)

Lavie P, Lord WJ, Frank RA (1974) Basic rest–activity cycle in the perception of the spiral after effect: a sensitive detector of a basic biological rhythm. Behav Biol 11:373–379

Lavie P, Kripke DF, Hiatt JF, Harrison J (1978) Gastric rhythms during sleep. Behav Biol 23:526–530

Lavie P, Oksenberg A, Zomer J (1979) It's time, you must wake now. Percept Mot Skills 49:447–450

Lavie P, Metanya Y, Yehuda S (1984a) Cognitive asymmetries after wakings from REM and NONREM sleep in right-handed females. Int J Neurosci 23:111–116

Lavie P, Pratt H, Sharf B, Peled R, Brown J (1984b) Localized pontine lesion: nearly total absence of REM sleep. Neurology 34:118–120

Manseau C, Broughton RJ (1984) Bilaterally synchronous ultradian EEG rhythms in awake adult humans. Psychophysiology 21:265–273

McCarley RW, Hobson AJ (1975) Neuronal excitability modulation over the sleep cycle: a structural and mathematical model. Science 189:58–60

McCarley RW, Massaquoi SG (1986) A limit cycle mathematical model of the REM sleep oscillator system. Am J Physiol 251:R1011–R1029

McPortland RJ, Kupfer DJ (1978) REM sleep cycle, clock time and sleep onset. Electroencephalogr Clin Neurophysiol 45:178–185

Mitler MM, Gujavarty KS, Sampson MG, Browman CP (1982) Multiple daytime nap approaches to evaluating the sleepy patient. Sleep 5:119–127

Moses J, Lubin A, Johnson L, Naitoh P (1977) Rapid eye movement cycle is a sleep-dependent rhythm. Nature 265:360–361

Ohlmeyer P, Brilmayer H, Hullstrung H (1944) Periodische Vorgange im Schlaf. Pflugers Arch 249:50–55

Orr W, Hoffman H, Hegge F (1976) Ultradian rhythms in extended performance. Aerosp Med 49:995–1000

Richardson DS, Carskadon MA, Flagg W, Van den Hoed J, Dement WC, Mitler MM (1978) Excessive daytime sleepiness in man: multiple sleep latency measurement in narcoleptic and control subjects. Electroencephalogr Clin Neurophysiol 45:621–627

Schulz H (1985) Ultradian rhythms in the nychthemeron of narcoleptic patients and normal subjects. In: Schulz H, Lavie P (eds) Ultradian rhythms in physiology and behavior. Springer, Berlin Heidelberg New York, pp 165–185

Schulz H, Dirlich G, Zulley J (1975) Phase shift in the REM sleep rhythm. Pflugers Arch 358:203–212

Snyder F (1966) Toward an evolutionary theory of dreaming. Am J Psychiatry 123:121–142

Ursin R, Moses J, Naitoh P, Johnson LC (1983) REM–NONREM cycle in the cat may be sleep-dependent. Sleep 6:1–9

Weitzman ED, Czeisler CA, Zimmerman JC, Ronda JM (1980) Timing of REM and 3 + 4 sleep during temporal isolation in man. Sleep 2:391–407

Zepelin H (1986) REM sleep and the timing of self-awakening. Bull Psychonom Soc 24:254–256

The Basic Rest–Activity Cycle – 32 Years Later: An Interview with Nathaniel Kleitman at 96

E. L. Rossi

Nathaniel Kleitman, the author of the pioneering volume *Sleep and wakefulness*, (1963)[1] is lively and well at 96, living alone in his apartment in Southern California where I have interviewed him a number of times over the past few years. His general appearance and élan is so reminiscent of the American humorist George Burns that it was all I could do to suppress a laugh each time I saw him. His daughter, Esther, who lives nearby, was present at some of the interviews and helped in the preparation of this material for publication. Kleitman has been in retirement for 30 years since he left his faculty position at the University of Chicago at the age of 65. Perhaps his personal experience with the ageing process reported here can serve as a necessary corrective to many of the misconceptions that currently distort our understanding of sleep, wakefulness and well-being in senior citizens.

Rossi (R): What was your most important contribution to science?

Kleitman (K): My idea is that the 90-min rest–activity cycle, manifested as REM–NREM[2] in sleep, is also present in the waking state. In other words, it operates around the clock. In wakefulness, this cycle is involved in the maintenance of the organism through the 90-min cycle of hunger contractions. It is also involved in the maintenance of the species through a similar cycle of sexual excitation. These two processes, the preservation of the individual and the preservation of the species, led to my designation of the ultradian periodicity as the basic rest–activity cycle (BRAC).

R: What were the important steps that led to the discovery of the REM–NREM cycle?

K: I began as a physiologist interested in the brain and the waking state of consciousness. But back in the 1920s when I began there were no well-known and accepted physiological approaches to the brain and wakefulness as there were to other systems. Dr Anton J. Carlson, who was the chairman of the physiology department at the University of Chicago, encouraged me to find my

Fig. 14.1. Nathaniel Kleitman at age 96, with five of the famous boxes of filecards for his book *Sleep and wakefulness*.

own way. I began with the study of sleep as a way of studying the waking state. When I came to write my book *Sleep and wakefulness* I put the word *Sleep* first because the library card catalogues did not have any heading for *Wakefulness* or *Awake*.

R: Your *Sleep and wakefulness* has truly been a remarkable volume and it is very fortunate that it has recently been reprinted in paperback. With its more than 4000 references it has been a standard for almost two generations of researchers.

K: I still have those references in their original cardboard files with all their coloured tags for cross referencing [see Fig. 14.1]. I spent more time assembling and typing up those reference cards one by one than in the actual writing of the book itself! Young researchers today should thank their lucky stars that they have computers to help them with such tedious tasks.

R: Your paper in 1953 on "Regularly occurring periods of eye motility, and concomitant phenomena, during sleep",[3] co-authored with your graduate student, Eugene Aserinsky, is credited with having started the current revolution of research into the nature of dreaming and sleep. Can you say exactly how that came about?

K: When I began my research in the 1920s there was some awareness of periods of altered respiration and heart rate in sleep. I also was aware of the slow eyeball movements in sleep but no one knew anything about what they meant. I asked Aserinsky to watch the eye movements of infants at the university hospitals while they were asleep. Like any graduate student he reported his findings back to me about rapid eye movements in sleep but we were not yet aware of their significance. Because REM accompanied an EEG resembling that of wakefulness, and increase in heart rate and respiration, it occurred to me that it may signify dreaming.

To learn more about these eye movements I suggested that we study a group of adults by waking them up when their eyes were moving rapidly during sleep – and when they were not – to see what they might report. We found that they reported dreaming when their eyes were moving rapidly but they reported no dreaming when their eyes were not moving. William Dement then became another of my graduate students to carry on the research, particularly in investigating sleep disorders, in which he is having a distinguished career.

R: You have also pioneered the idea that there is a continuity of the basic rest–activity cycles in wakefulness as well as in sleep.

K: Yes, as I mentioned, the basic rest–activity cycle runs through both wakefulness and sleep. This area remains controversial and a great deal of work remains to be done, but I have cited several research reports on animals and man that amply support my view.[4]

R: Is there any correspondence between the REM in dreaming and the performance peak in the BRAC during wakefulness?

K: Yes, but REM itself is of no physiological significance. It is possible that the REM phase is an atavistic phenomenon that had survival value in prehistoric times when it afforded an opportunity for the organism to awaken periodically to check the environment for predators. REM was probably gradually abolished during wakefulness, as it interfered with fixing the gaze on an object – friendly or hostile – in the visual field. There was apparently no need to abolish REM during sleep, as one couldn't see anyway with eyes closed.

Aside from the absence of REM during wakefulness, there is a great correspondence between the functions of the cerebral cortex during the activity phase of the BRAC in sleep and wakefulness. The main difference is the faulty performance of the cortex during the activity phase of the BRAC in sleep, resulting in dreaming.

R: How do you account for your own longevity and well-being?

K: I don't know how to answer that. However, I never smoked or drank alcohol and I eat sensibly.

R: Do you attend to your own BRAC on a daily basis? Do you use your understanding of the BRAC to enhance your life? Do you follow periods of activity, exercise and rest throughout the day?

K: At my age, it is not essential to utilize the activity phase of the BRAC to do creative work. Many individuals, however, without any knowledge of the BRAC during wakefulness, have repetitive outbursts of creative activity – painting, writing, composing, problem-solving, etc. – during the activity phases of the BRAC.

I do not follow any definite schedule with respect to activity, exercise and rest during the daytime hours. At night, I usually sleep less than I did years ago. Old people are often concerned about only getting 5 or 6 h of sleep during the night, compared to 7 or 8 h when they were younger. That accounts for their frequent recourse to sleeping pills. It should be remembered that in old age, when one no longer has to work, and especially, if living alone, there may be nothing to be awake for. Therefore, in addition to night sleep, old people often have one or more naps in the daytime, which they fail to add to the shortened night's sleep. They may actually be sleeping more than previously.

R: What research directions would you now like to see pursued? Any advice for young researchers?

K: We need to continue the excellent research on sleep disorders, to raise public awareness of the operation of the basic rest–activity cycle and the effects of sleep deprivation as they relate to accidents in transportation and industry. Really, research is just beginning in these areas.

In conclusion, it should be noted that the discovery of the BRAC in sleep led to the finding that the same cycle operates in wakefulness as well. This bears out the opinion, which I indicated earlier, that by studying sleep characteristics we may be able to add to our understanding of various features of wakefulness.

Notes

1. Kleitman N (1963) Sleep and wakefulness. University of Chicago Press, Chicago, IL.
2. Abbreviations: REM, rapid eye movement; NREM, non-rapid eye movement; EEG, electroencephalograph.
3. Aserinsky E, Kleitman N (1953) Regularly occurring periods of eye motility, and concomitant phenomena, during sleep. Science 118:273–274.
   Dement WC, Kleitman N (1957) Cyclic variations in EEG during sleep and their relation to eye movements, body motility, and dreaming. EEG Clin Neurophysiol 9:673–690.
4. Kleitman N (1982) The basic rest–activity cycle – 22 years later. Sleep 5:311–315.

Chapter 15

The Sleep–Wake Threshold in Human Circadian Rhythms as a Determinant of Ultradian Rhythms

R. A. Wever

Introduction

When considering human circadian rhythmicity, one must distinguish between rhythms of two types: first, steady rhythms of continuously measurable variables (example: body temperature) and, second, square waves from variables that alternate between two discrete states (example: sleep–wakefulness). It would facilitate all further analyses of the circadian system if both types of rhythm could be reduced to one type.

It has been shown that all types of time series analysed by a special periodogram algorithm (Wever 1979) can be performed consistently with both types of rhythm; the statements to be derived (concerning, for instance, dominant period, reliability, phases etc.) are independent of the character of the variable on which the rhythm is based. In the description of a rhythm by modelling procedures, the alternation between two discrete states (e.g. between wakefulness and sleep) has been reduced to threshold crossings of a steady rhythm (Wever 1960). As long as the rhythm runs above a fixed threshold it describes the one state (e.g. wakefulness), and as long as it runs below the threshold it describes the other state (e.g. sleep). The actual interval between rhythm and threshold may provide further information of interest which is not included in the original square wave (for instance: above the threshold, the degree of alertness; below the threshold, the depth of sleep).

Present data allow one to deduce quantitatively the position of the hypothetical threshold separating wakefulness and sleep. This chapter shows how this sleep–wake threshold may lead to a better understanding of the relationship between sleep and wakefulness and the evaluation of the nature of sleep disturbances. These considerations lead me to the view that the sleep–wake threshold determines the interaction between ultradian and circadian rhythms. This implies that displacements of the threshold (relative to the mean value of the rhythm) can release disorders of the ultradian sleep rhythm, independent of the actual circadian rhythmicities.

Determination of the Sleep–Wake Threshold

The steady rhythm hypothetically underlying the sleep–wakefulness alternation is defined by a mean value and the amplitude of this rhythm if the rhythm can be assumed to be more or less sinusoidal; when the relative fractions of wakefulness or sleep are considered (and not the absolute lengths of the episodes) the period of this rhythm is without meaning. Since these parameters of a hypothetical "activity rhythm" are unknown (because of a lack of any consistent data), in a first approximation the rhythm of body temperature (the parameters of which are well known for all individual subjects) may be taken for such a rhythm. The position of the sleep–wake threshold may then be given by temperature. The identification of the "activity rhythm" with the rhythm of body temperature (or better, the assumption of a constant relationship between the two rhythms) is justified by the observation that there are close correlations between the rhythms of body temperature and the sleep–wake cycle. It has been shown, for instance, that there is a very strong correlation between the amplitude of the body temperature rhythm and the fraction of sleep (or wakefulness) in the sleep–wake cycle (Wever 1988a; Fig. 15.1). In human subjects where the wake episodes are commonly longer than the sleep episodes, the position of the threshold is always, for geometric reasons, below the mean temperature level of the subject.

In the following, the procedure for deducing the position of the sleep–wake threshold will be illustrated with a homogeneous sample of long-term experiments of the free-running rhythms of male subjects. Free-running rhythms obtained under constant experimental conditions have been selected for this investigation because the reliability of such rhythms is commonly higher than that of (24-h) synchronized rhythms of everyday life. In free-running rhythms the lengths of the wake and sleep episodes are freely selected by the subject and are determined exclusively by biological necessities. In the natural 24-h day, however, the sleep–wake cycle is usually modulated by real or imagined social constraints. Moreover, long-term experiments are necessary because it may take up to about 2 weeks before a steady state is achieved (rhythm parameters can be determined only in a steady state; Wever 1979). In the following consideration the data are restricted to male subjects because of the well-known sex differences in most rhythm parameters (Wever 1984a,b). When data for both sexes are considered together, the correlations are frequently lost.

Fig. 15.1 shows the correlations for 16 male subjects for about 1 month under constant conditions; they all show free-running rhythms which are internally synchronized (i.e. the rhythms of all different variables run with identical periods, or in internal synchrony). The lower diagram presents data for the mean values versus the amplitudes, both of the body temperature rhythms. For both parameters the standard deviations (and twice these values, which about mark the range of the data) are drawn separately; and the calculated regression lines (in both directions, because a special direction of the relationship cannot be given) illustrate a possible relationship between the parameters; the coefficients of correlation (r, parametric; R, non-parametric) are indicated. It can be recognized that there is no correlation at all, or, in other words, the interindividual variations of these two parameters are

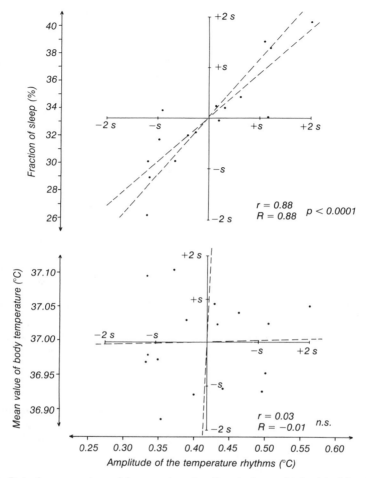

Fig. 15.1. Data for parameters of free-running circadian rhythms, obtained in 16 male subjects living each for about a month under temporally constant conditions. Against the amplitudes of the rhythms of body temperature are plotted, in the lower diagram, the mean values of these rhythms and, in the upper diagram, the fractions of sleep within the sleep–wake cycles. For all parameters the means, standard deviations ($\pm s$) and twice the standard deviations ($\pm 2s$) are indicated. Moreover, regression lines (in both directions of dependence) are drawn and the coefficients of correlation are indicated (r, parametric; R, non-parametric after Spearman). n.s., not significant.

completely independent of each other. The upper diagram presents the inter-relationship between the amplitudes of the body temperature rhythms and the sleep fractions within the sleep–wake cycles for the same subjects. In con-trast to the lower diagram, here very strong (positive) correlations can be recognized. This means the larger the temperature amplitude, the longer the sleep episode (or the shorter the wake episode), and vice versa.

Fig. 15.2 demonstrates the construction of the sleep–wake threshold by utilizing the data from Fig. 15.1 (mean values and amplitudes of the tempera-ture rhythms, and the fractions of sleep within the sleep–wake rhythms). In

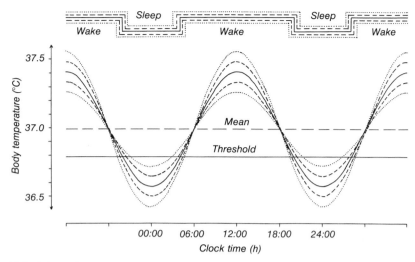

Fig. 15.2. Schematic construction of the sleep–wake threshold utilizing the data from Fig. 15.1. The lower diagram shows mean cycles of body temperature, drawn with the mean of all mean values, and with the mean amplitude (continuous line), the mean ±s.d. amplitude (dashed lines) and the mean ±2s.d. amplitude (dotted lines). The upper diagram shows mean sleep–wake cycles, drawn with the mean sleep–wake ratio (continuous line), the mean ±s.d. ratio (dashed lines) and the mean ±2s.d. ratio (dotted lines). The crossings of the projections from sleep-off (or wake-on) and sleep-on (or wake-off) with the mean temperature cycles define the threshold values; the thresholds defined by using the respective variations of the parameters are about the same as the (original) threshold defined by using the respective means.

the lower diagram, schematized (sinusoidal) courses of mean temperature cycles are drawn, by utilizing, on the one hand, the mean of all mean values (36.990 °C) and, on the other, the mean and also the mean ±1 and ±2 standard deviations (s.d.) of the amplitudes (0.418, 0.491 and 0.345 °C; 0.564 and 0.272 °C). Since mean value and amplitude have been shown to vary inter-individually independently of each other (cf. Fig. 15.1, lower diagram), the utilization of the same mean value for all representations of the mean tempera-ture cycle is justified. The upper diagram of Fig. 15.2 shows the schematic courses of the mean sleep–wake cycles, likewise utilizing the mean sleep fraction for all subjects and also the mean ±1 and ±2 standard deviations (33.3%, 37.2% and 29.4%, and 41.1% and 25.5%). The threshold is then defined by the temperature where the projections from sleep-on (or wake-off) and sleep-off (or wake-on) cross the course of the temperature cycle in the lower diagram. Originally, the crossings of the means of both parameters (continuous lines) constituted the threshold; remarkably enough, however, the crossings as constructed with the mean ± s.d. (dashed lines) and even the mean ±2 s.d. (dotted lines) of both parameters each constitute about the same threshold. This means the position of the sleep–wake threshold is independent of the interindividual variations of temperature amplitude and fraction of sleep (or wakefulness) because, according to Fig. 15.1 (upper diagram) both these parameters are always in strong correlation.

Meaning of Interindividual Variations of Sleep and Wakefulness

A consequence of the correlation between temperature amplitude and sleep fraction in Fig. 15.2, is that subjects who need longer sleep and who show, correspondingly, shorter wakefulness episodes also show higher maximum alertness. This means, in spite of their shorter wake episodes, that their total amount of activity per wake episode is certainly not smaller but rather greater than that of subjects with longer wake and shorter sleep episodes. Within the range of variability in the ratio between wakefulness and sleep in healthy subjects, therefore, subjects with wake episodes that are shorter than average (and sleep episodes that are longer than average) do not show a reduction in

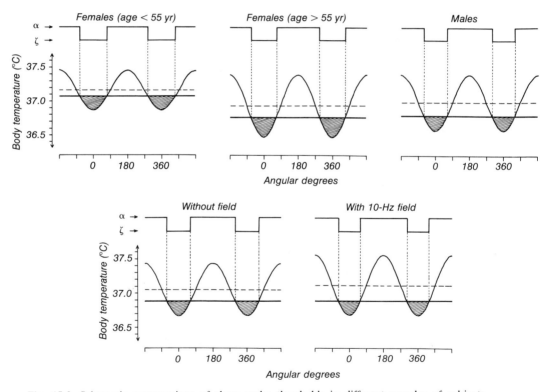

Fig. 15.3. Schematic constructions of sleep–wake thresholds in different samples of subjects, according to Fig. 15.2. Upper row of diagrams: thresholds by utilizing data from younger female, older female and male subjects (the last diagram corresponding to Fig. 15.2). Lower row of diagrams: thresholds by utilizing data from the same subjects but different sections of the same experiment for each; during the different sections the subjects were either exposed to a weak electric 10-Hz square-wave field, or protected from every natural or artificial electromagnetic field.

the total amount of activity but rather a temporal concentration of activity with an enhancement of its total amount (Wever 1988b).

With regard to sleep, however, a similar compensation between variations in the length of the sleep episodes and the depth of sleep is not present. On the contrary, according to Fig. 15.2 shorter sleep episodes seem to be accompanied by lower sleep efficacies. However, here the sole inspection of circadian rhythms is insufficient, and the additional consideration of superimposed ultradian sleep rhythms is indispensable. According to an established model (cf. Figs. 15.6 and 15.7), a decline of the circadian rhythm to positions below the sleep–wake threshold releases ultradian sleep cycles; the minimum values of these sleep cycles depend only marginally on the actual values of the original circadian rhythm (as long as it passes below the threshold). Only the minimum values of the ultradian sleep cycles but not those of the original circadian cycles determine the sleep efficacy, so that the efficacy of sleep does not depend on the minimum values of the underlying circadian rhythm (which is strongly related to the length of sleep). This means that subjects with shorter sleep episodes have about the same efficacy per hour of sleep than do subjects with longer sleep episodes (and not lower efficacies, as Fig. 15.2 seems to suggest). In summary, within the range of variabilities in the sleep–wake ratio in healthy subjects, shorter sleep episodes constitute a lower amount of total sleep, to an extent which is about proportional to the length of sleep.

Hence, the interindividual variabilities in the lengths of the wake and sleep episodes have differential consequences with respect to the variabilities of the overall levels of wakefulness and sleep. Whereas variations in the length of the sleep episodes reflect directly differences in the total amount of sleep, variations in the length of the wake episodes are more or less compensated (or even overcompensated) for in their effects on the total amount of activity. This result suggests that the need for sleep is different in different subjects, but the need for activity is similar in different subjects (only the distribution of activity within the sleep–wake cycle is different).

Properties of the Sleep–Wake Threshold

Fig. 15.2 suggests that the specific position of the sleep–wake threshold is a general feature of male subjects which does not depend on the individual values of single parameters of the different rhythms. Since in female subjects the values of most rhythm parameters systematically deviate from those of male subjects (the sleep episodes are longer by 1.56 (±1.09) h, the amplitudes of the temperature rhythm are smaller by 34 (±17)%, and the mean values of body temperature are higher by 0.176 (±0.078) °C in younger women than in men; Wever 1988a), a question arises about the position of the threshold in females. Since in most of these parameters (in particular, in those of the rhythm of body temperature) the deviations can be observed only in younger females (up to the age of the menopause), the calculations must be performed separately for younger and older female subjects. Fig. 15.3 shows the results of these calculations. In the upper row of diagrams the sleep–wake threshold is constructed separately for younger and older female and for male subjects, each in the same way as in Fig. 15.2 but only for the means of the correspond-

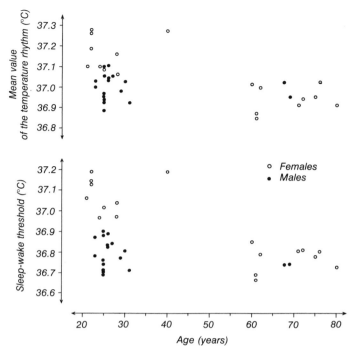

Fig. 15.4. Data for sleep–wake thresholds (lower diagram) and, for comparison, temperature mean values (upper diagram), originating from all subjects included in the upper three diagrams of Fig. 15.3; the data are plotted against the age of the subjects, independently of the affiliation of the subjects to the different samples. The diagrams justify the grouping into younger and older female subjects (unfortunately, there were no subjects with intermediate ages); the data of the two older male subjects do not deviate from data of younger males.

ing parameters. It has been shown in every single group (as it has been shown for males in Fig. 15.2) that interindividual fluctuations of the parameters leave the position of the threshold unchanged (Wever 1988a).

The sleep–wake thresholds as a function of age for all subjects are shown in the lower diagram of Fig. 15.4; for comparison, the mean values of the temperature rhythms are drawn in the same way in the upper diagram. Fig. 15.4 shows clearly that the sleep–wake thresholds are higher in younger female (37.077 ($\pm$0.086) °C) than in male (36.794 ($\pm$0.070) °C) subjects, without any overlap in the data; this difference is highly significant statistically, parametric ($t = 8.951$) and non-parametric ($u = 4.076$). The lower diagram of Fig. 15.4 shows, moreover, that the sleep–wake thresholds are lower in older (36.765 ($\pm$0.069) °C) than in younger female subjects (see above), again without any overlap; again, this difference is highly significant statistically ($t = 8.455$, $u = 3.576$). However, there is no significant difference between threshold positions of older female and male subjects, in spite of the significant differences in the sleep fractions (37.8 ($\pm$2.9)% versus 33.3 ($\pm$3.9)%): $t = 1.000$, $u = 0.792$. The mean values of the same groups of subjects (upper diagram of Fig. 15.4)

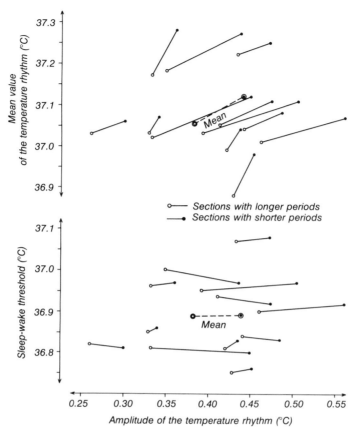

Fig. 15.5. Data for sleep–wake thresholds (lower diagram) and, for comparison, temperature mean values (upper diagram), plotted against the amplitudes of the temperature rhythms; the data originate from the subjects included in the lower two diagrams of Fig. 15.3; data from different sections of the same experiments (i.e. from sections with and without the weak electric 10-Hz field in operation) are combined by lines. In the sections with the field continuously in operation the periods were consistently shorter and the amplitudes were consistently larger than in the field-free sections. While the temperature mean value was consistently higher with than without the field, the sleep–wake threshold was independent of the field.

deviate from each other in the same directions as the thresholds but for slightly lower extents. In contrast to the thresholds (lower diagram) there is some overlap in the data; correspondingly, the statistical significances in the differences between the groups are slightly smaller (for instance, in the comparison between younger female and male subjects: $t = 5.574$, $u = 3.737$). In general, the sex difference in the position of the sleep–wake threshold is guaranteed at a higher level than that in any other rhythm parameter.

The next finding concerns the influence not of inter- but of intraindividual variations of rhythm parameters on the threshold position. Changes in the experimental conditions may affect rhythm parameters such as the amplitude of the temperature rhythm and the sleep fraction within the sleep–wake cycle in a systematic way. This finding is based on an experimental series where 12 subjects were exposed to a continuous weak electric 10-Hz square-wave field (2.5 V/m) in one section, while they were protected from any artificial and natural electromagnetic field in another section of the same experiment (Wever 1967, 1985d). Nearly all rhythm parameters showed highly significant differences in the sections with and without the field; the free-running period was shorter, the circadian temperature amplitude was larger, the mean temperature level was higher, and the fraction of sleep was smaller with the field. The special advantage of this study is that the subjects (in contrast, for instance, to experiments where the intensity of illumination was varied) cannot perceive consciously the controlling stimulus though it is effective; this suggests that behavioural effects cannot account for the results.

Fig. 15.5 (lower diagram) shows individual changes in the threshold position for all 12 subjects of this experimental study as a function of the amplitude of temperature rhythm; again, for comparison, the upper diagram shows changes in the mean value drawn in the same way. Data from sections with the shorter free-running periods (in all cases, these are sections with the 10-Hz field continuously in operation) are combined with data from sections of the same experiments with the longer periods (in all cases, these are sections under field-free conditions). Both diagrams of Fig. 15.5 show that the rhythms of all subjects have the larger amplitudes in the sections with the shorter periods. The lower diagram shows that the sleep–wake threshold is slightly higher in the sections with the shorter periods in some subjects while it is slightly lower in other subjects, without any systematic difference; on average, there is no relevant difference between the two sections. The mean value of the temperature rhythm (Fig. 15.5, upper diagram), however, is strongly correlated to other rhythm parameters; without exception, it is higher in the sections with the field in operation (i.e. with the significantly shorter periods, larger amplitudes and shorter sleep episodes) than in the sections under field-free conditions. Correspondingly, the mean of the temperature levels for all 12 subjects is significantly higher with the 10-Hz field continuously in operation than under field-free conditions.

In the lower row of diagrams in Fig. 15.3, mean sleep–wake thresholds are constructed separately for the sections without the field and for sections with the 10-Hz field continuously in operation. The diagrams show that, in spite of the significant differences in all rhythm parameters which underlie the threshold construction (difference in amplitudes, $0.057 (\pm 0.038)\,°C$; in the mean values, $0.066 (\pm 0.029)\,°C$; in the sleep fractions, $-1.73 (\pm 2.45)\%$), the sleep–wake threshold does not show a relevant difference between the two types of section $(0.002 (\pm 0.017)\,°C)$.

In summary, the position of the threshold separating sleep from wake is an individual constant that remains even when all rhythm parameters (e.g. amplitude and mean value of the temperature rhythm, fraction of sleep) vary due to chance variations or to systematic variations in the experimental conditions. On the other hand, the position of the threshold is subject to significant changes when the internal conditions within the subjects change; this has been shown with the difference between threshold positions in younger and older female subjects (with the menopause as the limit), which is apparently controlled hormonally. From these results it is plausible that also internal changes due to an illness may displace the sleep–wake threshold.

Meaning of the Sleep–Wake Threshold

The foregoing considerations show that the properties of the sleep–wake relationships depend on circadian rhythmicity and also on the position of a sleep–wake threshold (which is independent of circadian rhythmicity). This conclusion implies that sleep disorders may be due to irregular shiftings of the sleep–wake threshold (without any disorder in the circadian rhythmicity). Hence, in order to evaluate causes for sleep disorders one must investigate separately effects of changes in circadian rhythms as well as changes in the threshold position. While such a separation is useful theoretically, it does not mean, in practice, that sleep disorders are always exclusively one of these two types; frequently both types of disorder are present together.

The way to relate sleep disorders unambiguously to changes in circadian rhythmicity or threshold position is to use a sleep model where these systems can be manipulated separately. The model proposed here is based on a super-imposition of a circadian and an ultradian rhythm (Wever 1985a). According to the general model, all self-sustaining rhythms (also in the circadian and the ultradian domain) are manifest only within a limited "oscillatory range" (Wever 1964, 1984c, 1987). It is the special quality of this model that the (upper) limit of the ultradian oscillatory range is identical with the sleep–wake threshold (Wever 1985a). This means that the ultradian system exerts active (self-sustaining) deflections only during sleep; it is in rest during the awake state.

Other relationships between the ultradian oscillatory range and the sleep–wake threshold are possible. There is, for instance, the possibility that the ultradian system oscillates only during wakefulness and not during sleep (to describe ultradian rhythms of other types). Or there is the possibility of a symmetric ultradian oscillatory range describing self-sustainment all the time during wakefulness and sleep (or any relationship in between). In the latter case the parameters of the ultradian rhythm (e.g. its period) are different during the circadian sleep and wake episodes. This difference in the periods would be true even if the ultradian rhythm was stronger in one of the states (e.g. during sleep) than the other (e.g. during wakefulness), so that, with a rough inspection, the impression might arise that the ultradian rhythm is present only during one of the states (e.g. only during sleep). It is a general consequence of the modulation of ultradian rhythms by circadian rhythms that time series analyses in the ultradian domain over several circadian cycles are meaningless, independent of the type of modulation. If ultradian rhythms are present only during the circadian sleep episodes, ultradian rhythmicity starts anew at every circadian sleep onset independently of the previous sleep episode. so that there cannot be a continuation of the ultradian phase in successive circadian sleep episodes. If ultradian rhythms continue during the full circadian sleep–wake cycle (but with different rhythm parameters – e.g. period and amplitude – during sleep and wakefulness), there is, in principle, a phase shift between the ultradian rhythms in successive circadian sleep episodes of necessarily unknown extent, due to the different speeds of the ultradian system during the interposed wake episodes. Again, therefore, there is no continuation of the ultradian phase in successive circadian sleep episodes.

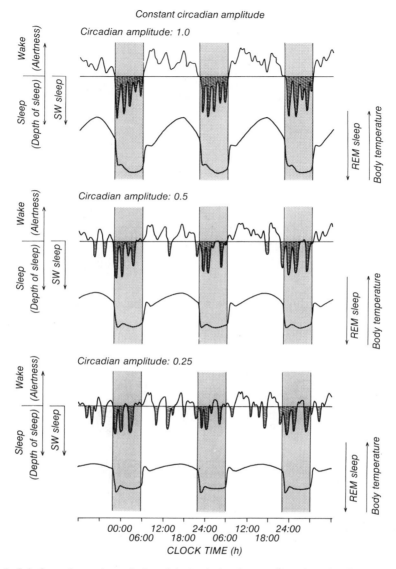

Fig. 15.6. Solutions of a mathematical model, simulating the coupling of an ultradian sleep rhythm on a circadian activity rhythm in the 24-h day; a coupled rhythm of body temperature (with the masking effect) is also shown. In all three diagrams, the sleep–wake thresholds and the parameters of the ultradian rhythms (and also the superimposed random fluctuations) are identical; only the amplitudes of the circadian rhythms are varied in three steps. The upper diagrams each ("sleep–wake") show (below the threshold) the efficacy of sleep (i.e. the actual fraction of slow-wave (*SW*) sleep of the total sleep); and the lower diagrams each ("body temperature") show the patterns of *REM sleep* (i.e. the lower the actual temperature, the longer are the respective REM episodes).

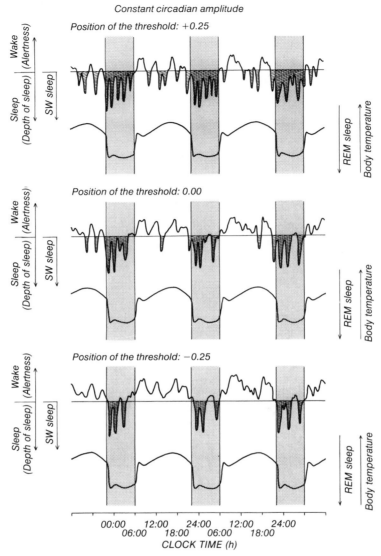

Fig. 15.7. Solutions of a mathematical model, simulating the coupling of an ultradian sleep rhythm on a circadian activity rhythm in the 24-h day; a coupled rhythm of body temperature (with the masking effect) is also shown. In all three diagrams the parameters of the circadian and the ultradian rhythms (and also the superimposed random fluctuations) are identical; only the position of the sleep–wake threshold is varied in three steps. For further explanation, see Fig. 15.6.

Figs. 15.6 and 15.7 show simulations calculated by utilizing this model for sleep–wake and body temperature during a 24-h day. In Fig. 15.6 the position of the sleep–wake threshold is held constant, and the amplitude of the circadian rhythm is varied in three steps. in Fig. 15.7, by way of contrast, the amplitude (and all other parameters) of the circadian rhythm is held constant, and the position of the sleep–wake threshold is varied in three steps (the middle diagrams of both these Figures show identical simulations). In both these Figures the hypothetical sleep–wake rhythm is first superimposed by random fluctuations, and then an ultradian system is coupled to the circadian system; the ultradian variations reflect the rhythmic variations of deep, or slow-wave sleep ("SW sleep"; i.e. stages 3 and 4). In addition, in Figs. 15.6 and 15.7 simulations of the rhythms of body temperature are included; these rhythms reflect simultaneously the rhythmic variations of REM (rapid eye movement) sleep; it has been shown that the single REM episodes within a sleep episode are the longer the lower is the actual body temperature (Wever 1985b). The rhythm of body temperature is superimposed by a masking effect, i.e. by a component which directly reflects the state of activity (Wever 1985c).

For a better comparability, the superimposed random fluctuations (in the sleep–wake rhythm) and the masking effect (in the temperature rhythm) are identical in all simulations of the two Figures; moreover, the parameters of the ultradian system are identical in all cases. These conclusions seem to be reasonable, since in disorders such as depression the circadian amplitude is reduced but not the superimposed regular and irregular fluctuations; rather, the relative extent of the masking effect and the random fluctuations seem to increase in depression and other disorders (Wever 1988a, 1992). To empha-size this, the simulation of the body temperature rhythm is drawn schematically, i.e. without superimposed random fluctuations, because in this rhythm particu-larly the influence of the masking effect is relevant.

Fig. 15.6 shows the influence of changes in the circadian amplitude. If the ultradian system is constant, the superimposed sleep rhythm changes its pattern dramatically when the circadian amplitude changes: whereas it shows a highly ordered pattern with large circadian amplitude (upper diagram, corresponding to the picture in really healthy subjects), the sleep pattern becomes nearly chaotic with decreasing circadian amplitude (lower diagrams, corresponding to increasing degrees of mental disorders) (Wever 1991). During this dramatic change of the sleep pattern, the total length of sleep shows only small changes. The rhythm of body temperature shows a great change in the relative portion of the masking effect. With large circadian amplitude (uppermost diagram) the masking effect does not influence the wave shape; in particular, the temporal positions of the extremum values are not affected by the masking effect. For instance, the temperature minimum is positioned close to the end of every sleep episode, as it would be without a masking effect; this means that the single REM episodes become longer during every sleep episode. With small circadian amplitude (lowermost diagram), on the other hand, the masking effect is dominant; in particular its overshoots (Wever 1985c) determine the positions of the extremum values. For instance, the temperature minimum is positioned immediately after sleep onset; this means the first REM episode is always the longest in every sleep episode, and there may be an increased tendency toward "sleep onset REMs" (such a sleep pattern resembles that of sleep in depressives). With the intermediate amplitude (middle diagram) there are two equivalent temperature minima, the first immediately after sleep onset ("masking minimum") and the second close to the end of sleep ("circadian minimum"); this probably means that sleep onset REMs (with a long first REM episode) are combined subsequently with a normal REM architecture, i.e. with lengthening REM episodes during every sleep episode.

Fig. 15.7 shows the influence of displacements of the sleep–wake threshold relative to the level of the rhythm (the picture would be the same if the threshold is held constant and the level of the rhythm is displaced). In spite of the unchanged circadian and ultradian systems, the sleep architecture again changes dramatically. Here the fraction of sleep changes more than the sleep pattern. With the higher threshold (uppermost diagram), any larger fluctuation leads to a fall below the threshold and, hence, an ultradian sleep cycle. The result is a greater sleep fraction as is characteristic of narcoleptics. With the lower sleep–wake threshold (lowermost diagram), only in rare cases will a fluctuation reach the threshold and lead to an ultradian sleep cycle. The result is an unusually small fraction of sleep as is characteristic of patients suffering from hyposomnia. In this type of simulation (Fig. 15.7) where the circadian rhythm is supposed to remain unchanged, the rhythm of body temperature and the REM architecture remains unchanged; it is always identical with that described for the middle diagram of Fig. 15.6. While in the simulations of Fig. 15.6 the entire sleep architecture (i.e. SW and REM sleep) changes considerably, in the simulations of Fig. 15.7 only the architecture of SW sleep changes but not that of the REM sleep.

In reality, it is unlikely that only one of the two systems – circadian rhythm and threshold – is subject to changes, while the other system remains unchanged. This is particularly clear if, instead of the absolute threshold position, its position relative to the rhythm is considered, i.e. the interval between threshold and mean value of the rhythm. Indeed, it has been shown that changes in the circadian amplitude (the actions of which are presented in Fig. 15.6) which are due to a change in the experimental conditions, are, with high probability, combined with changes in the mean value of the temperature rhythm (cf. Fig. 15.5, upper diagram), i.e. with changes in the interval between threshold and mean value; this means that the actions as presented in Fig. 15.7 must be considered as additive.

As an example of such combined changes of several rhythm parameters, the diagrams in the lower row of Fig. 15.3 may be considered. With the introduction of an electric 10-Hz field, the rhythm's amplitude enlarges and its mean value rises (and several more rhythm parameters change their values systematically); the sleep–wake threshold, however, remains unchanged. This means that the architecture of sleep alters simultaneously in the direction from the lower to the upper diagrams of Fig. 15.6 (due to the increase in the amplitude) and in the direction from the upper to the lower diagrams of Fig. 15.7 (due to an increase in the interval between threshold and mean value). This increase does not occur with a lowering of the sleep–wake threshold (as underlying Fig. 15.7, for clarity) but with a rise in the level of the rhythm (with fixed threshold) which is inevitably combined with the change in the amplitude (due to the non-linear qualities of the rhythms). In this example, therefore, the observed alteration in the length of sleep is due exclusively to a change in the rhythm itself, and not in the position of the sleep–wake threshold.

On the other hand, there are examples where sleep disorders are due only to displacements of the sleep–wake threshold, independent of changes in any rhythm parameter. In a series of patients suffering from profound sleep disorders there was a 27-year-old woman who did not show an irregularity in any parameter of her circadian rhythmicity. However, she showed a particularly low sleep–wake threshold (36.78 °C) which must be compared with the mean threshold in healthy younger female subjects (37.077 (±0.086) °C, see above; Wever 1992). According to Fig. 15.7 a low threshold (lowermost diagram) indicates a particularly low level of sleep. And a deviation in the threshold position from the corresponding mean for not less than 3.5 standard deviations excludes, with very high probability, a chance variation within the same sample of healthy subjects. The patient, therefore, must belong to another sample (i.e. ill patients). In other words, the sleep disorder of this patient is exclusively due to a displacement of the sleep–wake threshold and not to a virtual rhythm disorder.

Opposing contributions from both systems, circadian rhythm and sleep–wake threshold, in the alteration of sleep can be observed when young female subjects and male subjects of any age are compared (cf. Fig. 15.3, upper row of diagrams). All rhythm parameters are different in the two sexes. The lower mean value in males, when considered solely in combination with a fixed threshold, would result in a considerably longer sleep; likewise the larger amplitudes in the rhythm of body temperature would result in a longer sleep in men than in (young) women. Both, however, are in contrast to the experimental data. Only when in combination with a lowering of the threshold which clearly dominates that of the mean value are these effects of changes in rhythm parameters on the sleep length overcompensated. In summary, therefore, there should be shorter sleep episodes in males than in (young) females, as is observed. Sleep length in elderly women, however, equals that in younger women and is, correspondingly, likewise longer than in men, but elderly women and men have the same sleep–wake thresholds. Here, slightly lower mean values and slightly larger amplitudes of the temperature rhythm in elderly women than in men (both differences being opposite to the corresponding differences when young women are considered, and both differences being not significant statistically when considered separately) sum up to a significantly longer sleep in elderly women than in men of any age, in spite of the unchanged sleep–wake threshold. This means differences in the length of sleep between different groups of subjects can be described only by a common consideration of both systems, circadian rhythmicity and sleep–wake threshold.

Conclusions

The concept of a threshold separating wakefulness and sleep, though more than 30 years old, gains new importance when its influence on the structure of sleep is considered. Originally, the threshold was introduced only to describe changes in the ratio between the lengths of wake and sleep by changes in the relative position of threshold and circadian level. Recently, however, the considerable influence of the threshold on ultradian sleep rhythms has been recognized. Apart from the circadian rhythmicity itself, the sleep–wake threshold controls sleep, i.e. its length, its total amount and its architecture. Considerations of circadian rhythmicity alone, therefore, are frequently insufficient for understanding sleep disorders; the sleep–wake threshold is indispensable for analysing the basic nature of sleep disorders.

References

Wever R (1960) Possibilities of phase-control, demonstrated by an electronic model. Cold Spring Harb Symp Quant Biol 25:197–201

Wever R (1964) Zum Mechanismus der biologischen 24-Stunden-Periodik. III. Mitteilung: Anwendung der Modell-Gleichung. Kybernetik 2:127–144

Wever R (1967) Über die Beeinflussung der circadianen Periodik des Menschen durch schwache elektromagnetische Felder. Z Vergl Physiol 56:111–128

Wever RA (1979) The circadian system of man. Springer, Berlin Heidelberg New York

Wever RA (1984a) Properties of human sleep–wake cycles: parameters of internally synchronized freerunning rhythms. Sleep 7:27–51

Wever RA (1984b) Sex differences in human circadian rhythms: intrinsic periods and sleep fractions. Experientia 40:1226–1234

Wever RA (1984c) Toward a mathematical model of circadian rhythmicity. In: Moore-Ede MC, Czeisler CA (eds) Mathematical models of the circadian sleep–wake cycle. Raven Press, New York, pp 17–79

Wever RA (1985a) Modes of interaction between ultradian and circadian rhythms: toward a mathematical model of sleep. Exp Brain Res 12 [Suppl]:309–317

Wever RA (1985b) Circadian aspects of sleep. In: Kubicki S, Herrmann WM (eds) Methods of sleep research. Gustav Fischer Verlag, Stuttgart New York, pp 119–151

Wever RA (1985c) Internal interactions within the human circadian system: the masking effect. Experientia 41:332–342

Wever RA (1985d) The electromagnetic environment and the circadian rhythms of human subjects. In: Grandolfo M, Michaelson SM, Rindi A (eds) Biological effects and dosimetry of static and ELF electromagnetic fields. Plenum Press, New York London, pp 477–523

Wever RA (1987) Mathematical models of circadian one- and multioscillator systems. In: Carpenter GA (ed) Some mathematical questions in biology: circadian rhythms. American Mathematical Society, Providence, RI, pp 205–265

Wever RA (1988a) Order and disorder in human circadian rhythmicity: possible relations to mental disorders. In: Kupfer DJ, Monk TH, Barchas JD (eds) Biological rhythms and mental disorders. Guilford Press, New York, pp 253–346

Wever RA (1988b) Circadian control of vigilance. In: Leonard P (ed) Vigilance: methods, models and regulation. Peter Lang Verlag, Frankfurt-am-Main, pp 149–165

Wever RA (1991) Interactions between human circadian and (about 90-min) sleep rhythms: problems in the simulation and the analysis. In: Haken H, Koepchen HP (eds) Synergetics, Vol 55, Rhythms in physiological systems. Springer, Berlin Heidelberg New York, pp 235–253

Wever RA (1992) Possible relations between disorders in circadian rhythmicity and mental disorders. In: Emrich HM, Wiegand M (eds) Integrative biological psychiatry. Springer, Berlin Heidelberg New York, pp 159–180

Reality, Stress and Imagination in Temporal Isolation Experiments: An Interview with Rütger A. Wever

E. L. Rossi

Mathematical Models in Chronobiology: From Early Differential Equations to Current Chaos Theory

Rossi (R): I have been very interested in your efforts to develop mathematical models of psychobiological processes. Can you tell me something about the history and development of your thought in this area?

Wever (W): My first start in the field of biological rhythms in 1960 was on modelling procedures; I started with a simple electronic model. Due to the desired low-frequency output (period around 1 s) the electronic elements (e.g. the coil) had, inevitably, considerably non-linear features; it was only a few years later that this putative disadvantage turned out to be a great advantage, because the non-linearities established important biological features. The model was able to describe a remarkable multiplicity of biological results which were known at that time from a few animal experiments[1]. However, such a "hardware model" was limited in its flexibility, and it included the danger of confounding biological substrates with specific electronic elements (e.g. the coil or the capacitor). Very soon, therefore, I moved to a "software model", i.e. a mathematical differential equation[2].

When we came to Andechs, we performed a series of preliminary experiments with humans under temporal isolation in Munich. Only after we had finished these experiments successfully[3], did we start to build the special "bunker" for temporal isolation studies at the Max Planck Institute in Andechs in 1962. It was not ready for research until 1964. So I had the time to continue with some mathematical work before I could begin the main human experiments.

R: Did you have a strong mathematical background?

W: I am neither a biologist nor a physician. I was originally a physicist – specializing in nuclear physics. So I had some mathematical background. But I was certainly not an expert in differential equations when I started with a very simple oscillation equation to model biological rhythms. Therefore, I asked mathematicians for help. When I developed the differential equation which included some non-linearities, the mathematicians told me, "You cannot do that!" The normal way to deal with non-linearities at that time was to sterilize them by making them linear, at least stepwise. But I soon recognized that specific non-linearities were typical in describing apparently biological peculiarities. I recognized this long before the current emphasis on chaos theory in biology, which is all based on non-linearities, of course[4].

It is of some interest that in one of my early papers[5], I published a picture of a simulated biological rhythm just because of its curiosity; it appeared to me to be, on the one hand, so strange, that I couldn't do anything with it, but it appeared, on the other hand, to be of some interest. We can now recognize that it described a circadian rhythm with an ultradian rhythm superimposed on it. Only 25 years after I published that Figure, I measured how long the period was of the superimposed ultradian rhythm and found it was 90 min.

R: I believe this is fundamental for a new model of human consciousness and mind–body communication in general[6].

W: The model now is nearly 30 years old, i.e. it was developed before anything was known about human rhythms. During all this time it was not necessary to alter any term in the equation though nowadays it is to be applied mainly to human rhythms. In particular, at the time when the papers appeared (in the early 1960s) I had not heard anything about the 90-min ultradian cycles. Only about 20 years later this area became a bit more popular in mathematical congresses of rhythms in biology. I have some more recent publications wherein I summarize the model work[7,8]. In particular, I proceeded in modelling the interaction between circadian and ultradian rhythms, based on the unintended figure mentioned[9]; for instance, it could be shown that rather small changes in the circadian rhythm (without any change in the parameters of the ultradian rhythm) result in dramatic changes in the pattern of the superimposed ultradian rhythm, changing between very regular and nearly chaotic patterns[10]; and it could be shown that, with an unchanged actual circadian cycle but a change in a threshold separating sleep from wake, dramatic changes in the interacting ultradian cycle occur[11].

Paradoxical Performance and Mood with Desynchrony

R: I am interested in the paradoxical effects you describe where human performance apparently goes up and your subjects experience euphoria with the internal desynchronization of their psychobiological rhythms. You have summarized these results recently as follows[12]:

To summarize, in objectively measurable performance as well as in subjectively scored feeling, there are circadian rhythms with free-running components and with components entrained to a zeitgeber. It is the apparently paradoxical but significant result of many experiments with partial synchronization that both types of behavioral data, objective performance and subjective feeling, are better when the subject's rhythms are internally desynchronized than when they are synchronized (i.e., when the circadian system is in a state that contradicts the original biological meaning of the system). In addition, this apparently paradoxical state is combined with less sleep. These findings justify the performance of additional experiments with partial synchronization because such experiments could not be performed if there were any adverse effects. However, the present result is valid only for experiments lasting relatively short times (compared with a lifetime), and it will certainly be modified when the state of internal desynchronization lasts much longer. Experiments with flies (*Phormia terraenovae*) show a considerably shortened longevity when exposed to comparable rhythm disorders during total lifetime . . . The paradoxical result of the human experiments just discussed may be explained by a relative poverty of external stimulation during isolation wherein the internal stimulus of the rhythm disorder may optimally enhance the total amount of stimuli to be processed by the subject. Another explanation may be that subjects in the state of internal desynchronization are in a kind of euphoria. Because of the possible practical importance of this state, experiments are in progress to differentiate between these two explanations. In any case, the results discussed in the foregoing cannot be generalized to the statement that internal rhythm disorders lead to an enhanced state of well-being for the subjects.

R: What is your view about the biochemical basis of these paradoxical effects?

W: We have new confirmations of that apparently paradoxical effect, including those from other authors who performed similar experiments; however, I have no idea of its biochemical basis. Possibly, this effect is a kind of enhancement of the other effect which is likewise apparently paradoxical, namely that most subjects enjoyed generally their experience of an isolation experiment; at least for outsiders it sounds very strange that subjects agree voluntarily in performing long-term isolation experiments. The simplest way to show this enjoyment is to report that at least 80% of our subjects asked at the end, "Could I do another experiment?" Unfortunately, I could not do that with most of these subjects, because they did not have the time to "leave the world" for the term of another experiment. Nevertheless, we have still more than 50 subjects who participated in two, three, or even four temporal isolation studies in the bunker successively, with intervals of between 1 month and 19 years. Here, it must be emphasized that it is an indispensable pre-condition of this type of experiment that it is absolutely clear to all subjects, without any restriction, that they are free to terminate the experiment at any time; nothing is locked, so that they can leave the experimental unit even without informing the experimenters. As we know from many subjects, without the fulfilling of this pre-condition, the high motivation and the well-being of nearly all subjects cannot be guaranteed. As a result, not more than 10 out of all our subjects (more than 450, from 17 to 81 years old) made use of this possibility. About half of these subjects terminated the experiment for reasons which were independent of the special type of the experiments. The rate of subjects who stopped an experiment before the originally agreed time, therefore, was only in the range of 2%.

R: How do you account for that enjoyment? May I share a biochemical hypothesis with you? Would you agree that this desynchrony of biological rhythms is a stress?

W: Yes, I agree; we assume that internal desynchronization is a stress for the subjects, but we could not show this unambiguously, for instance by measuring an increase of special hormones. On the other hand, we have shown that stress can effect circadian rhythms by lengthening the free-running period as well as increasing tendency toward the occurrence of internal desynchronization of many physiological rhythms. To be sure, this effect of stress could be shown only under conditions which were unnatural and, hence, irrelevant under natural 24-h conditions.

R: Are you familiar with the research that documents how an ultradian peak in β-endorphin follows a peak of cortisol by about 20 min[13]? Could the general euphoria and sense of well-being your subjects experience in isolation be due to the fact that stress associated with internal desynchronization activated the cybernetic POMC–ACTH–cortisol–β-endorphin[14] cascade? Could heightened levels of these hormones lead to this enhanced level of performance and mood you observed in many of your subjects? I have hypothesized that this may be the biochemical basis of many of the similar paradoxical effects patients experience in emotional catharsis, hypnosis and psychotherapy[15]. Do you think this biochemical hypothesis could be explored with subjects living in your type of isolation experiments?

W: On the one hand, your idea has the great advantage that it is a testable hypothesis. And this means: yes, it could be done. On the other hand, there is a crucial difficulty. As far as I understand, the test of this hypothesis requires analyses of blood samples, and this means you have to take blood samples from the subjects in long-term experiments with intervals not longer than 20 min. It is well known that such an experimental procedure induces stress, and this means that the measuring procedure could itself account for the increase in blood levels of any stress-induced molecules you are measuring.

R: Recent research on salivary cortisol and a variety of other hormonal messenger molecules suggests a non-stress inducing methodology[16].

The Observer Effect

R: Do we have anything in chronobiology like the quantum effects in physics, where the presence of the observer interferes with what is being observed?

W: Yes, in a sense that at least the measuring procedure may interfere with the parameter to be measured. An example has just been given: to measure stress needs a procedure which, in itself, produces stress. What we cannot do, unfortunately, are double blind studies. The reason is that the experimental procedures (with humans as "experimental animals") are so subtle that a continuous control of the measuring data (e.g. body temperature or the activity of the subject) and the total technical equipment is indispensable, and such a control demands the experimenter's knowledge of the actual experimental protocol. For compensation of this disadvantage I never inform the subjects about the intention and the protocol of the experiment before the experiment, and I avoid telling the subjects anything about results of previous experiments (but, in addition, I tell the subjects that after the termination of their stay in the bunker, they can ask me whatever they want, and they can be sure to get right answers). Moreover, the protocol of every experiment always was fixed before the start of the experiment, and we followed this protocol disregarding the actual course. Apart from the mentioned continuous control of the health state and the mood of the subjects, we start analysing the experiment (e.g. with regard to period or amplitude of the rhythms) only after its termination. So, we avoid biassing ourselves by the actual results of running the experiment. Of course, after the completion of these analyses, the subjects were informed about all results of their experiment. In summary, we have the reasonable hope that we excluded feedback of the actual results during the experiment to the subject.

R: What did your subjects consistently find most difficult to believe when you questioned them afterwards?

W: The great majority of our subjects could not believe the deviation in the duration of their subjective days from the real clock time of 24 h. This is true particularly under constant conditions where the rhythms free-run; not one subject out of nearly 200 consciously experienced the deviation of their free-running subjective days from real clock time of 24 h, even when it was dramatic (between 12 and 65 h). In case of zeitgeber experiments with changing periods or phases, only a minority of more than 250 subjects experienced a change in the zeitgeber condition; and about half of these subjects estimated a wrong direction of this change. Surprisingly enough, all these wrong estimations were independent of whether or not the subjects had some (or even full) knowledge about the regularities of circadian rhythmicity.

R: Would you say it is easier to shift and entrain ultradian rhythms in humans?

W: I cannot say too much about that because we did not have continuous EEG-recordings of sleep (which only enabled us to establish ultradian rhythms) on more than about 10% of our subjects, most of them under constant conditions. And even these recordings constituted a tremendous expenditure (from every subject who was recorded completely and continuously, we got many miles of paper). Unfortunately, no external stimulus is known to be able to influence ultradian sleep rhythms, i.e. to entrain or phase-shift these rhythms. There is an interaction between circadian and ultradian rhythms, however. We can say that the circadian rhythm can be dramatically changed but the ultradian rhythm stays unchanged. For instance, changes in the length of circadian sleep episodes between about 5 and 25 h do not measurably change the period of the sleep rhythm (what is changing is only the number of sleep cycles per episode[17]).

The Mind–Body Connection

R: Can your studies say anything about mind–body communication?

W: The most important external stimuli that entrain and synchronize human circadian rhythms with all their physiological functions are of an informational and behavioural nature. It is not only the physiological effect of the light–dark cycle that modulates circadian physiology but also mainly its informational content. The light–dark cycle can be effective directly in shifting the circadian cycle, but only when the lights are bright enough (more than 2500 lux; i.e. more than any indoor illumination) and long enough (more than 6 h/day) at the right time. Therefore, normal people in our society usually cannot make use of the light stimulus alone. They usually need psychosocial cues to tell them when to sleep and when to wake up.

R: So what do you think of the new, popular light therapy that is used to deal with depression – particularly in seasonal affective disorder (SAD)[18]? Could that be a placebo effect?

W: It is clear that light therapy is very helpful though only in a small minority of depressives (suffering from SAD). It is not yet entirely clear to what extent this is a placebo effect. In some of the few studies which include effects of light of lower intensity the health effects of dim light (as an apparent placebo) and bright light (as the verum, true drug effect) are similar. The main problem is that a true placebo experiment testing the influence of light on SAD, is not possible in general because the subject can differentiate consciously between dim and bright light (at least when it is in operation only for short intervals as in light therapy). This is in great contrast to experiments testing drugs where the subject has no way of telling the difference between "placebo" and verum. This means that dim light is not a real placebo.

It is not yet clear whether the health effect of bright light in the case of SAD, on the one hand, and the strong zeitgeber effectiveness of bright light, on the other hand, are using identical pathways (though there are suggestions that the therapeutic effect in cases of SAD is based finally on a strengthening of the effective zeitgeber). In the case of the rhythmic effects of bright light it is obvious that it is not a placebo effect. We have a great number of different experiments where the reactions of the sleep–wake rhythm, i.e. of the only rhythm which is perceptible consciously, to dim and bright light are identical, while the reactions of the physiological rhythms (e.g. those of body temperature or cortisol secretion) are dramatically different; and no subject was able in any way to decide whether or not their physiological rhythms were synchronized to the zeitgeber[19].

We have found that information is the most important zeitgeber. An absolute light–dark zeitgeber (with normal light intensities) is effective only when the transition from light to dark cues the subject to rest, and the transition from dark to light enables the subject to start activities. This means information is a much stronger zeitgeber than the purely physical stimulus of the light–dark cycle; the same is true for social contacts between humans. In summary, for normal people living in normal situations, the psychological stimuli are the most important zeitgebers governing all the physiology of the body – the connection between mind and body[20].

R: That is why I ask you this question: is there proof that mind–body rhythms are a source, window, marker, or vehicle of the mind–body connection, as I and others are currently conceptualizing it[21–25]?

W: I said more than 20 years ago that one of the fascinations of the field of circadian rhythmicity is that it holds a boundary position between mind and body. The link between mind and body seems to include time as an inherent parameter.

R: That's your idea? I thought it was my idea! So, you have long had this view that the science of chronobiology, the rhythms of psychobiology, can say something about the essence of the mind–body connection. Where does research in this area need to go now? What is the next step in exploring the relationships between mind–body rhythms and mind–body communication?

W: You are speaking about humans?

R: Yes, of course.

W: After the properties of the circadian system in animals and humans are known to some extent, the way of research in the future will be the exploration of physiological pathways in the generation and control of circadian rhythmicity. In animals this means you have to do anatomical research with a knife, removing special organs, or separating these organs from others; so, you can look for the role of these organs in generating or controlling circadian rhythmicity. With humans, very fortunately, you cannot do that, so you have to think about more sophisticated experiments when you want to look for such physiological pathways. For instance, we are utilizing the technique of "fractional desynchronization" where we separate successively one rhythm after the other from a zeitgeber and, hence, also from each other[26]. This means we can investigate separated rhythms of different organs and also of different psychological performances, singly and independent of each other, and that even without any conscious perceptibility by the subjects. In this case, the rhythms are separated in time and not in space, as in the animal experiments mentioned, but the possibilities of getting insight into the functioning of the circadian system are similar with both types of separation (of course, the bloodless technique of fractional desynchronization can be utilized also in animal experiments).

R: What are practical implications of such chronobiology research?

W: Apart from the wide field of medical applications (you know, for instance, that mental disorders seem to be accompanied by rhythm disorders) there are the problems of "jet-lag" and shift work. In both cases we showed 7 years ago that bright light can alleviate the rhythm disorders which are inevitably associated with these alterations[27]. In simulated jet-lag studies (which constitute the more lucid problem) the duration of the re-entrainment to a shifted zeitgeber was halved when the subjects were exposed (during and after the shift) to a "bright-light zeitgeber" (with about 3000 lux during the light phase) instead of the "normal-light zeitgeber" (with about 300 lux during the light phase). Shift work, on the other hand, is more widespread and possibly more dramatic in its consequences. For instance, it is well documented that most atomic plant accidents happened at 3:00 or 4:00 a.m., when the human circadian rhythms do not adapt to the optimal performance that is required. Humans make more cognitive errors and have a longer reaction time at that time of day than at others. That is even true when they have slept before their night shift. The reason is the social nature of the relevant zeitgeber: these humans know very well that they have to stay awake at the wrong time, i.e. when most people are sleeping. And this knowledge apparently is sufficient to account for the fact that the rhythms of the shift workers usually remain unshifted, despite their shifted work schedule. Now, we have for the first time the opportunity to shift the rhythms of shift workers by exposing them to bright light. The workers feel better and they are more effective when they live and work in accordance with their circadian rhythms.

R: So you do recommend the use of bright lights to facilitate shift workers when necessary?

W: Yes, but many social factors are involved. You need, in addition to the bright light, great changes at the social level to optimize shift workers' performance.

R: You are referring to the important role of psychosocial stimuli in shifting the circadian clock?

W: Yes, but not exclusively. Nowadays, the shift worker knows a great deal that can help. *The information zeitgeber is paramount. For most people it is the most relevant stimulus*. Bright light can be effective and it can even override the information zeitgeber when applied in the right way and for long enough. To be sure, while most studies show that under the present lighting conditions a rapidly rotating shift system is optimal, under a bright-light zeitgeber a slowly rotating system should be preferred, to give the circadian system sufficient time to adapt to the shifted zeitgeber; it is clear that such a change in the rotation speed has social implications.

R: But in normal social living, information or psychosocial stimuli are usually more important than the physical light stimulus?

W: Normally, yes.

R: It seems that this is one of the more surprising and important conclusions of your lifetime of research.

W: Yes it is surprising – the extent of the interaction between information, social contact, and even imagination on the rhythms of mind and body.

What is Reality, Imagination and Social Contact?

W: When we say, "the most relevant zeitgeber for circadian rhythms is social contact", you should not ask me what is the meaning of "social contact". I don't know what that is. During our experiments, of course, there are no real social contacts between subject and environment, but more or less a simulation of social contacts (through our prearranged signals about how the experiments are to be conducted). So the subjects have an imagined social contact. But what is that?

R: Fantasized social contact? So there still is a mystery about what is real versus imagined psychological and social contact under such controlled living conditions?

W: Yes, what is the meaning of it all to the subjects? To give you an example. In group experiments, activities are synchronized; people eat together and sleep together – it is simpler to do things together. This is trivial – we know it well from observations of everyday life. But it is by no means trivial to find that when the individuals of the group are internally desynchronized (this state can be forced in every group by exposure to artificial days of 28 h or longer, or 22 h or shorter), the body temperature rhythms (and other imperceptible physiological rhythms) of the different individuals run in mutual synchrony – independent of their activity rhythms. We hypothesize that their mutual temperature synchronization can take place only through their social contacts. But what does "social contact" mean in case of the separated body temperature rhythms?

R: What is the informational pathway between social contact, the mind of the desynchronized individual, and the body rhythms that fall into synchrony in the group?

W: Yes, what is the pathway for that? One possibility would be the sense of smell. The only comparable example known to me is that women living in close proximity in a college will often find their menstrual cycles falling into synchronization. This mutual synchronization of menstrual cycles may take place via olfactory pathways by pheromones, unconsciously. But the daily temperature cycle – could that be synchronized by smell? That is hard to believe.

R: That leaves us with a profound mystery: how do psychosocial stimuli modulate the many circadian and ultradian mind–body rhythms?

W: Yes, we do not know that. There must be a psychobiological pathway, but we have no idea about what it is.

What is Stress? The Desynchronization Hypothesis

W: What is stress, for example. You need some stimulation for your well-being; only when this stimulation exceeds a special threshold, it acts as stress. It is clear that our subjects in an isolation experiment are socially deprived and, perhaps, they have a little bit less stimulation than they have in their social life and which is needed for optimal well-being. Then, the additional stress due to an internal desynchronization of their circadian rhythms may lift the level of stimulation just to the optimum – with the result that the well-being is improved under desynchronization. In the normal life, on the other hand, many people have enough stimulation beyond the optimum and, hence, are under stress even without an additional rhythm disorder; here, desynchronization would cause a troublesome increase of stress[28].

R: Hans Selye defined stress as a hypertrophy of the adrenals, a diminution of immune functions, and stomach ulcers. That triad defines something about the physiological pathways of stress. Could we introduce a time dimension into the concept of stress and say that stress itself is a result of internal desynchronization of mind–body rhythms? Could this be a new way of conceptualizing what stress is? We have both commented on the significance of Stroebel's finding in Rhesus monkeys[29], internal desynchronization led to a "stress condition" characterized by "psychosomatic" symptoms[22–24,30].

W: Stress leads to desynchronization – for that we have significant evidence. In addition, we have evidence that stress increases that what we call the "masking effect"[31].

R: But could you say that the desynchronization of mind–body rhythms is the essence of stress?

W: No.

R: Why not?

W: In normal life when you are under stress, you usually are not desynchronized. It could be that under special circumstances desynchronization is a trigger for stress – but only one possible trigger among many different triggers.

R: But what if normal life requires that you take a rest every hour and a half or so, as suggested by Kleitman's concept of the basic rest–activity cycle (BRAC). What if you do not take that rest because you are so compulsively driven by the outside world? Could that not lead to a desynchronization if you chronically ignored your natural ultradian rest–recovery periods? Are there any data that would support such an ultradian desynchronization hypothesis of stress?

W: What you are describing is, in my terminology, rather a deprivation from certain ultradian phases than a desynchronization. The statement of desynchronization requires the temporal relation to another rhythm of a deviating but similar frequency, either external or internal. However, we do not know any external rhythm that is able to control the ultradian rhythm, i.e. to act as a zeitgeber; and we do not know different ultradian rhythms which run with slightly different periods, i.e. which can desynchronize internally from each other. So the meaning of ultradian desynchronization is not really clear to me. On the other hand, during the wake episodes we have never seen an about 90-min rhythm in the state of activity or the rating of sleepiness, or in body temperature; this is true also in subjects who have no concrete programmes for their activities during the isolation and, hence, are certainly not under any temporal pressure. So I cannot see any indication of a sleep deprivation. But I have to say, my experience with ultradian rhythms is so small that I cannot reasonably comment on your hypothesis.

What I would like to know is to what extent in normal life the lack of social contact in older people leads to the internal desynchronization of their rhythms in the circadian domain. That is one of the great psychological problems of older people – they tend to lose their social contacts. What we do know now is that this is not only a psychological problem – it becomes, in addition, a physiological problem. They have a lack of social zeitgebers and, hence, they may desynchronize externally. The problem is that the sleep–wake rhythm, as the only one from which you can obviously see its course, is the rhythm with the closest connection to the zeitgeber. Consequently, the most critical danger is synchronization of the sleep–wake rhythm (pretending a fully synchronized rhythmicity) combined with an external desynchronization of the physiological rhythms. Such a partial external synchronization (which is hardly detectable without troublesome long-term measurements of physiological variables) is always combined with internal desynchronization. This could lead to periodic sleep disorders. In fact, many really old people in nursing homes have sleep–wake cycle disturbances of this type. We know only that they have 24-h sleep–wake rhythms, but we do not know about the status of their temperature rhythms (i.e. whether they are possibly free running); to clarify this we would need continuous temperature measurements over 2 or 3 weeks which, of course, are difficult to get in such persons.

R: This is research that needs to be done?

W: I can show you figures where, in healthy people, the periodic sleep disorders mentioned appear as a beat phenomenon. In these cases we have really measured differences in the periods of the sleep–wake and the temperature rhythms (in occasional cases where subjects became ill during a running experiment but insisted on continuing the experiment, this difference can be as small as the common difference between the 24-h day and the free-running period). Unfortunately, we do not know how frequently such different periods are present in old people.

So I would think it important to control the desynchronization of physiological rhythms in really old people who are separated from normal social zeitgebers. If such a desynchronization is suspected of being the cause of the sleep disorders you can try to restore the rhythms, for example, by bright light. We have sufficient evidence that bright light is effective as a strong zeitgeber in synchronizing the rhythms also of older people. It is technically not a simple problem to install bright-light devices because of heat problems. But I am sure that it is worth trying.

Psychological Implications of Temporal Isolation Experiments

W: The motivations of the subjects for participating in these experiments are very different. From our experience, money motivation is not sufficient to account for their interest and their feeling well in long-term isolation; we pay only small "pocket-money". Our subjects came for very different reasons. Many were students who needed an extended period to concentrate on their studies, preparing for examinations, writing their theses, or whatever. Other people said, "I want to learn about myself. I've never had a chance to be alone in my life. I want to think about me." At the end of their term in temporal isolation, a number of them said they felt better about themselves than before. It was nice to be with themselves.

R: Do you have any written accounts of these subjective reports?

W: We have some such reports, but we have published, as had been agreed with the subjects, only objective tests of some general items like "contentment", "alertness", "feeling of physical and psychical efficiency", and similar. In addition, we have tested performances of different complexity, memory and logical reasoning. The results of these tests indicate that there are different performance rhythms that can desynchronize from each other and, hence, are not functionally related to each other; rather, they are functionally independent of each other[32].

R: This reminds me of the functional autonomy of psychological processes that was a popular view a number of years ago in experimental psychology[33].

W: Normally these performance measures are in mutual synchrony; this is particularly true in the normal 24-h day. In the textbooks of several years ago, these were regarded as causal relationships. They certainly are not if they can be desynchronized from each other. The same is true for nearly all psychobiological variables. To emphasize this, we speak of a "multioscillatory system" to indicate their functional independence[34].

R: What does this do to the classical notion of homeostasis? Is it out-of-date now?

W: Certainly not! Rhythmicity and homeostasis are two sides of the same coin. Circadian as well as ultradian 90-min rhythms can only vary around a central tendency. Thirty years ago I was engaged in the study of control theory in biological systems and that led to my chronobiological studies. What we have learned is that, in the language of mathematics, typical biological homeostatic processes are characterized by rhythmic variations around constant means, and by non-linearities which are identical with the non-linearities governing circadian rhythmicity (which had been introduced originally for completely different reasons). The link between homeostasis and rhythmicity (established by the non-linearities mentioned) is the highly effective stabilization of all rhythm parameters (mainly period, mean value and amplitude); without these inherent stabilization mechanisms, slight external disturbances would bring the system out of control in the long run[35].

R: Did you find any change in the way your subjects experienced and recalled their dreams?

W: We found no change in the percentage of REM sleep (but in the pattern of REM sleep[12]). In a part of the experiments we asked the subjects to report their dreams. But we did not find systematic differences in comparison to their normal dreams.

R: Do your findings have implications for psychoanalysis or any of the psychotherapeutic arts that deal with the psychosomatic symptoms of stress?

W: I cannot rule that out, of course. But the connections are certainly not obvious to me because I am not competent in that field.

R: Did any of your subjects who wanted to get away from it all to learn about themselves actually come up with important insights? Did anyone say, "Gee, I found the answer to the meaning of my life!"?

W: Not specifically. We looked at personality data before and after every experiment: did the subjects become more neurotic, depressive, or better integrated? There were no negative effects. Rather, the subjects tended to become more stable and to show an improved IQ during a long-term isolation experiment (but not at significant levels). Certainly we never found any physical reason to stop the experiments, and that is even true in case of internal desynchronization, which is not a physiological state. According to all our knowledge, it is the biological meaning of the circadian system to avoid such a state of internal desynchronization. The fact that desynchronization led to the improvement in some mood and performance measures suggests that we might explore the possible therapeutic value of forced desynchronization in cases of mental disease, as a kind of stimulation therapy.

R: What are your projects for the future?

W: You may know that my former research unit was closed two years ago, i.e. some time after my own retirement. This means that I am no longer able to do further experiments. What I have to do for many more years, however, is to analyse data from former experiments (and to write the corresponding reports). No fear that I will ever run out of things to do or become bored!

However, I have the feeling now, in complete contradiction to what I felt 5 years ago – or even 3 years ago – that I have reached the state in my experiments that I can stop my experimental work without great frustrations. Obviously, the results all fit nicely together now.

R: There has been a natural completion to your experimental work.

W: Yes, at least to some degree. This does not mean that I lack further questions for experiments within the next 20 or even 50 years. But there are no questions the answer to which is essential for the congruity of my concept of the circadian system. Five, or even 3 years ago I still had to say, "This and that do not fit together – I have to look for missing links." At that time it would have been very frustrating to stop experimenting, but now it is not. I am speaking here primarily about the exploration of basic system properties of circadian rhythmicity; the utilization of the new knowledge for applications in many areas is only at its beginning.

R: That must be a wonderfully satisfying feeling for a scientist to have about his life's work.

W: My statement, of course, cannot exclude the possibility that 5 years from now the analyses will run into new discrepancies and, hence, I will have the feeling, "No, it was too early to stop after all."

R: Back to the laboratory?

W: Unfortunately, this will not be possible since there is no longer a laboratory available. I can go back only to my large pool of data, in the hope of finding answers to the new questions in the old data.

Notes

1. Wever R (1960) Possibilities of phase-control, demonstrated by an electronic model. Cold Spring Harb Symp Quant Biol 25:197–201.
2. Wever R (1962) Zum Mechanismus der biologischen 24-Studen-Periodik. Kybernetik 1:139–154.
3. Aschoff J, Wever R (1962) Spontanperiodik des Menschen bei Auschluss aller Zeitgeber. Naturwissenschaften 49:337–342.
4. Wever R (1964) Zum Mechanismus der biologischen 24-Studen-Periodik. III. Mitteilung: Anwendung der Modell-Gleichung. Kybernetik 2:127–144.
5. Wever R (1963) Zum Mechanismus der biologischen 24-Studen-Periodik. II. Mitteilung: Der Einfluss des Gleichwertes auf die Eigenschaften selbsterregter Schwingungen. Kybernetik 1:213–231.
6. Rossi E (1991) The wave nature of consciousness. Psychol Perspect 12:6–14.
7. Wever RA (1984a) Toward a mathematical model of circadian rhythmicity. In: Moore-Ede MC, Czeisler CA (eds) Mathematical models of the circadian sleep–wake cycle. Raven Press, New York, pp 17–79.
8. Wever RA (1987) Mathematical models of circadian one- and multioscillator systems. In: Carpenter GA (ed) Some mathematical questions in biology: circadian rhythms. American Mathematical Society, Providence, RI, pp 205–265.
9. Wever RA (1985b) Modes of interaction between ultradian and circadian rhythms: toward a mathematical model of sleep. Exp Brain Res 12 [Suppl]:309–317.
10. Wever RA (1991) Interactions between human circadian and (about 90-min) sleep rhythms: problems in the simulation and the analysis. In: Haken H, Koepchen HP (eds) Synergetics, Vol 55, Rhythms in physiological systems. Springer, Berlin Heidelberg New York, pp 235–253.
11. Wever, Chap. 15.
12. Wever RA (1982) Behavioral aspects of circadian rhythmicity. In: Brown FM, Graeber RC (eds) Rhythmic aspects of behavior. Lawrence Erlbaum Associates, Hillsdale, NJ, pp 105–171.
13. Iranmanesh A, Lizarradle G, Johnson M, Veldhuis J (1989) Circadian, ultradian, and episodic release of β-endorphin in men, and its temporal coupling with cortisol. J Clin Endocrinol Metab 68:1019–1025.
14. Abbreviations: POMC, proopiomelanocortin; ACTH, adrenocorticotrophic hormone; SAD, seasonal affective disorder; BRAC, basic rest–activity cycle.
15. Rossi E (1990) From mind to molecule: more than a metaphor. In: Zeig J, Gilligan E (eds) Brief therapy: myths, methods and metaphors. Brunner/Maz, New York, pp 445–472.
16. Kirschbaum C, Hellhammer D (1989) Salivary cortisol in psychobiological research: an overview. Neuropsychobiology 12:150–169.
17. Wever RA (1985d) Circadian aspects of sleep. In: Kubicki S, Herrmann WM (eds) Methods of sleep research. G. Fischer Verlag, Stuttgart, pp 119–151.

18. Rosenthal N (1989) Seasons of the mind: why you get the winter blues. Bantam, New York.
19. Wever RA (1986) Characteristics of circadian rhythms in human functions. J Neur Transmiss 21 [Suppl]:323–373.
20. Wever RA (1989) Light effects on human circadian rhythms: a review of recent Andechs experiments. J Biol Rhythms 4:161–185.
21. Rapp P (1987) Why are so many biological systems periodic? Prog Neurobiol 29:261–273.
22. Rossi E (1982) Hypnosis and ultradian cycles: a new state theory of hypnosis? Am J Clin Hypn 25:21–32.
23. Rossi E (1986a) Altered states of consciousness in everyday life: the ultradian rhythms. In: Wolman B, Ullman M (eds) Handbook of altered states of consciousness. Van Nostrand, New York, pp 97–132.
24. Rossi E (1986b) Hypnosis and ultradian rhythms. In: Zilbergeld B, Edelstien M, Araoz D (eds), Hypnosis: questions and answers, W. W. Norton, New York, pp 17–21.
25. Rossi E (1986c) The psychobiology of mind–body healing: new concepts of therapeutic hypnosis. Karger, New York.
26. Wever RA (1983) Fractional desynchronization of human circadian rhythms: a method for evaluating entrainment limits and functional interdependencies. Pflügers Arch 396:128–137.
27. Wever RA (1985c) Use of light to treat jet lag: differential effects of normal and bright artificial light on human circadian rhythms. Ann NY Acad Sci 453:282–304.
28. Wever RA (1988) Order and disorder in human circadian rhythmicity: possible relations to mental disorders. In: Kupfer DJ, Monk TH, Barchas JD (eds) Biological rhythms and mental disorders. Guilford Press, New York, pp 253–346.
29. Stroebel C (1969) Biologic rhythms correlates of disturbed behavior in the Rhesus monkey. In: Rohles FH (ed) Circadian rhythms in non-human primates. S. Karger, New York, pp 91–105.
30. Wever RA (1979) The circadian system of man. Springer, Berlin Heidelberg New York.
31. Wever RA (1985a) Internal interactions within the human circadian system: the masking effect. Experientia 41:332–342.
32. Folkard S, Wever RA, Wildgruber CM (1983) Multi-oscillatory control of circadian rhythms in human performance. Nature 305:223–226.
33. Boring E (1950) A history of experimental psychology. Appleton-Century-Crofts, New York.
34. Wever R (1975) The circadian multi-oscillator system of man. Int J Chronobiol 3:19–55.
35. Wever RA (1984b) Properties of human sleep–wake cycles: parameters of internally synchronized freerunning rhythms. Sleep 7, 27–51.

Chapter 17

Human Biological Rhythms: The Search for Ultradians in Movement Activity Behaviour

Helen C. Sing

Introduction

Human biological rhythms are the expressions of orderliness governing the proper functioning of the organism throughout its existence. The most familiar of these rhythms is the circadian (cycling about once a day) by virtue of its immediacy in generating the pattern of daily life. It is centred around a period of approximately 24 h, give or take an hour or two in either direction. It is the easiest to track in that under normal circumstances humans dutifully follow the sleep–activity cycling within each 24-h day. When this circadian rhythm is disrupted by whatever reason (i.e. continuous wakefulness (or sleep), transmeridian flight across several time zones, illness, etc.), the individual feels "out of sorts" or desynchronized from his or her ordinarily constant patterns. This is because many of the major physiological measures in humans also follow the same circadian rhythmicity as the sleep–activity cycle. These include body temperature, heart rate, movement activity, brain activity, cognitive performance, levels of certain hormones, enzymes, and blood constituents. When the circadian rhythms of these major regulators of normal functioning for humans are perturbed, then a system accustomed to running "like clockwork" may be on the verge of disorder. But the human is an eminently adaptable entity and unless the perturbation is extraordinarily severe and prolonged, circadian rhythms are sufficiently labile to allow adjustment to the new circumstances over time.

What underlies the circadian pattern and its adaptability to change? If all that is evident is the cycling once a day (in which for each measure there is a peak and a trough) then there is not much information to determine what happens for the rest of the 24-h period. Might there not be other, faster rhythms within the framework of the circadian contributing vital support, or, alternatively, an ensemble of harmonics of which the circadian is the fundamental? The literature cited within this volume by other authors attests to the existence of such rhythms. What are some of these faster rhythms? Collectively they have been called *ultradian* rhythms, i.e. cycling faster than once per 24 h, ranging from 12 h to cycles of a minute or less. Examples of ultradian rhythms abound in the literature including cellular (Lloyd and Kippert 1987), neuronal (Stevens et al. 1971; Naitoh et al. 1973), hormonal (Weitzman et al. 1974; Veldhuis and Johnson 1988), cognitive (Globus et al. 1971; Orr et al. 1974), hypnotic (Rossi 1982), motor (Globus et al. 1973), physiological (Cugini et al. 1987; Meneses and Corsi 1990), metabolic (Brandenberger et al. 1985), sensory and psychomotor perception (Lovett Doust and Podnieks 1975), visual vigilance reaction time (Lovett Doust et al. 1978; Lavie 1979), respiratory (Horne and Whitehead 1976) and biochemical (Mejean et al. 1988) processes. In addition, there are at least two excellent review articles (Broughton 1982; Stupfel and Pavely 1990). Among those who study such rhythms, the goal is to characterize (and possibly categorize) all aspects of human rhythmicity and thus provide the capacity for intervention when rhythms go awry. Implicit in this aspiration is the ability to measure, analyse, and define these rhythms. To this end we (see Acknowledgements, p. 369) would like to add to this pool of collective knowledge an exhaustive study of two sets of human movement activity data collected with non-invasive measurement devices. Analytical techniques are presented for the determination of the ultradian rhythms that may be inherent components of the data.

Search for Ultradian Rhythms – A Case Study with Movement Activity

Source of Data

The Walter Reed Army Institute of Research has had a long and abiding interest in the use of movement activity monitors (commonly referred to as actigraphs). The Walter Reed actigraph (see Redmond and Hegge 1985) is a small rectangular device (6.5 cm × 4 cm × 1.5 cm) worn like a watch on the individual's non-dominant wrist. Among the many appealing attributes of the actigraph are: (a) programmability in setting the epoch length of movement recording ranging from 1 s to several minutes; (b) capacity to store the movement activity within its self-contained memory over several days; and (c) ruggedness. It allows the gathering of a vast amount of data, and does so in such a very safe and non-invasive manner that the wearer is oblivious to its presence and can pursue his or her daily activity unrestrained. Actigraphs have been fielded during training exercises, wartime, transmeridian flights, etc., to gauge sleep and rest accurately, quantitatively and qualitatively in military personnel. In our laboratory, where the effects of sleep deprivation are studied, the actigraphs are worn regularly by human subjects while continuously awake over 64 h and while in recovery sleep of 8–10 h. During the 64 h of wakefulness, subjects take a battery of computerized cognitive performance tasks (Thorne et al. 1985) every 2 h (lasting approximately 20 min) and are otherwise allowed normal activities, i.e. reading, walking, watching TV, etc. Normal sleep/rest activity of each subject is also recorded during the week before the subject enters into a sleep-deprivation study. All activity records along with their identification and recording parameters are transferable from the actigraph readout unit to a personal computer (PC) for further analysis.

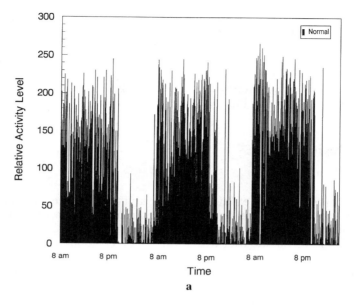

a

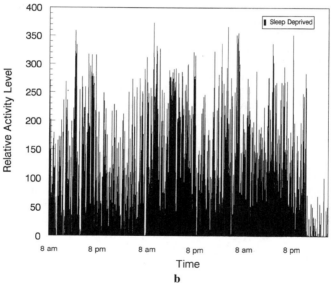

b

Fig. 17.1. a Record of raw activity data from an actigraph worn on the non-dominant wrist of an individual pursuing his normal daily routine. Each column represents an epoch length of 2 min of cumulation of movement activity. The total time displayed for this record is 68.27 h. **b** Record of raw activity data from an actigraph worn on the non-dominant wrist of an individual confined to a laboratory suite and kept continuously awake for at least 62 h. The epoch length and total time of the record are the same as in **a**.

Our case study consists of activity measures taken from two male subjects – aged 36 years (normal) and aged 37 years (sleep deprived). The length of each activity epoch in the normal record was set at 2 min to accommodate 7 days of recording. In the sleep-deprived case, epoch length was set at 30 s for the 4 days of the experiment and were transformed to 2-min epochs, allowing for direct comparisons with the normal activity records. Examples of these motor activity records are shown in Fig. 17.1a and b.

Data Analysis
Fast Fourier Transform

The activity record may be viewed as a signal varying over time and in this case because of its epoch-by-epoch recording, a discrete-time sampled series. The standard treatment for a discrete-time series is initially to determine its power spectrum through the fast Fourier transform (FFT) technique so that an idea of the signal content is obtained. This is a method for taking time-dependent information and transforming it into what is known as the frequency domain. In this sense the varying values of the time series may be viewed in terms of the relative magnitudes of the frequencies that are constituents of the signal. For example, a data set containing only a pure sine or cosine function of a single frequency will have a spectrum consisting of only one value, representing the estimate of power or magnitude (the square root of power) of that single frequency. For a signal consisting of mixed frequencies, the spectral representation will consist of estimates of the relative magnitudes of these frequencies. A limitation of the FFT is its requirement that the number of input values must be a power of 2. This is not a serious limitation and most data sets can be trimmed or padded with zeros to satisfy this requirement.

FFT Output

In our FFT analysis, we used a software package called ANAGRAPH, specifically designed and written for PCs by Dr Alan Fridlund (personal communication). This software enabled speedy and efficient FFT processing of our two sample data sets with options of various data transformations, detrending methods (i.e. removing the trend component which in this case is the mean) and filters. We selected the options of detrending by mean subtraction and spectral analysis by periodographs. The results of these analyses are displayed in Fig. 17.2. In interpreting the output of the FFT, results are in terms of frequencies, i.e. cycles which are the reciprocals of period lengths. For example, the circadian rhythm is defined in frequency as one cycle per day (1 cpd) with a period of 24 h. Harmonics of it are given as cycles per day or period lengths of fewer than 24 h. Table 17.1 gives the relationship of representative cycles and periods with the circadian as the fundamental rhythm.

The FFT output consists of one-half of the number of the original data points which are indexed sequentially starting at zero. The zero index refers to zero frequency (also known as the "direct current" or DC frequency) and is usually

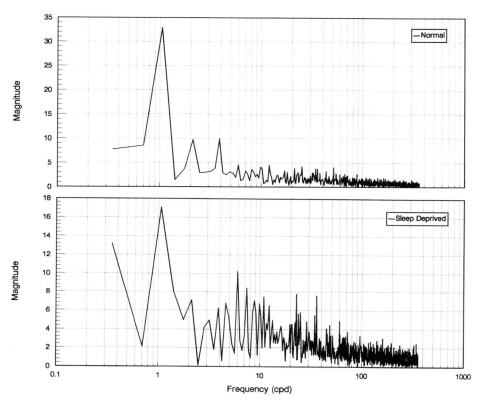

Fig. 17.2. Spectra of the frequency components from Fourier transform of the raw movement activity data shown in Fig. 17.1a and b for the normal subject (top) and for the sleep-deprived subject (bottom).

the mean value of the data set. Each subsequent index is linked with a frequency and corresponding period value. The frequency is calculated from division of its index by the total number of data points used in the FFT, while the period is its reciprocal. These calculations presume unit epoch lengths so that the period output of the FFT may not be the same in real time as for the data set in question and it may not even be in integer multiples of the period of interest, which in our case is 24 h or 1440 min. Hence, some further calculation is needed to scale the frequency to relate to the actual periods or frequencies of interest (cycles per day). In Fig. 17.2, this adjustment has been made, as explained below.

Interpretation of FFT of Data Sets

Our data sets consist of 2048 (a power of 2) values, each with an epoch length of 2 min. The actual period for the entire data set then is 4096 min (equal to 68.27 h) and is the period of index 1. Subsequent periods are multiples of the

Table 17.1. Integer periods and frequencies based on 1440 min/day

Frequency (cpd)	Period (min)	Frequency (cpd)	Period (min)	Frequency (cpd)	Period (min)
1	1440	25	58	53	27
2	720	26	55	55	26
3	480	27	53	58	25
4	360	28	51	60	24
5	288	29	50	63	23
6	240	30	48	65	22
7	206	31	46	68	21
8	180	32	45	72	20
9	160	33	44	76	19
10	144	34	42	80	18
11	131	35	41	85	17
12	120	36	40	90	16
13	111	37	39	96	15
14	103	38	38	103	14
15	96	39	37	111	13
16	90	40	36	120	12
17	85	41	35	131	11
18	80	42	34	144	10
19	76	43	33	160	9
20	72	45	32	180	8
21	69	46	31	206	7
22	65	48	30	240	6
23	63	50	29	288	5
24	60	51	28	360	4

reciprocal of the indices. Thus, under these circumstances, we do not obtain an exact 24-h period, the closest being 22.76 h or 1365 min for index 3. Nevertheless, the spectral output of the normal and sleep-deprived data sets have distinctly different characteristics. We can immediately observe that the relative power of the 68.27-h period is almost as large as that of the 22.76-h period in the sleep-deprived subject's spectra, reflecting the continued wakefulness/ activity (64 h) over much of that time. On the other hand, the spectra for the subject with normal daily cycle of sleep and activity have the bulk of the power located at the near circadian period of 22.76 h. Its magnitude is almost twice that of the sleep-deprived subject's, in this instance reflecting the dominance of regularity within the circadian period. Differences in power for other frequencies are also noted between the two subjects' spectra and these are discussed in greater detail in later sections when another analytical technique is introduced.

In a statistical sense, the power of these frequencies may be thought of as the variance accounted for by each frequency component contained in the time series. Thus, the greater the power, the more dominant is that frequency in the contribution to the waveform of the time series, or, alternatively, the more confident we may be that the frequency is "real" and not random noise.

Complex Demodulation Technique

Although the FFT provides a quantitative sense of the magnitude of each constituent frequency within the realm of possible frequencies for the data set, it does not provide the temporal relationship of these frequencies and so we cannot discern the behaviour of individual frequencies over the entire sampling time. To uncover the temporal aspect, we make use of another mathematical analysis technique called complex demodulation (Sing et al. 1980), which has been regularly employed in our laboratory for the analysis of various time-based recordings such as actigraphy, electrocardiography (ECG), cognitive performance tasks, mood assessments, body temperature, etc. This analytical method not only determines the relative amplitude of constituent frequencies but also provides a smoothed function for each frequency that can be displayed over the same time interval as the original data. From this smoothed function, which is termed a remodulate and which oscillates around zero amplitude, characteristic properties can be determined such as actual period lengths, peak and trough amplitudes, and zero crossover periods (period length calculated between the crossings of the waveform at zero and encompassing a peak and a trough within these limits). The real time of occurrences for each property can be determined as well.

Methodology

A very brief description of the methodology for complex demodulation will be presented without the usual reference to mathematical notations. Further details and some applications may be obtained from the following: Bloomfield 1976; Sing et al. 1985; Babkoff et al. 1991.

The first step in the method is the detrending of the data set by subtracting the mean, calculated from all the values in the series, from each datum. Each resulting value in our activity time series is then transformed into two values by multiplication with a cosine function for one value and a sine function for the other. The argument for the sine and cosine is the same and is equivalent to the period (or frequency) of interest and its sequential index in time. Since the frequency of the circadian rhythm is considered to be the fundamental and all other frequencies in the series to be harmonics of it, the initial transformation of the data results in mapping values with implicit circadian rhythm onto the complex plane and essentially shifting this desired frequency into the zero frequency position. The complex plane simply means that the real and imaginary, or respectively cosine and sine, are now represented in what are x and y axes in Euclidean geometry. Mathematically, this has great utility in that amplitudes and phases are easily derived using these representations. Suffice to say that multiplication with sines and cosines produces sums and products of other frequencies which are considered to be "noise" in light of the desired frequency and these frequencies must be eliminated. The process for doing this is digital filtering which in practice requires fast computers.

Filter

The key to the capability and success of the complex demodulation technique in extracting a single frequency from a mixture of frequencies lies in the characteristics of its filter. Our original exponential filter (Sing et al. 1980) was successful in extracting relatively pure frequencies from data of 10 min or longer epochs and where the interest was in the ultradian range of frequencies of fewer than about 36 cpd. It was found not to be as successful with data sampled at shorter epoch lengths where the potential for faster ultradian rhythms may be determined. Consequently, a low pass filter consisting of 128 terms (coefficients) was designed and then tested with a series of cosine functions of known frequencies (i.e. from 1 cpd to 360 cpd, as given in Table 17.1) generated from the same ANAGRAPH software used above in the FFT analysis. The output of the operation with this new filter on each of the tested frequencies provided proof that the filter not only succeeded in extracting the desired frequency centred at the zero frequency position, but also preserved the original amplitude value. By folding out the data series at either end by 128 values each before the filtering process, and by filtering with forward and reverse passes twice, the usual phase shifting inherent to filtering processes was eliminated. The actual filtering procedure involved multiplication and summation of the 128 values of the filter with a corresponding set of 128 values of the data set at each instance so that each pass of the filter involves 2176 of these 128 multiplication and summation steps (557 056 operations!). Some end effects of amplitude diminution at the very end of the data set are observed for the very fastest frequencies tested, i.e. 240 and 360 cpd. One might be well advised in movement activity studies to extend recording to a few hours beyond that required if very fast frequencies are to be studied. (We note that this portion in our data sets involved activity measures where both subjects were asleep.) The incorporation of this filter into our standard complex demodulation technique enables the analysis of data sampled at any epoch length.

Harmonics of the Circadian Rhythm

Once the circadian remodulate has been determined, its values are subtracted from the original detrended data set. The reason for doing this is because the magnitude of the frequency of the circadian rhythm is usually so much greater than the other frequencies that its position adjacent to zero frequency would cause leakage of power from it to the frequency being extracted and may give erroneously high power values. All harmonics of the circadian frequency are complex demodulated from this new data set.

Our interest is focussed mainly on human biological rhythms and, as such, the circadian rhythm whether known as a frequency of about 1 cpd or a period in the range of 22 to 26 h is fundamental. Rhythms with frequencies slower than that of the circadian are categorized as infradian and those faster as ultradian. Previously (Sing et al. 1980), we arbitrarily, but descriptively, categorized frequencies as intermediate (2–5 cpd); subultradian (6–11 cpd); ultradian (12–20 cpd); supra-ultradian (21–29 cpd); and very high (30–36 cpd). We need to extend these categories as we progress to determination of even higher frequencies.

We are theoretically able to derive ultradian rhythms up to a frequency of 360 cpd (period of 4 min) from the 2-min epochs of the activity record. This is the Nyquist limit of discrete time sampling theory in which the maximum period derivable from a time series is twice the sampling rate. Prudence dictates the selection of 72 frequencies to represent the range of 360 possible frequencies contained within each actigraph data set. These frequencies with their associated periods are those presented in Table 17.1. Note that this choice was deliberate to provide a correspondence of integer periods.

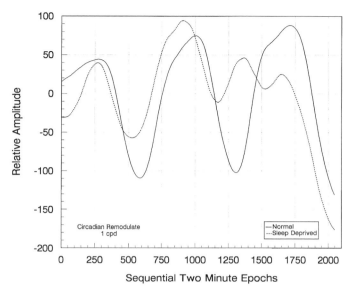

Fig. 17.3. Circadian functions (remodulates) of the normal and sleep-deprived subjects extracted from the activity data by complex demodulation. Note that the normal subject maintains stable cycling while the sleep-deprived subject shows bifurcation of his/her circadian peak on the third day of continuous wakefulness.

Remodulates

One of the features of complex demodulation is the generation of phase and amplitude values for each frequency extracted at each epoch in the time series. The remodulate function for each frequency is reconstructed from these phase and amplitude values of each epoch of the demodulated frequency. Fig. 17.3 illustrates the circadian remodulates, and other representative frequencies are shown in Figs. 17.4–17.8. Strictly speaking, these waveforms should be represented as their discrete epochs; however, we have connected the epochs with a continuous line to aid in visualizing the undulation of the waveforms. In contrast with spectral analysis, the remodulate depicts exact peaking and troughing in the time of a frequency's cycling so that comparison of the amplitudes from one cycling to the next may be observed. One can note how each cycle in a particular frequency waxes and wanes over a time interval, either in or out of phase with other frequencies over the same time interval. One can also imagine the difficulty in comparing and grasping the significance of each frequency over the entire sampled time interval and across all frequencies! The next section provides one way in which these may be visualized simultaneously.

Comparison of Normal Versus Sleep Deprived

It was indicated above in the discussion of spectral output from the FFT that the frequency structure of our subject with normal rest/activity is notably

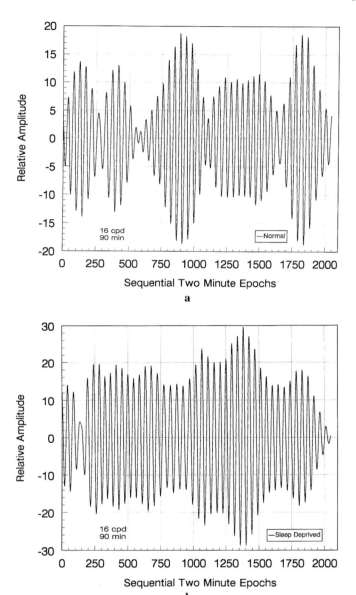

Fig. 17.4. a Remodulate of the 16 cpd (90-min period) activity pattern of the normal subject. The rhythm appears to cycle in bursts of waxing and waning. **b** Remodulate of the 16 cpd (90-min period) activity pattern of the sleep-deprived subject. Unlike the normal subject, there is less evidence here of cycling in bursts.

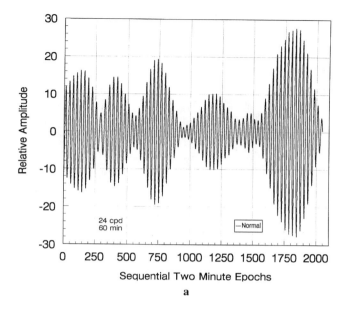

a

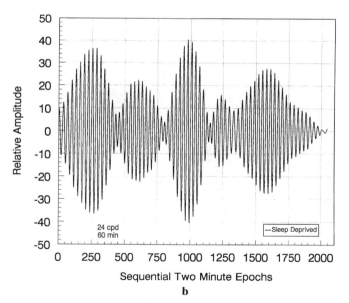

b

Fig. 17.5. a Remodulate of the 24 cpd (60-min period) activity pattern of the normal subject. Note the high amplitude of the rhythm on the third day (centred around epoch 1750) indicating greater movement activity than in the previous 2 days, although this is not immediately obvious in the raw activity record. **b** Remodulate of the 24 cpd (60-min period) activity pattern of the sleep-deprived subject.

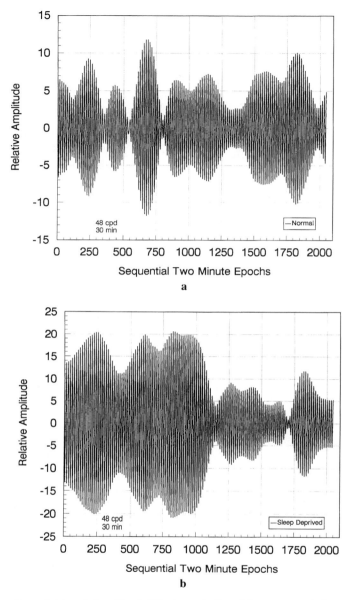

Fig. 17.6. a Remodulate of the 48 cpd (30-min period) activity pattern of the normal subject.
b Remodulate of the 48 cpd (30-min period) activity pattern of the sleep-deprived subject.

different from our sleep-deprived subject. This difference is revealed in
greater detail by intersubject comparison of the remodulates. In Fig. 17.3, the
remodulate of the circadian rhythm for our normal subject has a set of well-
defined peaks and troughs within each 24-h period for the entire 3 days of the

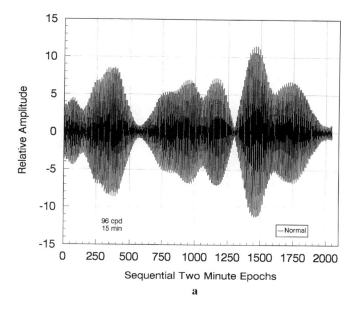

a

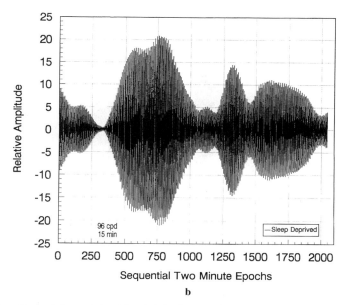

b

Fig. 17.7. a Remodulate of the 96 cpd (15-min period) activity pattern of the normal subject. **b** Remodulate of the 96 cpd (15-min period) activity pattern of the sleep-deprived subject.

data record, with the peaks of almost equal amplitudes. The remodulate of our sleep-deprived subject, on the other hand, shows evidence of bifurcation of the circadian peak after 40+ h of continued wakefulness, an indication of incipient disruption of the circadian rhythm. Peaks and troughs of the next frequency of

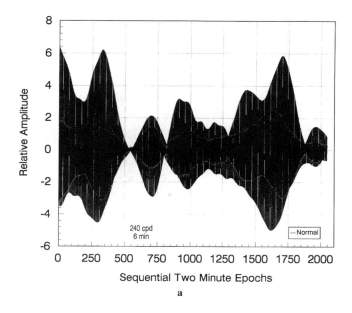

a

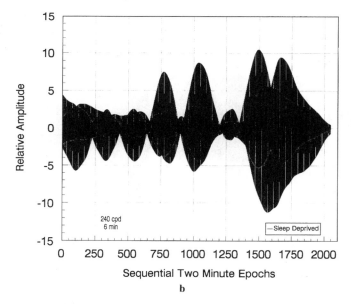

b

Fig. 17.8. a Remodulate of the 240 cpd (6-min period) activity pattern of the normal subject. **b** Remodulate of the 240 cpd (6-min period) activity pattern of the sleep-deprived subject.

2 cpd (period of 12 h) are double the number of circadian peaks and troughs within each 24-h period and similarly increase proportionately with each higher frequency.

The enormous difficulty in envisaging simultaneously all the demodulated frequencies, their respective peak amplitudes and their temporal relationships is made easier by a surface plot of the type presented for each subject in Fig. 17.9a and b. These plots are a three-dimensional representation of frequency on the abscissa, epoch number (time in 2-min intervals) on the ordinate, and amplitude of the peaks on the vertical axis. The striking difference between the normal and sleep-deprived activity records over 68 h is immediately apparent. At the same time, for both records, the waxing and waning of peak amplitudes within and across frequencies are readily observable, reinforcing the notion of non-stationarity of human rhythms. Bands of frequencies are seen to be in phase, with approximately equal peak amplitudes at one moment, and then are both diminished in amplitude and out of phase at another moment in time. Thus, the distribution of energy across time and across the frequency spectrum may be easily compared from record to record.

Ultradian Rhythms of Note

An important consequence of knowledge of the remodulate parameters determined from complex demodulation is the ability to derive a variety of graphical plots which provide further enlightenment of not only the relative contribution of each of the 72 individual frequencies but also the intrinsic relationships of these parameters. Information conveyed from illustrative plots as shown in Fig. 17.10a and b of the frequency versus peak amplitude and in Fig. 17.11a and b of calculated zero crossover period versus peak amplitude indicate that there is indeed a major contribution of energy in the ultradian range of 30 to 45 cpd (period of 48 min to 32 min). Even the very fastest frequencies, i.e. 240 to 360 cpd (periods of 6 and 4 min), appear not to be noise components in the activity record, but may in fact be minor contributors of energy at key moments of time.

Non-linear Dynamics Aspects

The remodulate functions obtained from complex demodulation show very clearly that human biological rhythms are not symmetrical stationary waveforms, i.e. they are not sine or cosine functions which have equal peak and trough amplitudes and cycle consistently with the same period length. Rather, the circadian remodulate of humans under normal conditions is asymmetrical, with a longer interval of positive values than of negative values; that is to say more time is spent in activity than in the rest phase, and its exact period from one day to the next may be very variable. Hence, in a real sense, predictability of exact circadian periodicity is not possible – one can give only probable periods between a narrow range of values. If this is true of the circadian rhythm which is the most robust of all the frequencies, then one would infer that higher frequencies are equally variable over time. This is evident in the

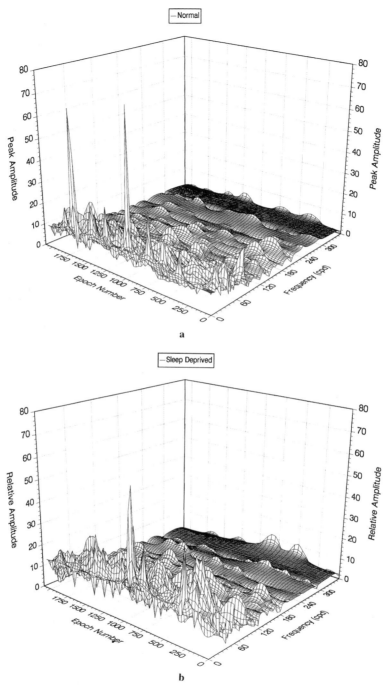

Fig. 17.9. a Surface plot of the peak amplitudes of remodulates derived from complex demodulation for the normal subject. Frequency (cpd) is displayed on the abscissa [*x*]; epoch number on the ordinate [*y*]; and relative amplitude on the vertical [*z*] axis. Visual comparison of peak amplitudes of all the frequencies may be observed simultaneously. Note that frequencies between 288 cpd and 360 cpd are not extracted for this analysis and are seen as the band of flat surface in the plot. **b** Surface plot of the peak amplitudes of remodulates derived from complex demodulation for the sleep-deprived subject. The axes are oriented as in **a** and the same explanations apply.

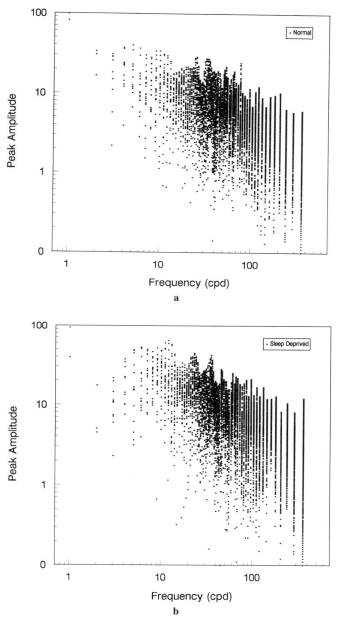

Fig. 17.10. a Scatter diagram of frequency versus peak amplitudes of the remodulates for the normal subject plotted on logarithmic scale. Peak amplitude values of each frequency's remodulate is mapped on that frequency. Note that the higher the frequency, the greater the number of peaks plotted. The span of peak amplitudes of each frequency and across frequencies may be compared. **b** Scatter diagram of frequency versus peak amplitudes of the remodulates for the sleep-deprived subject plotted on logarithmic scale. See **a** for explanation.

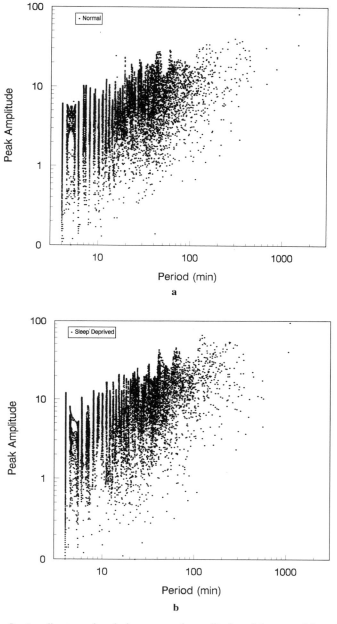

Fig. 17.11. a Scatter diagram of period versus peak amplitudes of the remodulates for the normal subject plotted on logarithmic scale. The peak amplitude value within each calculated zero cross-over period of a frequency's remodulate is mapped on that period. Unlike the peak amplitude versus frequency scattergram, where we denote fixed cycles, the calculated zero crossover period can vary; hence, the arrangement of the scatter is not as orderly as in Fig. 17.10a and b. **b** Scatter diagram of period versus peak amplitudes of the remodulates for the sleep-deprived subject plotted on logarithmic scale. See **a** for explanation.

changing amplitudes and period lengths observed for each frequency. In the Figures, it appears that each frequency's cycling is in dynamic flux.

Phase-Plane Plots

The field of non-linear dynamics systems has generated a multitude of innovative ways to look at data that are oscillatory but not with exact periodicity. Graphical depictions borrowed from non-linear dynamics systems are a natural way to envisage the pseudo-periodic nature of these waveforms. These are graphs called phase-plane plots in which the time series is plotted with ordinate value as the next sequential value (or some other integer lag) relative to the abscissa (for a very instructive treatise on methodology for non-linear dynamics, see Denton et al. 1990), for example, in the two-dimensional xy plane: $(x,y) = (x_t, x_{t+1})$. Raw data phase-plane plots of activity records show chaotic topology (Fig. 17.12a and b); even so, the partition or clustering of activity into two separate areas of activity and rest is detectable for the normal, but not for the sleep-deprived, record. However, phase-plane plots of the isolated frequencies show a dynamic orderliness as illustrated for the circadian in Figs. 17.13a and 17.14a for normal and sleep deprived, respectively. Here the circadian remodulates are plotted with a 120 epoch (240 min) lag in order to depict the cycling orbits more distinctly. These plots are similar to the limit cycles mentioned frequently in connection with chaotic processes, in which the system after some external (or also in the case of human systems, internal) perturbation always reverts back to cycling around the same inherently set limits. If the original data plotted in phase-plane plots indicate inherent chaotic behaviour, then the phase-plane plots of the remodulates indicate that there is an orderliness embedded in this chaos. For the normal record, the cycles are well formed and regularly traverse the same paths, i.e. pathways between rest and activity. For the sleep-deprived record, the cycling appears chaotic, is centred at high activity level (with circadian troughs apparent at slightly lower activity during the continued wakefulness) until finally sleep is allowed, as indicated in the "tailspin" towards the deep trough. A sense of time progression for these rhythms is shown in three-dimensional aspect plots given in Figs. 17.13b and 17.14b. Other representative examples of phase plots of individual frequencies and their respective time plots are shown for each activity record in Figs. 17.15 to 17.18. Note that the waxing and waning of the amplitudes of the frequencies are expressed in larger and smaller cycles around the origin and their amplitude changes with each cycling are clearly visible in the three-dimensional plots. The most startling and aesthetically pleasing of the phase plots are those for the higher frequencies, in particular that for the 240 cpd (6-min period). Here again, there is a distinct difference in the nature of the rhythm between the normal and the sleep deprived.

Inter-subject Phase-Plane Plots

How does one then gain a sense of the real differences between any two sets of remodulate frequencies each of which consists of 2048 data points? Even more daunting is to try to capture the relationship among all 72 frequencies. We

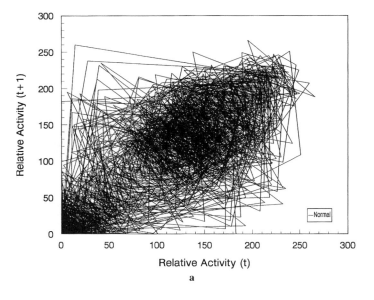

a

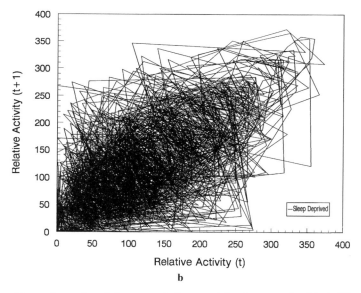

b

Fig. 17.12. a Phase-plane plot of the raw activity record for the normal subject. The first epoch value of the raw data is plotted versus the second epoch value and this ordering (lag of one epoch) is repeated until the end of the data set. There is a density of values in two locations, one at high activity and the other towards zero activity, indicating the circadian peaking and troughing of the inherent rhythm. **b** Phase-plane plot of the raw activity for the sleep-deprived subject. The pattern here is different from the normal subject with higher density values at medium to high activity levels, reflecting the lack of sleep/rest times.

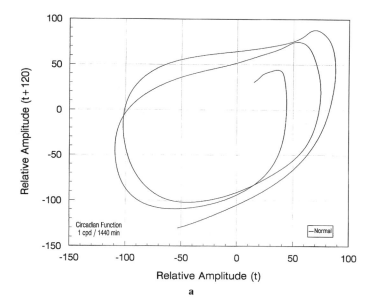

a

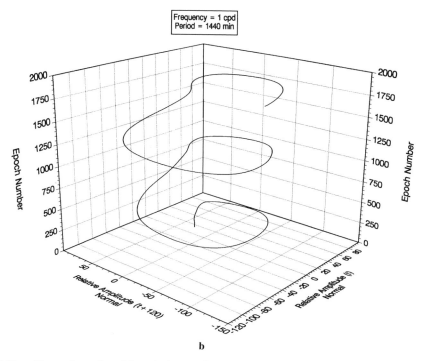

b

Fig. 17.13. a Phase-plane plot of the relative amplitude of the circadian remodulate for the normal subject in two dimensions. A lag of 120 epochs (240 min) is used to show the cycling in more circular orbits (otherwise very flat orbits result). Note that the orbits trace almost the same paths, indicating the regularity of the circadian rhythm for this subject. **b** Phase-plane plot of the circadian function for the normal subject as in **a**, plotted in three dimensions to show the temporal relationship of the cycles over 3 days.

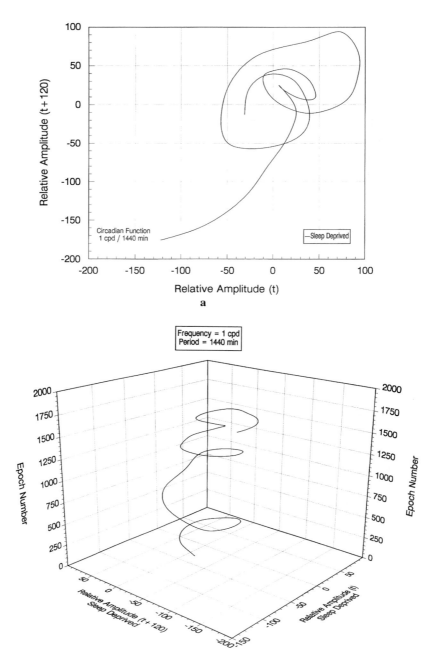

Fig. 17.14. a Phase-plane plot of the relative amplitude of the circadian remodulate for the sleep-deprived subject in two dimensions (plotted as in Fig. 13a). Here the orbits trace different paths over the 3 days and show the bifurcation (frequency doubling) of the rhythm as an indentation within the last orbit between 0 and 50 on both axes. The long "tail" traces the path to recovery sleep on the third night. **b** Three-dimensional phase-plane plot of the circadian remodulate for the sleep-deprived subject, illustrating that the rhythm is contained mainly in the positive amplitude range of activity over the 3 days.

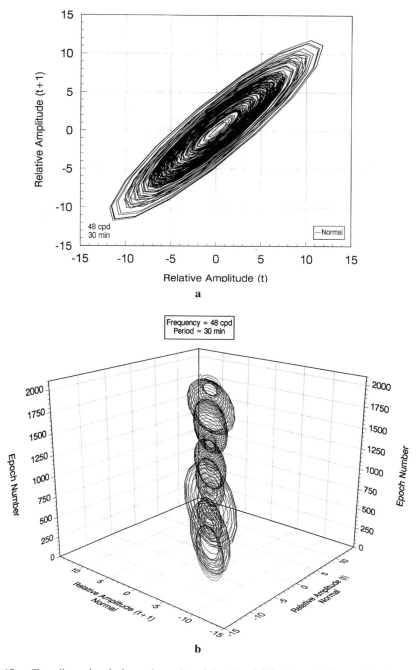

Fig. 17.15. a Two-dimensional phase-plane plot of the 48 cpd (30-min period) remodulate for the normal subject. Amplitude changes of the orbits are evident. **b** Three-dimensional phase-plane plot of the 48 cpd remodulate for the normal subject. The pattern of amplitude changes of the orbits over time are clearly seen in this type of plot.

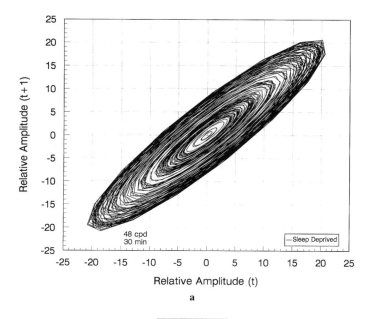

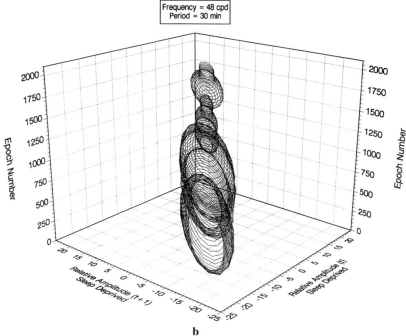

Fig. 17.16. a Two-dimensional phase-plane plot of the 48 cpd remodulate for the sleep-deprived subject. The amplitude changes of the orbits appear similar to those in the normal subjects. **b** Three-dimensional phase-plane plot of the 48 cpd remodulate for the sleep-deprived subject. "Stretching out" the orbits over time reveals how very different the pattern of amplitude changes is from that of the normal subject.

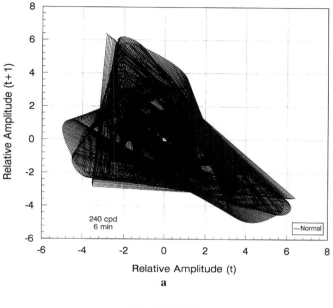

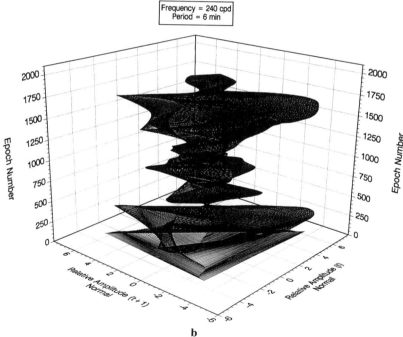

Fig. 17.17. a Two-dimensional phase-plane plot of the 240 cpd remodulate for the normal subject. Since the activity epochs are each 2 min in length, the orbits of this frequency (6-min period) are triangular in shape. **b** Three-dimensional phase-plane plot of the 240 cpd remodulate for the normal subject. The waxing and waning of groups of orbits over time are noted.

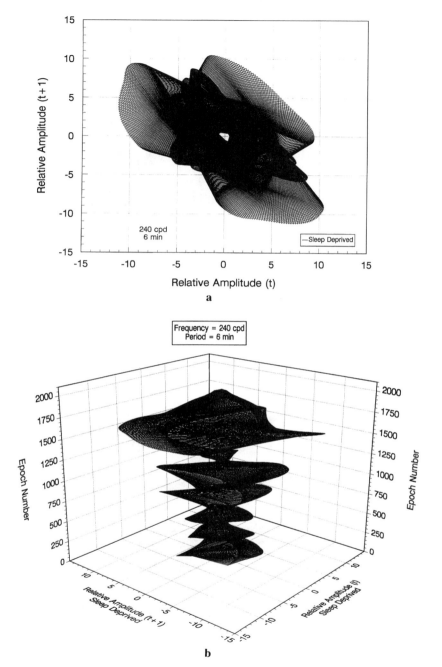

Fig. 17.18. a Two-dimensional phase-plane plot of the 240 cpd (6-min period) remodulate for the sleep-deprived subject. The plot for this frequency is very different in appearance to that of the normal subject. **b** Three-dimensional phase-plane plot of the 240 cpd remodulate for the sleep-deprived subject. Note that the orbital amplitudes and their change over time are distinctly different in their patterns from the normal subject.

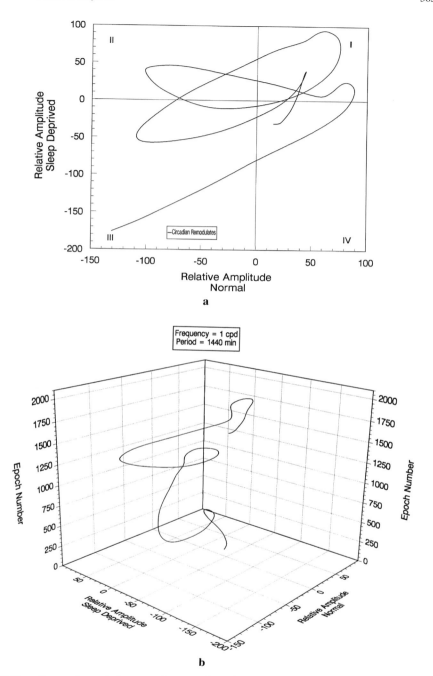

Fig. 17.19. a Comparison of the normal versus sleep-deprived circadian remodulates. When the joint functions are located in quadrants I and III, the two circadian rhythms are in phase; while in quadrants II and IV, they are out of phase. **b** Addition of temporal dimension to **a**. The three-dimensional plot shows when (in epoch sequence) the normal and sleep-deprived circadian rhythms are not in phase.

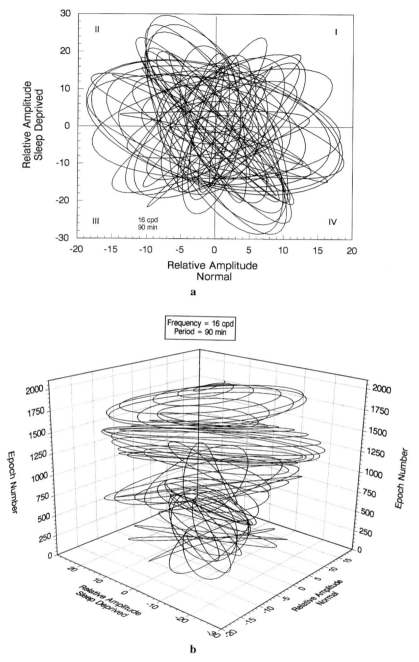

Fig. 17.20. a Comparison of the normal versus sleep-deprived 16 cpd (90-min period) remodulates as in Fig. 17.19a. The rapidly changing nature of phase and amplitude of this rhythm for each subject indicates that the subjects are more often out of than in synchrony. **b** Three-dimensional plot of **a**. This plot shows that the apparent desynchrony of the two subjects' 16 cpd rhythm appears mainly at the beginning. By the second day, the orbits trace more orderly paths.

have made an initial effort to set forth a simple way of showing basic relationships consisting of plotting amplitudes (at one remodulate frequency) against each other. An example is Fig. 17.19a, showing circadian amplitude of the normal subject on the abscissa and sleep-deprived values on the ordinate which compresses 2048 pairs of data points into a direct relationship between the two. When the resulting function is found in quadrants I and III (when both are either positive or negative in value), the two remodulate frequencies are in phase; while in quadrants II and IV (values of mixed signs), they are out of phase. Within each quadrant, the relative magnitude of each is easily determined and overall differences and similarities assessed. A sense of the changing orbits over time may be viewed in a three-dimensional plot, as shown in Fig. 17.19b. In a similar fashion, any pair of remodulate frequencies may be plotted together and evaluated. This is particularly useful in viewing ultradian rhythms where a linear plot over time may be extremely dense depending on the frequency. Another example is given in Fig. 17.20a and b for the 16 cpd rhythm, both in two and three dimensions.

Fractal Dimensions of Waveforms

The fractal dimension was originally devised by mathematician B. B. Mandelbrot (1983) to describe the dimension of irregularly shaped objects, coastlines for example, that seem to change depending on one's view or distance from the object and so cannot be described in a fixed Euclidean dimension, i.e. that cannot be said to be 0 (for a point), or 1 (for a length), or 2 (for a plane), or 3 (for a volume). This different dimension is a measure of the ratio of the logarithm of total length of the object being measured to the logarithm of its span. The idea of a fractal dimension has been extended from its original usage to many areas of study where different methods of calculating it have also been devised. Its utility in aiding understanding or comparison of data in a new light has expedited its general usage.

A very enlightening publication by Katz (1988) in the use of fractal dimensions in the analysis of waveforms shows that one may not only determine whether a waveform is randomly generated but may also make quantitative comparisons of waveforms which are not random. According to Katz, a waveform's dimensionality "is a measure of its convolutedness", so that a dimension of one would indicate a straight line across the span measured, and increasingly complex waveforms would have increasingly larger values. It therefore is also a measure of the information content contained in the waveform.

We have used the software from Katz's paper to generate the fractal dimension values of our two original activity records as well as for all the remodulates. All of these values proved to be above the significant statistical limit at the $p < 0.01$ level for the random-walks table generated by Katz and so are not randomly generated waveforms. The fractal dimensions of the original data sets confirm that the sleep-deprived activity record with a dimension of 2.05147 is richer in detail than the normal record with a dimension of 1.84405. Fig. 17.21 is a plot of the fractal dimensions of the remodulates (as given in Table 17.1) determined from the two activity records. The fractal dimensions of lower frequency rhythms are very similar for the normal and sleep deprived, but as the fre-

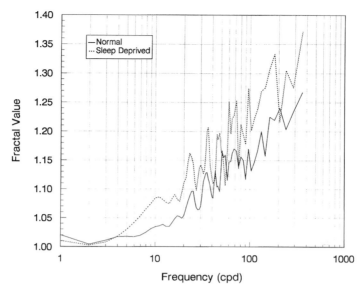

Fig. 17.21. Calculated fractal dimension values for the normal and sleep-deprived subjects. Frequency is given in logarithmic scale on the abscissa and the calculated fractal value on the ordinate. Note that in general the fractal dimension (information content) of the sleep-deprived subject is greater than that of the normal subject. One of the notable exceptions is the circadian (1 cpd) fractal value, which is greater for the normal subject. This bears out both the greater spectral power of the circadian rhythm for the normal subject and its well-defined circadian remodulate shaping.

quency increases, the values of the dimensions diverge and are higher for the sleep-deprived remodulates. Clearly, the pattern of activity, in particular in the ultradian frequency range, is distinctively different; whether this is characteristic of the individuals or of conditions under which they are functioning cannot be determined until the hundreds more activity records we have collected are analysed in the same manner. We are in the process of so doing.

Applications

How can the wealth of information that is elicited from the non-invasive monitoring of human movement be applied in helpful and meaningful ways? What practical use might be made of ultradian and supra-ultradian rhythms from activity records? In particular, what do they signify and how may they be applied to benefit the human beings from whom they are derived. The very unobstrusive nature of the measuring device lends itself to a multitude of applications. Lovett Doust et al. (1978), in their study of reaction time measurements of a perceptual task presented at 60 s intervals, found periods of cycling of between 4 and 15 min. They theorized that these are endogenous oscillations and "may be related to metabolic turnover of chemical transmitter substances in synapses of cell assemblies and their local circuits". Since movement or motor activation is the ultimate extrinsic response of neuronal processes, then it is conceivable that supra-ultradian rhythms from activity measures are correlates to the most fundamental of mental processes. Might it not be possible to distinguish between health and pathology, if one were to characterize profiles of normal ultradian/supra-ultradian/very high frequency rhythms against which abnormal ones would be compared. Examples of beneficial applications would be in tracking the progress of drug therapy in hyperkinetic children, manic-depressives, Parkinsonian patients and others where differences in movement may signify therapeutic or pathological changes.

One might even track these rhythms in healthy humans such as athletes wielding tennis racquets or pitchers throwing fast balls at over 140 km/h!

Future Developments

We are excited about the development of a new generation of activity monitors that will be about half the size (4.5 cm × 3.2 cm × 1.1 cm) of the old one, have a wide range of programmability and are capable of sampling at 1-s epochs. We look forward to the gathering of data with this new actigraph and to the application of the analytical techniques described here. There is the expectation of discovery of even faster ultradian rhythms of movement activity in daily life!

Acknowledgements

The editorial "we" throughout this chapter acknowledges the invaluable contribution of Dr Frederick W. Hegge, who has been a prime advocate of actigraphy and its applications within the Walter Reed Army Institute of Research. He has provided the impetus for improvement of actigraphs spanning the last 15 years. Just as importantly, he has also devoted much thought to the nature of actigraph data. This is reflected in his active participation in analysis of the movement activity data presented here and in devising the digital filter that allows extraction of each frequency component in the complex demodulation procedure.

Continued development of the actigraph would not have been possible without the technical expertise of Col. Daniel P. Redmond, MD, who has taken the actigraph progressively from the "dark age" to state-of-the-art technology and has at each iteration proved its worth in countless military field studies. That future actigraphs will provide a wealth of unique information will be due entirely to Col. Redmond.

References

Babkoff H, Caspy T, Mikulincer M, Sing HC (1991) Monotonic and rhythmic influences: a challenge for sleep deprivation research. Psychol Bull 109:411–428
Bloomfield P (1976) Fourier analysis of time series: an introduction. Academic Press, New York

Brandenberger G, Follenius M, Muzet A, Ehrhart J, Schieber JP (1985) Ultradian oscillations in plasma renin activity: their relationships to meals and sleep stages. J Clin Endocrinol Metab 61:280–284

Broughton R (1982) Human consciousness and sleep/waking rhythms: a review. J Clin Neuropsychol 4:193–218

Cugini P, Kawasaki T, Leone G, DiPalma L, Letizia C, Scavo D (1987) Spectral resolution of ultradian blood pressure in man. Jpn Circ J 51:1296–1304

Denton TA, Diamond GA, Helfant RH, Khan S, Karagueuzian H (1990) Fascinating rhythm: a primer on chaos theory and its application to cardiology. Am Heart J 120:1419–1440

Globus GG, Drury RL, Phoebus EC, Boyd R (1971) Ultradian rhythms in human performance. Percept Mot Skills 33:1171–1174

Globus GG, Phoebus EC, Humphries J, Boyd R, Sharp R (1973) Ultradian rhythms in human telemetered gross motor activity. Aerosp Med 44:882–887

Horne JA, Whitehead M (1976) Ultradian and other rhythms in human respiration rate. Experientia 32:1165–1167

Katz MJ (1988) Fractals and the analysis of waveforms. Comput Biol Med 18:145–156

Lavie P (1979) Ultradian rhythms in alertness. A pupillometric study. Biol Psychol 9:49–62

Lloyd D, Kippert F (1987) A temperature-compensated ultradian clock explains temperature-dependent quantal cell cycle times. Symp Soc Exp Biol 41:135–155

Lovett Doust JW, Podnieks I (1975) Comparison between some biological clocks regulating sensory and psychomotor aspects of perception in man. Neuropsychobiology 1:261–266

Lovett Doust JW, Payne WD, Podnieks I (1978) An ultradian rhythm of reaction time measurements in man. Neuropsychobiology 4:983–988

Mandelbrot BB (1983) The fractal geometry of nature. Freeman, New York

Mejean L, Bicakova-Rocher A, Kolopp M, Villaume C, Levi F, Debry G, Reinberg A, Drouin P (1988) Circadian and ultradian rhythms in blood glucose and plasma insulin of healthy adults. Chronobiol Int 5:227–236

Meneses OS, Corsi CM (1990) Ultradian rhythms in the EEG and task performance. Chronobiologia 17:183–194

Naitoh P, Johnson LC, Lubin A, Nute C (1973) Computer extraction of an ultradian cycle in sleep from manually scored sleep stages. Int J Chronobiol 1:223–234

Orr WC, Hoffman HJ, Hegge FW (1974) Ultradian rhythms in extended performance. Aerosp Med 45:995–1000

Redmond DP, Hegge FW (1985) Observations on the design and specification of a wrist-worn human activity monitoring system. Behav Res Methods Instr Comput 17:659–669

Rossi EL (1982) Hypnosis and ultradian cycles: a new state(s) theory of hypnosis? Am J Clin Hypn 25:21–32

Sing HC, Redmond DP, Hegge FW (1980) Multiple complex demodulation: a method for rhythmic analysis of physiological and biological data. In: O'Neill JT (ed) Proceedings of the 4th Annual Symposium on Computer Applications in Medical Care. IEEE, New York, pp 151–158

Sing HC, Thorne DR, Hegge FW, Babkoff H (1985) Trend and rhythm analysis of time-series data using complex demodulation. Behav Res Methods Instr Comput 17:623–629

Stevens JR, Kodama H, Lonsbury B, Mills L (1971) Ultradian characteristics of spontaneous seizure discharges recorded by radio telemetry in man. Electroencephalogr Clin Neurophysiol 31:313–325

Stupfel M, Pavely A (1990) Ultradian, circahoral and circadian structures in endothermic vertebrates and humans. Comp Biochem Physiol [A] 96:1–11

Thorne DR, Genser SG, Sing HC, Hegge FW (1985) The Walter Reed performance assessment battery. Neurobehav Toxicol Teratol 7:415–418

Veldhuis JD, Johnson ML (1988) Operating characteristics of the hypothalamo-pituitary – gonadal axis in men: circadian, ultradian, and pulsatile release of prolactin and its temporal coupling with luteinizing hormone. J Clin Endocrinol Metab 67:116–123

Weitzman ED, Nogiere C, Perlow M et al. (1974) Effects of prolonged 3-hour sleep–wake cycle on sleep stages, plasma cortisol, growth hormone and body temperature in man. J Clin Endocrinol 38:1018–1030

Chapter 18

The Wave Nature of Being: Ultradian Rhythms and Mind–Body Communication

E. L. Rossi and B. M. Lippincott

Introduction

A major ideal of science is to integrate apparently different phenomena into a general theory of nature. The more widely divergent the phenomena that we can bring together in a meaningful way, the greater the beauty, scope and potential utility of the theory. We admire the simple elegance whereby Newton's theory of gravitation proposed a grand unification of celestial mechanics – the orbits of the sun, moon and planets – with the humble fall of an apocryphal apple here on earth.

Newton's accomplishment was made possible by (a) centuries of careful observation of planetary motion and (b) the creation of a new mathematics, calculus, that related time to changes in position and motion. The introduction of time as an independent variable could be regarded as a profound turning point in the creation of the scientific world view. Introducing time led to an understanding of "rates of change" so that predictions could be made accurately and easily. The experimental confirmation of such predictions led, in turn, to the scientific method of verifying theory.

Are we in a similar place today in our study of ultradian rhythms made possible by: (a) more than two centuries of observations since the astronomer De Mairan verified experimentally that living plants had their own endogenous rhythms; and (b) the current creation of a new mathematics, chaos theory, that seems capable of relating time to the deterministic yet bewilderingly complex rhythms we find characteristic of life at all levels from the molecular–genetic to the brain–mind? This chapter may be regarded as a prolegomenon to the unification hypothesis of chronobiology: there are cybernetic relationships between ultradian rhythms on the epigenetic, neuroendocrine, behavioural and psychosocial levels that facilitate the general life processes of adaptive homeostasis.

We begin with an overview of the nature–nurture controversy from this perspective and then summarize recent research that leads to an ultradian theory of adaptive homeostasis, stress and healing in psychosomatic medicine.

Nature–Nurture and Ultradian Mind–Body Communication

The question of what determines human behaviour in health and illness has been conceptualized as the so-called "nature versus nurture controversy". Is the source of a behaviour to be found in endogenous, in-built genetic factors or is it better understood as an adaptive process to outer environmental circumstances? In the field of chronobiology this question takes a more precise form: is the vast range of ultradian, circadian and seasonal rhythms that all of life is heir to primarily an expression of endogenous or environmental signals? More than two centuries ago when the astronomer De Mairan first established the endogenous source of circadian rhythms in plants for chronobiology, a similar development was taking place in the field of psychology. Mesmerism, the theory that the source of human behaviour, health and illness was to be found in the influence of magnetic forces emanating from the heavenly bodies, was being transformed into the field of hypnosis, depth psychology and psychosomatic medicine, where it was recognized that the sources of health, stress and psychopathology could be traced to more potent, proximal, psychosocial sources acting within the person.

In this primitive, polarized form the nature–nurture controversy misses what is most exciting at the leading edge of research today: it is the interaction between nature and nurture at the molecular–genetic level that is most informative in the emerging fields of psychobiology. In particular, it is our currently evolving knowledge of the molecular pathways of information transduction between environmental signals and the expression of certain constitutive ("housekeeping") genes that is throwing a new light on the dynamics of biology and behaviour at all levels.

This chapter surveys a wide variety of clinical and experimental observations that lead to an integrative view of communication and information transduction as the basic common denominator of biology and psychology. In particular, we focus on how ultradian rhythms may be explored as the parameters of information transduction in the cybernetic process of *adaptive homeostasis*. Adaptive homeostasis may be defined as the communication process whereby the genetic informational matrix of life, as expressed in endogenous epigenetic oscillations at the molecular–genetic level, is entrained and adaptively integrated with environmental signals. The signals from the environment include the availability of water and nutrients, temperature, light, toxins, physical stress, performance demands and that general class of behavioural stimuli usually called "psychosocial variables".

The Unification Hypothesis of Chronobiology: An Evolutionary View of Mind–Body Rhythms, Stress and Healing

The unification hypothesis of chronobiology implies an evolutionary view of the cybernetic loop of information transduction between the epigenetic level and environmental signals that may be surveyed by tracing three interacting levels of adaptive homeostasis in health and illness.

Cellular–Genetic Level

The epigenetic cycle is hypothesized as the source of life rhythms – the basic biological clock (Lloyd and Edwards 1984; Lloyd, Chap. 1). In more technical terms it is the "kinetic rate limiting factor"; the time required for genes to express themselves in the basic processes of cell metabolism, division, growth and repair (Lloyd and Edwards 1987; Todorov 1990).

Prokaryotes, the simplest and oldest forms of unicellular life that have no nucleus, require as little as 20 min for cell division and complete self-replication (Alberts et al. 1989). The M phase of cell division in the more highly evolved eukaryotes and in the cells that make up all complex organisms where chromosomal replication takes place before mitosis, however, usually requires 90–120 min even though it may take many hours, days or longer for the entire cell to replicate itself. A 20-min building-up period of mediating factors (e.g. maturation-promoting factor and H1 kinase) is required to trigger the 90–120-min process of genetic replication in the complete 24-h circadian cell cycle in eukaryotes (Murray et al. 1989; Murray and Kirschner 1989, 1991).

This emerging data on the molecular–genetic level suggests that there has been an evolution from 20 min to 90–120-min ultradian rhythms to 24-h circadian rhythms of cell cycle activity, replication and entrainment by environmental signals (Klevecz and Braly 1987; Emunds 1988; Rosbash and Hall 1989). It may be hypothesized that the original 20-min rhythm is most reflective of the basic rate-limiting factors on the thermodynamic–molecular level in prokaryotes while the 90–120-min ultradian, circadian and seasonal rhythms reflect progressively more inclusive cybernetic processes of information transduction between genetic and environmental signals in eukaryotes. This leads to the following interpretation of how ultradian and circadian rhythms at the molecular–genetic level may be reflected at the progressively more evolved brain–body (neuroendocrinal axis) and mind–brain levels (sensations, memory, learning etc.) in adaptive homeostasis.

Brain–Body Level

All the major systems of mind–body communication, such as the central nervous system (CNS), autonomic, endocrine and immune systems, have evolved ultradian and circadian rhythms that apparently integrate the molecular–epigenetic cycles within cells with the organism's external and internal environments for "adaptive homeostasis" (Rossi 1986a; Rossi and Cheek 1988). These include:

1. The pontine reticular activating system of the CNS that is associated with the 90–120-min ultradian rhythm of dreaming. The neurophysiologist Jouvet (1962, 1973) believed that the function of dreams was to exercise genetic patterns of behaviour associated with adaptation, emotion and cognition in humans as well as animals. It is noteworthy in our context that the phases of

rapid eye movement sleep (REM state) that are associated with dreaming range from 15 to 40 min, with an average of 20–30 min. Kleitman (1969) cites evidence that these alternating phases of sleep and dreaming correspond to his 90–120-min basic rest–activity cycle (BRAC) in the daytime that are the essence of genetically based processes of adaptive homeostasis such as food intake and sexuality.

2. The neuroendocrine system is now well recognized as having prominent ultradian and circadian components related to a variety of psychobiological behaviours associated with mental and physical activity, nutrition, metabolism and reproduction. There are experimentally verifiable 20-min couplings between peaks of associated hormones that are released in approximately 90–120-min ultradian rhythms: luteinizing hormone peaks lead prolactin and testosterone peaks by 10 to 20 min (Veldhuis et al. 1987; Veldhuis and Johnson 1988; Veldhuis, Chap. 8); glucose leads insulin by 15 to 20 min (Van Cauter et al. 1989); cortisol leads β-endorphin by 20–30 min (Iranmanesh et al. 1989). These associations that extend from the molecular–genetic generation of these hormones at the cellular level to their expression at the neuroendocrinal level and their interaction with the mind–brain processes of memory, learning and behaviour described below can hardly be accidental. They must play a significant role in the cybernetic process of information transduction between gene and adaptive behaviour in health and illness. They are the most vivid illustration of the conjecture of an evolutionary relationship between the 20-min, 90–120-min and 24-h rhythms of adaptive homeostasis at all levels.

Mind–Brain Level

Ultradian relationships whereby short-term memory is converted to long-term memory over a 90-min cycle, with a 20-min latency period capable of aborting the process have been found in organisms ranging from the marine mollusc *Aplysia* to humans (Kandel and Schwartz 1985). The molecular pathways between environmental stimuli and the genetic–cellular basis of virtually all classes of memory and learning associated with the mind–brain processes of psychotherapy have been discussed by Kandel. He carefully differentiates between changes in genetic structure and the regulation of gene expression by environmental factors as follows (Kandel 1989, pp. 122–123, our italics).

The genetic data on schizophrenia and on depression indicate that these diseases involve alteration *in the structure of genes*. By contrast, the data now emerging on learning suggest that neurotic illnesses acquired by learning, which can often respond to psychotherapy, might involve alterations *in the regulation of gene expression*. In this context, it is important to realize, as I have emphasized earlier, that genes have two regions: a regulatory region and a coding region. The *regulatory region* usually lies upstream of the coding region and consists of two types of DNA elements. One type of DNA element is called the promoter. This is the site where the enzyme RNA polymerase binds before it reads and transcribes the gene into messenger RNA. The second type of DNA region is called the *enhancer region*. It recognizes protein signals that determine in which cells, and when, the coding region will be transcribed by the polymerase. Whether the RNA polymerase binds and transcribes the gene and how often it does so in any given period of time is determined by a small number of proteins, transcriptional regulators, that bind to different segments of the upstream enhancer region. Development, hormones, stress, and learning are all factors that can alter the binding of the transcriptional regulator proteins to the regulatory regions of genes. I suggest that at least certain neurotic illnesses (or components of them) represent a reversible defect in gene regulation, which is produced by learning and which may be due to altered binding of specific proteins to certain upstream regions that control the expression of certain genes.

According to this view, schizophrenia and depression would be due primarily to heritable genetic changes in neuronal and synaptic function in a population carrying one or more mutations. By contrast, *neurotic illnesses might represent alterations in neuronal and synaptic function produced by environmentally induced modulation of gene expression. Insofar as psychotherapy works and produces long-term learned changes in behavior, it may do so by producing alterations in gene expression.* Needless to say, psychotic illness, although primarily caused by inherited alterations in gene structure, may also involve a secondary disturbance in environmentally acquired gene expression.

Clinical–experimental data that are consistent with Kandel's view have been cited by Rossi, who has hypothesized that chronic disruptions of ultradian rhythms of activity and rest may lead to the breakdown of adaptive homeostasis between environmental stimuli, the neuroendocrine axis and the epigenetic level that are manifest as stress and psychosomatic illness (Rossi 1982, 1986a,b; Rossi and Cheek 1988). It has been hypothesized that many forms of psychological and holistic healing utilizing hypnosis, the relaxation response, psychotherapy and meditation can facilitate stress reduction and healing right down to the molecular–genetic level by simply providing a therapeutic context for rest and recovery that can optimize ultradian rhythms (Rossi 1990b, 1991a).

One of the most intriguing areas of recent research exploring the ultradian interface between the mind–brain level, stress, psychosomatics and personality is the so-called nasal rhythm. The German rhinologist Kayser (1895) is credited with recognizing and measuring the widely varying ultradian shifts in "nasal dominance" in humans whereby the left and right chambers of the nose alternate in their size and shape to change the degree of air flow through each every few hours. Table 18.1 outlines some of the major studies in a century of research in this still highly controversial area. The most significant of these studies for understanding mind–body communication are those of Werntz (1981), who reported a contralateral relationship between cerebral hemispheric activity (electroencephalographic, EEG) and the ultradian rhythm of the nasal cycle. She found that relatively greater integrated EEG values in the right hemisphere are positively correlated with a predominant airflow in the left nostril and vice versa.

In a wide ranging series of studies Werntz et al. (1982a,b) found that subjects could voluntarily shift their nasal dominance by forced uni-nostril

Table 18.1. A century of research on the duration of the nasal cycle

References	No. of Ss	Frequency of measurements (min)	Duration of measurements	% of Ss with nasal cycle	Range of cycle (h)	Average duration of cycle (h)	Method of measurements
Kayser 1895		[Many clinical/experimental case studies]			[0.5–several hours]		Rhinomanometer
Heetderks 1927	60	10	2+ h	80	1–4 h	2.5	Direct observation
Stoksted 1953	26	30	3–4 h	38	1.5–5 h	2.5	Rhinomanometer
Soubeyrand 1964	50	15	NA	100	2–5 h	3.5	Rhinomanometer
Keuning 1968	17	15	3–4 h	71	2–7 h	4.5	—
Hasegawa and Kem 1978	50	15	7 h	72	1–6 h	2.9	Pneumotachograph
Eccles 1978	2	30	3–7 days	100	1–2.5 h	2	Pneumotachograph
Clarke 1980	5	30	30 days	100	1.5–6 h	4	Mirror condensation
Werntz 1981	43	15	0.66–3.6 h	51	0.5–3.3 h	NA	Thermistors
Gilbert 1989	9	5	8 h	44.4	3.5–6 h	4.5	Rhinomanometer

S, case-study.

breathing through the closed nostril. Further, this shift in nasal dominance was associated with an accompanying shift in cerebral dominance to the contralateral hemisphere and autonomic nervous system balance throughout the body (Klein et al. 1986). The ultradian nasal cycle is not only a marker for cerebral hemispheric activity, but it also could be used to change voluntarily the loci of activity in the highest centres of the brain and autonomic system that are involved in cybernetic loops of communication with most organ systems, tissues and cells of the body. Some of these investigators hypothesize that this nasal–brain–mind link may be the essential path by which the ancient practice of breath regulation in yoga led to the voluntary control of many autonomic nervous system functions for which the Eastern adepts are noted (Brown 1991; Rossi 1991b).

These relationships inspired a recent PhD dissertation by Osowiec (1991), who assessed hypothesized associations between the nasal ultradian rhythm, anxiety, symptoms of stress and the personality process of self-actualization. She found that: "(1) there is a significant positive correlation between self-actualizing individuals having low trait anxiety and stress related symptoms and a regular nasal cycle . . . and (2) non-self-actualizing individuals with high levels of trait anxiety and stress-related symptoms exhibit significantly greater irregularity in the nasal cycle . . . ". These results are reminiscent of the ancient texts that emphasize that an irregular nasal cycle, particularly one in which the person remains dominant in one nostril or the other for an excessively long period of time, is associated with illness and mental disorder (Rama et al. 1976).

Osowiec's findings with the ultradian nasal rhythm are 'similar to the types of association that are found between stress, symptoms, personality and responsiveness to therapeutic hypnosis. Since hypnosis, like chronobiology, was a historically significant turning point in the study of these cybernetic pathways of mind–body communication and healing, it will be used here as a paradigm of how psychology and biology may find a common denominator in ultradian time, rhythm and information transduction (Rapp 1987).

Matching Chronobiological Rhythms and Hypnotic Phenomena

It now appears that most of the known psychological and physiological processes that manifest a natural variability during ultradian and circadian rhythms are modifiable by hypnosis. A matching of the psychobiological processes that vary during ultradian rhythms with so-called "hypnotic phenomena" is outlined in Table 18.2. Table 18.2, like Table 18.1, is not complete in any sense; it is only an introduction to the wide range of research topics that chronobiology and hypnosis share in common. As can be seen, most of the psychobiological processes that manifest a natural variability during ultradian rhythms are modifiable by hypnosis. This matching provides an empirical data base for a more comprehensive assessment of the hypothesis that what has been tradition-ally called "clinical hypnosis" or "therapeutic suggestion" may be, in essence, the accessing and utilization of the natural variability of ultradian and circadian processes that respond to psychosocial cues. Within this framework, the classic phenomena of hypnosis may be conceptualized as extreme manifestations and/or perseverations of time-dependent psychobiological processes that are

Table 18.2. A matching of ultradian processes with classical phenomena of therapeutic hypnosis

Ultradian processes	Hypnotherapeutic phenomena
Entrainability Winfree 1980, 1987; Brown 1982; Schulz and Lavie 1985	*Suggestibility* Fromm and Shor 1972; Hilgard 1965; Rossi 1980
90-min sleep–dream rhythm Aserinsky and Kleitman 1953; Jouvet 1973; Hobson 1988	*Somnambulism/hypnoidal* Breuer and Freud 1957; Rossi and Smith 1990
Waking BRAC Globus 1966; Hartmann 1968; Kleitman 1982	*Common everyday trance* Fischer 1971a,b,c; Erickson and Rossi 1979;
Hemispheric laterality Klein and Armitage 1979; Gordon et al. 1982	*Hemispheric laterality* Frumkin et al. 1978; Carter et al. 1982; Sabourin 1982; Gabel 1988
Social variables Hayes and Cobb 1979; Lavie and Kripke 1981	*Rapport* Sarbin 1976; Spanos et al. 1985
Sensory/perceptual Lavie 1976, 1977; Lovett 1976; Elsmore and Hursch 1982	*Illusions/hallucinations* Bowers 1977; Rossi 1980; Migaly 1987
Rorschach: Globus 1966 *Visual illusions*: Gopher and Lavie 1980	Barber 1972; Orne 1972 Orne 1972; Rossi 1980
	Age regression Perry et al. 1988; Pettinati 1988
	Analgesia/anaesthesia Hilgard and Hilgard 1983; Spanos 1986
Cognition/memory Klein and Armitage 1979; Folkard 1982	*Amnesia/hypermnesia* Spanos 1986; Pettinati 1988
Fantasy Cartwright and Monroe 1968; Kripke and Sonnenschein 1978	*Imagination* Wilson and Barber 1978; Sheehan and McConkey 1982
Time sense Tepas 1982; Rose 1988; Young 1988	*Time distortion* Cooper and Erickson 1959; Aaronson 1969b; Zimbardo et al. 1972
Affective behaviour Friedman 1978; Poirel 1982; Wehr 1982	*Catharsis/mood* Breuer and Freud 1957; Gill and Brenman 1959
Transpersonal sense Broughton 1975; Eccles 1978; Funk and Clarke 1980	*Transpersonal experience* Aaronson 1969a; Erickson and Rossi 1980b; Tart 1983

responsive to psychosocial cues. Because the literature in hypnosis and psychobiological rhythms has developed independently until now, a certain amount of translation will be necessary to understand the proposed matchings in Table 18.2.

Even though the history of hypnosis indicates that many of the major phenomena of hypnosis were discovered originally as curious manifestations of normal and abnormal states (Ellenberger 1970; Tinterow 1970), the salient characteristic that made them particularly interesting from a social/psychological point of view was that they were "suggestible". The first row in Table 18.2 proposes that what is called "suggestible" in hypnosis is called psychosocially "entrainable" in the literature of chronobiology.

Table 18.2. *Continued*

Ultradian processes	Hypnotherapeutic phenomena
Motor behaviour	*Ideodynamic responses*
Body activity: Luce 1970; Clements et al. 1976; Naitoh 1982	Hilgard 1965; Rossi and Cheek 1988
Muscle tonicity	*Catalepsy*
Lovett et al. 1978; Katz 1980; Rasmussen and Malven 1981	Erickson and Rossi 1981; Hilgard 1965
Response latency	*Ideomotor retardation*
Lovett and Podnieks 1975; Bossom et al. 1983	Hilgard 1965; Erickson and Rossi 1980a
Eye behaviour	*Eye behaviour*
Ullner 1974; Krynicki 1975; Lavie 1979	Weitzenhoffer 1971; Spiegel and Spiegel 1978
Autonomic nervous system	
Cardiovascular	*Cardiovascular*
Lovett 1980; Dalton et al. 1986; Lydic 1987	Gorton 1957, 1958; Sarbin and Slagle 1972; Crasilneck and Hall 1985
Peripheral blood flow	*Peripheral blood flow*
Lovett 1980; Romano and Gizdulich 1980	Barber 1984; Olness and Conroy 1985; Rossi and Cheek 1988
Respiration	*Respiration*
Horne and Whitehead 1976; Feldman 1986; Lydic 1987	Edwards 1960; Rossi 1980
Thermoregulatory	*Thermoregulatory*
Hunsaker et al. 1977; Lloyd and Edwards 1987	Reid and Curtsinger 1968; Timney and Barber 1969
Urine flow	*Urine flow*
Lavie and Kripke 1977; Gordon and Lavie 1986; Brandenberger et al. 1987	Brown 1959; Freeman and Baxby 1982; Crasilneck and Hall 1985
Gastrointestinal–enteric	*Gastrointestinal–enteric*
Friedman and Fisher 1967; Kripke 1972; Lewis et al. 1977	Eichorn and Tracktir 1955; Gorton 1957, 1958; Dias 1963; Rossi 1980
Endocrine system	
Corticosterone	*Corticosterone*
Shiotsuka et al. 1974; Follenius et al. 1987; Simon et al. 1987	Levitt et al. 1960; Kosunen et al. 1977; Carli et al. 1979
Gonadotrophins	*Sexual processes*
Filicori et al. 1979; Knobil and Hotchkiss 1985	Araoz 1982; Rossi and Cheek 1988

A major theoretical and methodological common denominator between the ultradian literature on the *entrainability* of psychobiological processes and the hypnosis literature on the *suggestibility* of hypnotic phenomena is to determine the degree to which they are a function of endogenous (organismic) and exogenous (environmental or social) influences. A "two-factor hypothesis", for example, was used by Brown (1982) to account for the interaction between endogenous organismic states and external illumination on a variety of ultradian and circadian processes. Likewise, the literature of hypnosis abounds with "two-factor theories" to account for the interactions between hypnotizability as an innate, organismic trait versus its suggestibility and responsiveness to psychosocial cues (Fromm and Shor 1972; Sheehan and Perry 1976; Erickson and Rossi 1981). Table 18.2 cites only a few of the major approaches to

Table 18.2. *Continued*

Ultradian processes	Hypnotherapeutic phenomena
Leutenizing hormone & testosterone Steiner et al. 1980; Rasmussen 1986; Veldhuis and Johnson 1988	*Menstrual cycle* Erickson 1980a; Rossi and Cheek 1988
Growth hormone Millard et al. 1981; Bernardis and Tannenbaum 1987	*Breast growth* Barber 1984; Erickson 1980b
Neuroendocrine (ACTH, etc.) Carnes et al. 1986; Rasmussen 1986; Lydic 1987	*Stress* Sachar et al. 1965; Barber 1984; Rossi and Cheek 1988
Plasma adrenaline Levin et al. 1978	*Memory* Naish 1986; Rossi and Cheek 1988
Thyroid Bykov and Katinas 1979	*Metabolism* Rossi 1980; Rossi and Ryan 1986

Immune system

Macrophage activation Smolensky and Reinberg 1977; Lydic 1987	*Macrophage activation* Black et al. 1963; Goldberg 1985
Killer cell activity Williams et al. 1981; Lissoni et al. 1986	*Killer cell activity* Mason and Black 1958; Chapman et al. 1959; Bowers and Kelly 1979

Psychosomatic response

General Friedman 1978; Moore-Ede et al. 1983; Reinberg and Smolensky 1983	*General* Breuer and Freud 1957; Bowers and Kelly 1979; Rossi 1980; Rossi and Cheek 1988
Stress Stroebel 1969; Orr et al. 1974; Broughton 1975	*Stress* Putnam 1985; Wickramasekera 1987; Spiegel et al. 1988
Affective illness/depression Wehr and Goodwin 1981, 1983; Kripke 1984	*Affective illness/depression* Breuer and Freud 1957; Bower 1981
Habits and symptoms Friedman and Fisher 1967; Friedman et al. 1978; Poirel 1982	*Habits and symptoms* Rossi 1980; Spiegel et al. 1982; Rossi and Cheek 1988

Accidents

Wolcott et al. 1977; Shaffer et al. 1978; Holley et al. 1981	Wester and Smith 1984; Rossi and Cheek 1988

Social issues

Bowden et al. 1978; Gerkema and Daan 1985	Ritterman 1983; Wester and Smith 1984; Simons et al. 1988

studying the interactions between organismic and psychosocial variables in the literature on ultradian rhythms and their corresponding variables in the literature on hypnosis.

The proposed matchings of Table 18.2 outline how many varieties of sensation, perception, memory, cognition, motor and affective behaviour modifiable by hypnosis exhibit natural ultradian variations. A basic hypothesis for integrating chronobiology with psychology is that these natural ultradian variations of cognition, mood and behaviour are the "basic stuff" of hypnotic performance (Balthazard and Woody 1985) and the art of therapeutic suggestion in general. The responsivity of many autonomic, endocrine and immune system parameters to ultradian variation and hypnotic suggestion reinforces the view that they share a common denominator in the unification hypothesis of chronobiology. This hypothesis is supported further by the voluminous literature on stress, psychosomatics, accidents and social issues outlined in Table 18.2 that the researchers into ultradian rhythms and hypnotherapeutics have produced independently of one another until this time.

This ultradian hypothesis of therapeutic suggestion is entirely consistent with a generation of research in experimental hypnosis that firmly established that, contrary to popular belief, there is no transcendence of normal abilities in hypnosis (Wagstaff 1986). What seems to be an extension of the normal parameters of a wide range of mind–body performance skills via hypnosis is actually the optimization of the individual's natural range of abilities. These researchers openly acknowledge, however, that they have no adequate theory of the source and parameters of hypnotic performance. A prominent researcher, for example, has recently summarized the current situation as follows: "As [hypnotic] susceptibility is normally assessed, a high scorer is one who *produces* the behavior, the *reason* for its production remains unknown . . . [T]he claim was frequently made that cognitive processes are involved in the production of 'hypnotic' effects. However, the exact nature of these processes generally remained obscure" (Naish 1986). The ultradian theory of hypnotic suggestion proposed here is the first that provides a comprehensive view that can account for all the known historical, biological, psychological and sociological facts about the source and parameters of hypnotic behaviour (Rossi, 1986a,b, 1989; Rossi and Cheek 1988). Moreover, as will be seen in the following sections, the ultradian theory of hypnotic suggestion has generated and is in the process of verifying an entirely new set of predications that could not have been made within the world view of traditional hypnosis and psychobiology.

Winfree (1980, 1987) has succinctly defined some of the important concepts that are central to further research in this proposed integration of ultradian psychobiology with the mind–body healing arts as follows (Winfree 1980, p. 1, our italics):

In living systems, as in much of mankind's energy-handling machinery, rhythmic return through a cycle of change is an ubiquitous principle of organization . . . The word *phase* is used . . . to signify position on a circle, on a cycle of states. Phase provides us with a banner around which to rally a welter of diverse rhythmic (temporal) or periodic (spatial) patterns that lie close at hand all around us in the natural world. I will draw your attention in particular to *"phase singularities": peculiar states or places where phase is ambiguous but plays some kind of a seminal, organizing role.* For example in a chemical solution a phase singularity may become the source of waves that organize reactions in space and time.

Winfree (1980, 1987) discussed phase singularities that characterize a wide variety of ultradian rhythms ranging from the purely biological (e.g. photosynthesis, cell division) to the behavioural (e.g. activity rhythms, sleep–wake cycles) and the phenomenological (e.g. the experiences of colour and jet lag). He has not discussed hypnosis, but one would be hard pressed to find a more apt description of hypnosis than "peculiar states or places where [behaviour] is ambiguous but [suggestion] plays some kind of a seminal, organizing role". In its most general form, the ultradian hypothesis would predict that all the hypnotic phenomena listed in Table 18.2 would be most readily evoked during the "phase singularity" portion of those psychobiological processes that are proposed here as the "basic stuff" of each hypnotic phenomenon. Let us now examine a series of recent studies that explore the ultradian psychobiological parameters of what we shall call "naturalistic hypnosis", "the common every-day trance" or the "ultradian healing response".

Current Clinical–Experimental Studies
Ultradian Rhythms of Hypnotic Susceptibility

The first independent experimental assessment of the ultradian theory of hypnotic suggestion was carried out by Aldrich and Bernstein (1987), who found that "time of day" was a statistically significant factor in hypnotic susceptibility. They reported a bimodal distribution of scores on the Harvard Group Scale of Hypnotic Susceptibility (HGSHS) in college students with a sharp major peak at 12:00 h and a secondary, broader plateau around 17:00 to 18:00 h. The limitations of this study were that the subjects were tested in groups at hourly intervals during regular daytime class periods. The authors acknowledged that this group testing may have cancelled out the more individual 90–120 min ultradian rhythms in hypnotic susceptibility hypothesized by Rossi (1982), and they recommended that further assessment be made with individuals rather than groups.

Rossi (1990a) therefore designed a pilot study whereby individual subjects could keep diaries that might identify periodicity in their daily patterns of self-hypnosis and ultradian rest. A *Hypnosis Diary Group* comprised individuals who had expressed an interest in learning self-hypnosis and were led through at least one classical hypnotic induction involving eye fixation, imagery and relaxation to facilitate "mind–body healing". They were then encouraged to keep a "self-hypnosis diary" for 2 weeks in which daily they recorded three items: (a) the time of day when they did self-hypnosis, (b) how much time they remained in self-hypnosis and (c) anything about their healing experience of self-hypnosis that they found interesting.

The *Ultradian Diary Group* consisted of people who attended one of Rossi's lectures on "The ultradian healing response" (Rossi and Cheek 1988) as an approach to optimizing mind–body healing. To facilitate this new approach to mind–body healing they were to record the same items as the self-hypnosis group. Both groups were given the same purposely vague and non-directive instruction about when and how often they should do their "inner healing work" and diary recording.

Figure 18.1 presents an overview of the data for the total group of 16 subjects used in this pilot study. The jagged line with dark filled circles represents the original data of 292 diary reports by the total group. The symmetrical curve of open circles is the result of the analysis on the original data with the computer technique of multiple complex demodulation (MCD) carried out by Helen Sing of the Behavioral Biology Department of the Walter Reed Army

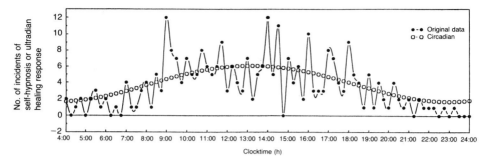

Fig. 18.1. An overview of the circadian rhythm in the "ultradian healing response" and "self-hypnosis" in 292 diary recordings of 16 subjects over a 1-week period. (Reproduced, with permission, from Rossi 1992a.)

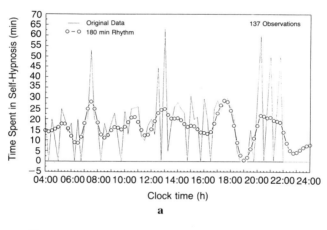

a

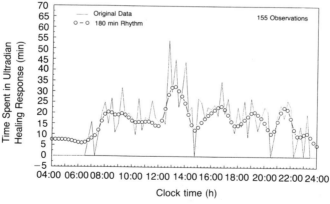

b

Fig. 18.2. a The predominant 180-min rhythm of the self-hypnosis diary group ($n = 9$). (Reproduced, with permission, from Rossi 1992a.) **b** The predominant 180 rhythm of the ultradian healing response diary group ($n = 7$). (Reproduced, with permission, from Rossi 1992a.)

Institute of Research (see Chap. 17). This symmetrical curve respresents a very prominent circadian rhythm with a peak between 12:00 and 13:00 h in the "number of incidents" of self-hypnosis and the ultradian healing response in the total group. This result is consistent with Aldrich and Bernstein's (1987) finding of a peak in hypnotic susceptibility at 12:00 h but the secondary plateau they reported at between 17:00 and 18:00 h is only marginally evident by inspection of the jagged original data. The contrast between the smoothed computer-generated circadian curve and the jagged original data which are more suggestive of an ultradian periodicity (with peaks at 09:00, 12:00, and perhaps at 14:00, 16:00 and 18:00 h) provides some empirical support for the view that the circadian cycle is a composite of many ultradian rhythms (Lloyd and Edwards 1987; Edmunds 1988).

While Rossi (1991a) hypothesized that there would be a 90–120-min BRAC rhythm in the data, Fig. 18.2a and b illustrates a 180-min component isolated by the MCD analysis that was a more prominent ultradian rhythm in both groups. Although the small number of observations of this pilot study do not permit us to make any statement about the significance of the differences between the two groups in the shape of the curves, it is evident that there are ultradian rhythms in the proclivities of these subjects to do "inner healing" whether it is identified as "self-hypnosis" or an "ultradian healing response".

Another observation of significance in Fig. 18.2a and b is that both groups of subjects tended to remain in hypnosis for about 15 or 20 min just as ultradian theory predicts. Further, when the self-descriptions of their inner experiences written in their diaries were examined, it was found that, while there were wide variations in what each individual reported, attenuated forms of all the classical hypnotic phenomena were described to some degree or other in the total group even though the therapist did not suggest them directly. While these observations are still qualitative, they are consistent with the basic hypothesis that the classical forms of hypnotic experience are all expressions of natural ultradian variations of a variety of the psychobiological processes of adaptive homeostasis.

Ultradian Rhythms of Self-Hypnosis and Healing

The striking confirmations of the predictions of the ultradian theory of hypnotic suggestion in the pilot study reported above led to another experimental design with two independent replications. Rossi (1992b) developed a variety of "naturalistic," permissive and non-directive approaches to therapeutic hypnosis, where individuals were encouraged to remain in a "naturalistic hypnosis", for as long as they wished, with the following words: "You will remain in hypnosis as long as necessary to resolve the issue you have been dealing with in as satisfactory a manner as is possible at this time. Your unconscious (or 'inner mind') will then allow you to awaken entirely on your own feeling refreshed and alert."

No further cues or statements were made while the patients remained in this self-guided state of naturalistic hypnosis. After the patients indicated that they had spontaneously awakened by opening their eyes, talking and moving normally, the therapist non-directively encouraged them to report something about their trance experience with questions such as: what was that experience like? what did you actually experience? can you tell me more about it? As a check on the reliability of the amount of time spent on this "natural therapeutic trance" the entire procedure was repeated with each person usually within a week or two.

The results of this experiment supported the prediction of ultradian theory: patients remain in a naturalistic therapeutic hypnosis for about 20 min. This naturalistic trance time was reliable in the sense that the correlation of 0.49 between time spent in the first and second trance was significant at the $p = 0.02$ level (d.f. = 28). The large standard deviations (± 10 min) and wide range of naturalistic trance times (between 2 and 67 min), however, does not recommend it as a useful predictor in practical therapeutic work.

This wide variability is highly characteristic of many ultradian behaviours because they are highly sensitive and adaptive to changing environmental circumstances. Kleitman (1970) and Wever (1989) have emphasized that most ultradian–circadian behaviours are easily modifiable by psychosocial stimuli. It is precisely this sensitivity and responsiveness to psychosocial stimuli that supports the view that hypnotic suggestion (certainly a class of psychosocial stimuli) may achieve its mind–body effects by entraining or synchronizing the naturally wide variability of ultradian behaviour in the process of adaptive homeostasis.

As was found in the previous study, when the subjects were later questioned about their experience they all volunteered at least one description implying that they had experienced a hypnotic phenomenon even though none was directly suggested. When a subject remarked, for example, "A lot was going on during the trance but now I don't remember any of it", it implies that *hypnotic amnesia* was experienced. When a subject commented, "I felt I was outside my body", it implies that a *hypnotic dissociation* took place. When a subject recalled, "At one point I felt I was a baby screaming", it was taken to imply that at least a partial *hypnotic age regression* was experienced even though it was not in any way suggested by the experimenter. This aspect of the study was limited to the qualitative level because we have no standardized scales for quantifying such spontaneous hypnotic phenomena (for an unstandardized scale, see Rossi 1986d).

Replications and the 20-Min Ultradian Trance Curve

Sommer (1990) and Lippincott (1990) each reported independent replications of the above study with (a) minor modifications of procedure, (b) subjects assessed in different age ranges and parts of the country and (c) scales for measuring hypnotic susceptibility. They both found essentially similar results: Sommer reported the mean naturalistic trance time for her 32 subjects to be 18.02 min (s.d. 9.62) when she controlled for a variety of variables such as sex (half of her subjects were male and half were female) and amount of previous trance experience. Lippincott reported a mean naturalistic trance time of 18.55 min (standard deviation, 14.11). Sommer and Lippincott each carefully documented that while there were the typically wide variations in the nature of what each subject reported about the subjective aspects of their naturalistic trance experience, within their total groups virtually all the classical phenom-

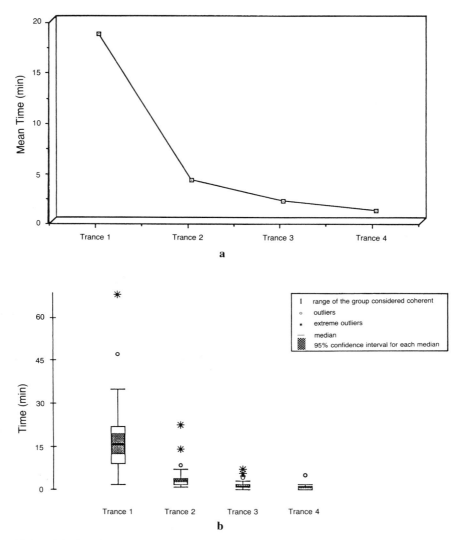

Fig. 18.3. a The lengths of successive naturalistic trances in a single 90-min ultradian period. (Median values from Lippincott 1990.) **b** The lengths of successive naturalistic trances within a single 90-min period, illustrating the 95% confidence interval and outliers. (Reproduced, with permission, from Lippincott 1990.)

ena of hypnosis were experienced by implication as described above even though they were not directly suggested.

In another study Lippincott (1990) tested a unique hypothesis derived from ultradian theory. If the 20-min naturalistic trance is, in fact, an important psychobiological period ("marker") for some endogenous psychobiological rhythm, one would expect that after subjects have experienced it, they would no longer have a need to remain in a second naturalistic trance for another

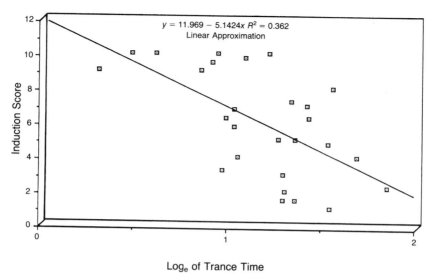

$y = 11.969 - 5.1424x \quad R^2 = 0.362$
Linear Approximation

Fig. 18.4. The distribution of hypnotic induction score versus $\log_e$ trance length. (Reproduced, with permission, from Lippincott 1990.)

20 min within the same 90–120-min BRAC ultradian period. This leads to the prediction that subjects ($n = 30$) will remain in a permissive naturalistic trance for much shorter periods if they are asked to experience a second, third and fourth trance in rapid succession immediately after they have awakened from their first 20-min naturalistic trance. This prediction was well supported by the results illustrated in the "successive trances in 90 min" shown in Fig. 18.3a and b, where the first trance had a mean length of 18.55 min while the second, third and fourth experienced in rapid succession were 4.03, 1.96 and 0.98 min, respectively.

A surprising aspect of this study was the inverse relationship between hypnotic susceptibility as measured by the hypnotic induction profile (Spiegel and Spiegel 1978) and naturalistic trance time. Fig. 18.4 illustrates this inverse relationship as the distribution of hypnotic induction scores versus $\log_e$ of naturalistic trance length with a Pearson product moment correlation coefficient of $r = -0.602$ ($p < 0.001$) and a coefficient of determination (R^2) of 0.362. One way of interpreting these results is to say that, whatever the psycho-biological basis of the ultradian–naturalistic trance time relationship may be, the more highly susceptible hypnotic subjects apparently are better at it since they require less time to do it. Replication and verification of these findings with a broader range of subjects, circumstances and measures of hypnotic suggestibility are required before we can accept their implications for an ultradian theory of mind–body communication and healing.

A recent review of the literature of experimental hypnosis prompted by these findings turned up a number of earlier studies that could be interpreted as providing further support for a naturalistic 20-min trance time. In an early methodological study Dorcus et al. (1941) compared the amount of time a group of 20 deeply hypnotizable subjects remained in trance, after the hypnotist left the room, with a control group who were told to simply lie down and relax. In both groups the majority of the subjects got up and left the room within 20 min. In two studies using subjects simulating hypnosis as control groups, it was found that highly hypnotizable subjects remained in trance when they believed they were left unobserved for 10.7 and 16.5 min (Orne and Evans 1966; Evans and Orne 1971) while the simulating low hypnotizable subjects acted as if they were in trance for 25.2 min. In a more clinically oriented recent survey Sanders (1991) mailed a self-hypnosis questionnaire to 1000 members of the American Society of Clinical Hypnosis. In the 233 responses she received it was found that a 15–20 min period was most typical in the use of self-hypnosis. It is interesting that, while none of these researchers set out to test the ultradian prediction that there is a naturalistic 20-min trance time, all their data support it.

Ultradian Owls, Larks and Hypnotic Susceptibility

The peaks in hypnotic susceptibility found in the morning (noon) and evening by Aldrich and Bernstein (1987) led Rossi (noted in Lippincott 1992) to hypothesize that it could be a reflection of the differential performance of a mixed population of "larks" (people who claim to be more alert in the morning) and "owls" (people who claim to be more alert in the evening). Lippincott (1991, 1992) then designed an experiment to test the hypothesis that (a) larks would have higher hypnotic susceptibility in the later afternoon (16:00 to 18:00 h), (b) owls would have a higher susceptibility in the morning (08:00 to 10:00 h) and (c) there would be no difference in hypnotizability between midnight and 02:00 h when both groups would normally be asleep.

Colman's "Owl and Lark Questionnaire" was used to select 42 subjects that could be identified as owls (n = 21) or larks (n = 21) using a median split within a larger group of college students who had taken the Harvard Group

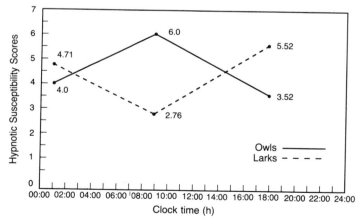

Fig. 18.5. The distribution of hypnotic susceptibility scores in "owls" and "larks" throughout the day. (Reproduced, with permission, from Lippincott 1991.)

Scale of Hypnotic Susceptibility, Form A (HGSHS:A). Fig. 18.5 illustrates a confirmation of all three hypotheses with a multivariate analysis of variance (MANOVA) using the Wilk's Criterion, a one-way univariate analysis of variance and Tukey's Honestly Significant Difference as a post hoc analysis. As can be seen in Fig. 18.5 owls have significantly higher scores in the morning (mean = 6, s.d. = 2.22) than larks (mean = 2.76, s.d. = 2.20; $p < 0.0001$). Larks, on the other hand, have significantly higher scores in the late afternoon (mean = 5.52, s.d. = 2.68) than owls (mean = 3.52, s.d. = 2.11; $p < 0.01$). There was no significant difference between owls (mean = 4, s.d. = 1.17) and larks (mean = 4.71, s.d. = 2.49; $p < 1.19$) between midnight and 02:00 h.

What is most significant about Fig. 18.5 is that if the owls and larks had not been separated there apparently would have be no significant differences in hypnotic susceptibility over time because the inverse periodic patterns of owls and larks would have cancelled each other out. If future studies confirm such ultradian and circadian performance shifts in owls and larks it will require a profound re-evaluation of many prevous studies on psychobiological variables in general as well as hypnosis in particular. These results challenge previous conceptions of hypnotic susceptibility as a relatively fixed, unvarying trait of the individual (Hilgard 1982) and suggest new studies to assess the hypothesis that ultradian rhythms are a significant aspect of the mind–body processes of adaptive homeostasis that are accessed and utilized by hypnotherapeutic suggestion and perhaps holistic medicine in general.

Suggestions for Future Research

The speculative nature of the unification hypothesis of chronobiology proposed in this chapter for a new integration of biology and psychology via ultradian time, rhythm and adaptive homeostasis at all levels from mind to gene can be justified only if it leads to further research generating novel facts for resolving old dilemmas. The sampling of studies proposed here are a supplement to the "Sixty-four research projects in search of a graduate student" published earlier (Rossi 1986b; Rossi and Cheek 1988).

Non-Linear Models in Ultradian Psychobiology

A difficulty with many of the studies cited in Table 18.2 is that their data are often statistically marginal and not easy to reproduce; conclusions drawn from them are consequently controversial. A portion of this problem may be accounted for by the currently evolving view that two distinct signalling systems are involved in mind–body communication: (a) the neural/synaptic basis of the central and autonomic nervous systems, and (b) the ligand–cell receptor system of molecular communication that characterizes the endocrine, immune and neuropeptide systems. It is now believed that molecular information substances (including hormones, growth factors, neuropeptides, neuromodulators, cytokines etc. (Schmitt 1984, 1986) are the older, original communication system that evolved in single-cell life forms while the neural/synaptic system evolved later to facilitate communication in multicellular organisms (Roth et al. 1985). It has been proposed that it is the complex and mutually modulating homeostatic interaction of the neural/synaptic and the molecular informational cell-receptor systems that are responsible for the state-dependent features of physiology (Lydic 1987) as well as memory, learning, and behaviour associated with hypnosis (Rossi 1986c, 1990c; Rossi and Cheek 1988) that give rise to the ever-changing rhythms of psychobiological variables.

An illuminating illustration of the difficulty in planning and interpreting these psychobiological studies is provided by the ultradian/circadian and hypnotic research literature on endorphins. The most simplistic approach has sought a direct correlation between one or more informational substances, psychological and behavioural states (Bergland 1985). Although not stated explicitly, this is the model that has guided most researchers within the past decade who have tried to determine whether endorphins were involved in hypnotic analgesia.

Many of these early studies used naloxone as an antagonist of the endogenous endorphins; the oversimplified view was that if hypnotic analgesia was blocked by naloxone, then some involvement of endorphins would be implied. However, naloxone can have paradoxical effects. In low doses it can relieve pain, while in high doses it can make pain worse (Grevert and Goldstein 1985). Placebo effects, emotional stress, and circadian rhythms are only a few of the factors that complicate the picture in ways that are not easily controllable. The initial studies by Barber and Mayer (1977) and Goldstein and Hilgard (1975) reported that naloxone did not affect hypnotic analgesia, while Stephenson (1978) reported that hypnotic analgesia was reversed by naloxone with one subject in very deep somnambulistic trance. Frid and Singer (1979) then reported that hypnotic analgesia is partially reversed by naloxone when *stress* is introduced before pain. More recent work by Domangue et al. (1985) has directly measured increases in β-endorphin-like immunoreactive molecules following hypnotherapy. Domangue et al. (1985) reported that "there were clinically and statistically significant decreases in *pain*, *anxiety* and *depression* and increases in β-endorphin-like immunoreactive material".

An intriguing but typical feature of these seemingly quixotic findings is that conflicting results are often reported when psychological variables such as *stress*, *pain*, *anxiety*, and *depression* are taken into account in evaluating somatic/molecular variables. Very slight changes in the initial conditions of either the psychological or somatic/molecular variables in such studies frequently give rise to dramatically different experimental outcomes. While this has been a little-understood and deeply disturbing feature of the psychobiological studies in the literature on ultradian/circadian rhythms as well as that on hypnosis, it is exactly what one would expect from the mathematical models of non-linear dynamics and chaos theory where it is now well known that slight differences in the initial conditions of any self-referential (i.e. homeostatic, cybernetic) system give rise to deterministic but, practically speaking, non-predictable outcomes (Ford 1986, 1988). These non-linear dynamics are currently being used to create mathematical models of the typical features of self-reproducing systems in genetics, the immune system, autocatalytic proteins, and neural networks (Campbell 1987). It is being proposed here that these same non-linear mathematical models are now required for uniting the previously unrelated fields of psychobiological rhythms, hypnosis and psychosomatic medicine.

A crucial link for studies in mind–body communication is to relate phenomenological variables (memory, learning, emotions, imagery, hypnosis, psychotherapy, meditation, the ultradian healing response, etc.) to the molecular messengers of the neuroendocrinal axis and the epigenetic level (Pert et al. 1985, 1989). An emerging methodology utilizing non-invasive tests for molecular messengers in saliva during stress, agression, depression, postpartum blues and a variety of other psychobiological states may be a generally useful approach. Relationships between salivary cortisol, cortisone, oestriol, progesterone, testosterone (Kirschbaum and Hellhammer 1989) and a variety of immunological parameters (Olness and Conroy 1985; Olness et al. 1980, 1989) may be the wave of the future for the assessment of the molecular messengers between psychosocial variables and cellular–genetic processes of information transduction in adaptive homeostasis.

Ultradians as the Basic Parameters of Therapeutic Hypnosis

The natural ultradian parameters of psychobiological processes that are entrainable by psychosocial cues are hypothesized as defining the range of performance and experience that can be modulated and optimized by hypnotic suggestion. This leads to the testable hypothesis that the ultradian processes of Table 18.2 that are most entrainable to external cues (variously named entraining agent, synchronizer, or zeitgeber) will be the most modifiable by hypnosis. Further, the parameters of ultradian and circadian rhythms entrainable by psychosocial cues set the limits of hypnotic suggestibility. This leads to the testable hypothesis that hypnotic suggestion will not be able to modify any of the sensory/perceptual and behavioural processes listed in the left-hand column of Table 18.2 beyond the range of their typical ultradian parameters.

Table 18.2 suggests many other direct and indirect tests of the ultradian theory of therapeutic hypnosis. Since the hypnotic phenomena of age regression and analgesia/anaesthesia have *not* been reported as being associated with the ultradian rhythms in any of the literature cited in Table 18.2, a direct way of assessing the ultradian hypothesis of hypnosis would be to test the prediction that age regression and analgesia would be most readily evoked during a specific phase of the BRAC, first described by Kleitman (1963, 1969). Three studies have already reported qualitative observations that support this hypothesis (Lippincott 1990; Rossi 1990a; Sommer 1990); we now need well-controlled quantitative studies.

Intuitively, one would predict that age regression and analgesia would be most evident during the rest phase of BRAC, when the subject is approaching some of the critical parameters of sleep. This may not be the case with the other hypnotic phenomena. For example, the experimental (Blum 1972; Evans 1972) and clinical (Erickson and Rossi 1979, 1981; Erickson et al. 1976) literature concur in finding that various states of activation are required for the optimal experience of many hypnotic phenomena. A conceptually sophisticated approach to this issue may be provided by the recent applications of geometrical dynamics and chaos theory to the analysis of periodic behaviour (Abraham and Shaw 1983a,b,c; Glass and Mackey 1988).

The suprachiasmatic nucleus of the hypothalamus is empirically well established as the circadian pacemaker entraining the neuroendocrinal system to the daily light–dark cycle in a manner intimately associated with a wide range of biological rhythms associated with optimum performance, health and illness (Klevecz and Braly 1987; Kupfer et al. 1988). Does naturalistic hypnotic suggestion and the ultradian healing response modulate these circadian rhythms to the same degree as light (Jewett et al. 1991)? This may be a useful way of determining the degree to which hypnosis can modulate circadian rhythms versus the ultradian rhythms that have been emphasized in this chapter.

The immune system appears to have a wide range of ultradian and circadian rhythms that are integrated with similar rhythms at all mind–body levels. Many of the ultradian relationships within the immune system (such as the commitment to T lymphocyte activation wherein 2 h are sufficient for DNA synthesis and morphological changes; Crabtree 1989) are also related to the more global mind–body processes of cancer and psychoneuroimmunology (Kiecolt-Glaser and Glaser 1986). Does naturalistic hypnotherapy modulate precisely those ultradian parameters of the immune system that respond to other types of psychosocial stimuli such as stress? Can such relationships account for the recent literature that finds a relationship between amelioration of cancer and psychosocial variables such a group psychotherapy (Spiegel et al. 1989) and hypnosis (Hall 1982–1983)? What, precisely, is the psycho-neuro-endocranial–epigenetic pathway of information transduction between stress, naturalistic trance, the ultradian healing response and any of the 80 oncogenes that are involved in cancer?

Ultradian Rhythms, Stress and Healing

It has been proposed that chronic stress engendered by individuals who override and disrupt their own ultradian rhythms (by ignoring their natural periodic needs for rest in any extended performance situation, for example) are thereby setting in motion a basic psychosocial process of psychosomatic illness (Rossi 1982). The stress related mind–body problems are viewed as an expression of distorted loops of cybernetic information transduction ranging from the epigenetic to the mind–brain level. Further, a naturalistically oriented therapeutic hypnosis that utilizes the 20-min ultradian rest period provides a comfortable, "healing state" during which disrupted ultradian parameters can normalize (reset or synchronize) themselves and thus undercut the process of psychosomatic illness at their psychobiological source. This permissive receptivity to our natural need for rest and recovery is called the "ultradian healing response" (Rossi 1982, 1986a,b, 1991b; Rossi and Cheek 1988). It now appears that most holistic methods of mind–body therapy (psychotherapy, shamanism, biofeedback, the relaxation response, progressive relaxation, autogenic training, imagery, etc.) may utilize the natural 20-min ultradian healing response as an unrecognized, general factor facilitating healing. This also may be the unrecognized psychobiological basis of the "therapeutic benefits" experienced in many meditation methods (Zen, Transcendentalism, Yoga etc.) that traditionally enjoin beginners to practise for 20 min. Advanced meditators typically practise for a full 90–120-min ultradian cycle to utilize the entire range of their psychobiological potentials for personal/spiritual development.

Most research studies in these areas of holistic healing cite the use of a 20-min period for therapeutic healing (Feher et al. 1989; Green and Green 1987). However, none of these papers ever cites any rationale for using this 20-min period. The reader should survey a representative sample of such studies to determine the degree to which their results are consistent with the ultradian theory of naturalistic healing (Sanders 1991).

One should design controlled experimental studies to assess further the hypothesis that during the natural 20-min ultradian rest-restoration period of the BRAC (Kleitman 1963, 1969, 1970, 1982) consciousness is in a mildly dissociated state (mildly independent of control by outer world stimuli). How would you test the hypothesis that this mildly dissociated (hypnoid) state apparently permits internal information transduction to proceed optimally with relatively less interference from competing outer world stimuli? Most generally, how would you assess the theory that the active phase of the BRAC enhances sensory-perceptual functioning for interacting optimally with the outside world while, by contrast, the rest phase of the BRAC shifts this focus to the inner world to optimize the processing of internal data and all the attendant systems of information transduction between the epigenetic source (genetic transcription, translation and processing of proteins), the neuro-endocrinal axis and the mind–brain processing of state-dependent memory, learning and behaviour in adaptive homeostasis (Rossi 1987, 1990d,e)?

The grand unification of the many diverse forms of holistic healing across cultures and historical time periods via their common denominator in the unification hypothesis of chronobiology may be a useful approach to resolving subtle aspects of the mind–body problem and the nature–nurture controversy. The nasal–brain link that has been proposed as the mechanism or pathway of information transduction whereby many workers in the healing arts achieve their effects is one of the most intriguing research avenues for exploring novel psychobiological parameters (Rossi 1990b). A century of studies on the ultradian nasal cycle rhythm outlined in Table 18.1 suggests that the verification and further exploration of the ultradian breath–brain–personality relationships pioneered by Werntz (1981; Werntz et al. 1982a,b), Klein (Klein and Armitage 1979; Klein et al. 1986), and Osowiec (1991) will be a major highway of future research on mind–body communication and the wave nature of being.

Summary

This chapter builds a bridge between biology and psychosocial processes via ultradian time, rhythm and information transduction. It begins by exploring the unification hypothesis of chronobiology: ultradian rhythms in the 20- to 120-min range on the cellular–genetic level are reflected in similar periodicities on the neuroendocrinal, behavioural and psychosocial levels. These ultradian rhythms are viewed as coordinators of information transduction between the expression of genes and environmental signals such as temperature, food and psychosocial variables in the general process of adaptive homeostasis. Failures in this process of adaptive homeostasis are reflected in what is commonly called "stress" and "psychosomatic problems". An extensive matching of the clinical-experimental data of chronobiology and psychology suggests that what the biologist calls the "*entrainment* of ultradian and circadian rhythms by psycho-social stimuli" is the psychobiological basis of what psychotherapists call "hypnotic suggestion to facilitate mind–body healing". Research paradigms and mathematical models now needed to explore this ultradian interface between biology and psychosocial processes are outlined.

References

Aaronson B (1969a) The hypnotic induction of the void. Paper presented at the American Society of Clinical Hypnosis, San Francisco, CA

Aaronson B (1969b) Time, time stance, and existence. Paper presented at the meetings of the International Society for the Study of Time, Freibourg, Germany

Abraham R, Shaw C (1983a) Dynamics – the geometry of behavior. Part 1: periodic behavior. Vismath vol 1. Aerial Press, Santa Cruz, CA

Abraham R, Shaw C (1983b) Dynamics – the geometry of behavior. Part 2: Chaotic behavior. Vismath vol 2. Aerial Press, Santa Cruz, CA

Abraham R, Shaw C (1983c) Dynamics – the geometry of behavior. Part 3: Global behavior. Vismath vol 3. Aerial Press, Santa Cruz, CA

Alberts B, Bray D, Lewis J, Raff M, Roberts K, Watson J (1989) Molecular biology of the cell, 2nd edn. Garland Publishing Inc., New York

Aldrich K, Bernstein D (1987) The effect of time of day on hypnotizability. Int J Clin Exp Hypn 35:141–145

Araoz D (1982) Hypnosis and sex therapy. Brunner/Mazel, New York

Aserinsky E, Kleitman N (1953) Regularly occurring periods of eye motility and concomitant phenomena during sleep. Science 118:273–274

Balthazard C, Woody E (1985) The "stuff" of hypnotic performance: a review of psychometric approaches. Psychol Bull 98:283–296

Barber J, Mayer D (1977) Evaluation of the efficacy and neural mechanism of a hypnotic analgesia procedure in experimental and clinical dental pain. Pain 4:41–48

Barber TX (1972) Suggested ("hypnotic") behavior: the trance paradigm versus an alternative paradigm. In: Fromm E, Shor R (eds) Hypnosis: research developments and perspectives, 2nd edn. Aldine, Chicago, IL, pp 115–182

Barber TX (1984) Changing unchangeable bodily processes by (hypnotic) suggestions: a new look at hypnosis, cognitions, imagining, and the mind–body problem. Advances 1(2):7–40

Bergland R (1985) The fabric of mind. Viking Press, New York

Bernardis L, Tannenbaum G (1987) Failure to demonstrate disruption of ultradian growth hormone rhythm and insulin secretion by dorsomedial hypothalamic nucleus lesions that cause reduced body weight, linear growth, and food intake. Exp Brain Res 66:572–576

Black S, Humphrey J, Niven J (1963) Inhibition of Mantoux reaction by direct suggestion under hypnosis. Br Med J 5346:1649–1652

Blum G (1972) Hypnotic programming techniques in psychological experiments. In: Fromm E, Shor R (eds) Hypnosis: research developments and perspectives. Aldine, Chicago, IL, pp 359–385

Bossom J, Natelson B, Levin B (1983) Ultradian rhythms in cognitive functions and their relationship to visceral processes. Physiol Behav 31:119–123

Bowden D, Kripke D, Wyborney V (1978) Ultradian rhythms in waking behavior of rhesus monkeys. Physiol Behav 21:929–933

Bower G (1981) Mood and memory. Am Psychol 36:129–148

Bowers K (1977) Hypnosis: an informational approach. Ann NY Acad Sci 296:222–237

Bowers K, Kelly P (1979) Stress, disease, psychotherapy, and hypnosis. J Abnorm Psychol 88:490–505

Brandenberger G, Simon C, Follenius M (1987) Night–day differences in the ultradian rhythmicity of plasma renin activity. Life Sci 40:2325–2330

Breuer J, Freud S (1957) Studies on hysteria. In: Strachey J (ed and transl), Freud A (ed) The standard edition of the complete psychological works of Sigmund Freud, vol II. W. W. Norton, New York [First published 1895]

Broughton R (1975) Biorhythmic variations in consciousness and psychological functions. Can Psychol Rev: Psychol Can 16:217–239

Brown F (1982) Rhythmicity as an emerging variable for psychology. In: Brown F, Graeber R (eds) Rhythmic aspects of behavior. Lawrence Erlbaum Associates, Hillsdale, NJ, pp 35–38

Brown P (1991) The hypnotic brain. Yale University Press, New Haven, CN

Brown T (1959) Hypnosis in genito-urinary diseases. Am J Clin Hypn 1:165–168

Bykov V, Katinas G (1979) Temporal organization of the thyroid in the A/He mice (morphometric investigation). Biol Bull Acad Sci USSR 6:247–249

Campbell D (1987) Nonlinear science: from paradigms to practicalities. Los Alamos Sci no. 15:218–262

Carli G, Garabollini F, Lopo di Prisco C (1979) Plasma corticosterone and its relation to susceptibility to animal hypnosis in rabbits. Neurosci Lett 11:271–274

Carnes M, Brownfield M, Kalin N, Lent S, Barksdale C (1986) Episodic secretion of ACTH in rats. Peptides 7:219–223

Carter B, Elkins G, Kraft S (1982) Hemispheric asymmetry as a model for hypnotic phenomena: a review and analysis. Am J Clin Hypn 24:204–210

Cartwright R, Monroe L (1968) Relation of dreaming and REM sleep: the effects of REM deprivation under two conditions. J Pers Soc Psychol 10:69–74

Chapman L, Goodell H, Wolff H (1959) Changes in tissue vulnerability induced during hypnotic suggestion. J Psychosom Res 4:99–115

Clarke J (1980) The nasal cycle II: a quantitative analysis of nostril dominance. Res Bull Himalayan Int Inst 2:3–7

Clements P, Hafer M, Vermillion M (1976) Psychometric, diurnal, and electrophysiological correlates of activation. J Pers 33:387–395

Cooper L, Erickson M (1959) Time distortion in hypnosis. Williams & Wilkins, Baltimore, MD

Crabtree G (1989) Contingent genetic regulatory vents in T lymphocyte activation. Science 243: 355–361

Crasilneck H, Hall J (1985) Clinical hypnosis. Grune & Stratton, New York

Dalton K, Denman D, Dawson A, Hoffman H (1986) Ultradian rhythms in human fetal heart rate: a computerized time series analysis. Int J Bio-Med Comput 18:45–60

Dias M (1963) Hypnosis and prolonged suggested sleep in gastroenterology. Hospital 64:983–993

Domangue B, Margolis C, Lieberman D, Kaji H (1985) Biochemical correlates of hypnoanalgesia in arthritic pain patients. J Clin Psychiatry 46:235–238

Dorcus R, Britnall A, Case H (1941) Control experiments and their relation to theories of hypnotism. J Gen Psychol 24:217–221

Eccles R (1978) The central rhythm of the nasal cycle. Acta Otolaryngol 86:464–468

Edmunds L Jr (1988) Cellular and molecular bases of biological clocks. Springer, Berlin Heidelberg New York

Edwards G (1960) Hypnotic treatment of asthma. Br Med J 1:492–497

Eichorn R, Tracktir J (1955) The effects of hypnotically induced emotions upon gastric secretions. Gastroenterology 29:432–438

Ellenberger H (1970) The discovery of the unconscious. Basic Books, New York

Elsmore T, Hursh S (1982) Circadian rhythms in operant behavior of animals under laboratory conditions. In: Brown F, Graeber R (eds) Rhythmic aspects of behavior. Lawrence Erlbaum Associates, Hillsdale, NJ, pp 273–310

Erickson M (1980a) Psychogenic alteration of menstrual functioning: three instances. In: Rossi E (ed) The collected papers of Milton H. Erickson on hypnosis, vol II, Hypnotic investigation of sensory, perceptual, and psychophysical processes. Irvington, New York, pp 207–212 [First published 1960]

Erickson M (1980b) Breast development possibly influenced by hypnosis: two instances and the psychotherapeutic results. In: Rossi E (ed) The collected papers of Milton H. Erickson on hypnosis, vol II, Hypnotic alteration of sensory, perceptual, and psychophysical processes. Irvington, New York, pp 203–206 [First published 1960]

Erickson M, Rossi E (1979) Hypnotherapy: an exploratory casebook. Irvington, New York

Erickson M, Rossi E (1980a) Two-level communication and the microdynamics of trance. In: Rossi E (ed) The collected papers of Milton H. Erickson on hypnosis, vol I, The nature of hypnosis and suggestion. Irvington, New York, pp 430–451 [First published 1976]

Erickson M, Rossi E (1980b) Autohypnotic experiences of Milton H. Erickson. In: Rossi E (ed) The collected papers of Milton H. Erickson on hypnosis, vol I, The nature of hypnosis and suggestion. Irvington, New York, pp 108–132 [First published 1977]

Erickson M, Rossi E (1981) Experiencing hypnosis: therapeutic approaches to altered states. Irvington, New York

Erickson M, Rossi E, Rossi S (1976) Hypnotic realities. Irvington, New York

Evans F (1972) Hypnosis and sleep: techniques for exploring cognitive activity during sleep. In: Fromm E, Shor R (eds) Hypnosis: research developments and perspectives. Aldine, Chicago, IL, pp 43–83

Evans F, Orne M (1971) The disappearing hypnotist: the use of simulating subjects to evaluate how subjects perceive experimental procedures. Int J Clin Exp Hypn 19:277–296

Feher S, Berger L, Johnson J, Wilde J (1989) Increasing breast milk production for premature infants with a relaxation/imagery audiotape. Pediatrics 83:57–60

Feldman J (1986) Neurophysiology of breathing in mammals. In: Mountcastle V, Bloom F, Geiger S (eds) Handbook of physiology, section I, The nervous system, vol IV, Intrinsic regulatory systems of the brain. American Physiological Society, Bethesda, MD, pp 463–524

Filicori M, Bolelli G, Franceschetti F, Lafisca S (1979) The ultradian pulsatile release of gonadotropins in normal female subjects. Acta Eur Fertil 10:29–33

Fischer R (1971a) Arousal-statebound recall of experience. Dis Nerv Syst 32:373–382

Fischer R (1971b) The "flashback": arousal-statebound recall of experience. J Psychedel Drugs 3:31–39

Fischer R (1971c) A cartography of ecstatic and meditative states. Science 174:897–904

Folkard S (1982) Circadian rhythms and human memory. In: Brown F, Graeber R (eds) Rhythmic aspects of behavior. Lawrence Erlbaum Associates, Hillsdale, NJ, pp 313–344

Follenius M, Simon C, Bradenberger G, Lenzi P (1987) Ultradian plasma corticotropin and cortisol rhythms: time-series analyses. J Endocrinol Invest 10:261–266

Ford J (1986) Chaos: solving the unsolvable, predicting the unpredictable! In: Barnsley M, Demko S (eds) Chaotic dynamics and fractals. Academic Press, New York, pp 1–52

Ford J (1988) What is chaos, that we should be mindful if it? In: Capelin S, Davies P (eds) The New Physics. Cambridge University Press, Cambridge, pp 248–372

Freeman R, Baxby K (1982) Hypnotherapy for incontinence caused by the unstable detrusor. Br Med J [Clin Res] 284:1831–1834

Frid M, Singer G (1979) Synoptic analgesia in conditions of stress is partially reversed by naloxone. Psychopharmacology 63:211–215

Friedman S (1978) A psychophysiological model for the chemotherapy of psychosomatic illness. J Nerv Ment Dis 166:110–116

Friedman S, Fisher C (1967) On the presence of a rhythmic diurnal, oral instinctual drive cycle in man: a preliminary report. J Am Psychoanal Assoc 15:317–343

Friedman S, Kantor I, Sobel S, Miller R (1978) On the treatment of neurodermatitis with a monomine oxidase inhibitor. J Nerv Ment Dis 166:117–125

Fromm E, Shor R (1972) Hypnosis: research developments and perspectives. Aldine, Chicago, IL

Frumkin L, Ripley H, Cox G (1978) Changes in cerebral hemispheric lateralization with hypnosis. Biol Psychiatry 13:741–750

Funk F, Clarke J (1980) The nasal cycle observations over prolonged periods of time. Res Bull Himalayan Int Inst Winter, 1–4

Gabel S (1988) The right hemisphere in imagery, hypnosis, REM sleep and dreaming. J Nerv Ment Dis 176:323–331

Gerkema M, Daan S (1985) Ultradian rhythms in behavior: the case of the common vole (*Microtus arvalis*). In: Schulz H, Lavie P (eds) Ultradian rhythms in physiology and behavior. Springer, Berlin Heidelberg New York, pp 11–31

Gilbert A (1989) Reciprocity versus rhythmicity in spontaneous alternations of nasal airflow. Chronobiol Int 6:251–257

Gill M, Brenman M (1959) Hypnosis and related states. International Universities Press, New York

Glass L, Mackey M (1988) From clocks to chaos: the rhythms of life. Princeton University Press, Princeton, NJ

Globus G (1966) Rapid eye movement cycle in real time. Arch Gen Psychophysiol 15:654–669

Goldberg B (1985) Hypnosis and the immune response. Int J Psychosom 32(3):34–36

Goldstein A, Hilgard E (1975) Failure of the opiate antagonist naloxone to modify hypnotic analgesia. Proc Natl Acad Sci USA 72:2041–2043

Gopher D, Lavie P (1980) Short-term rhythms in the performance of a simple motor task. J Motor Behav 12:207–221

Gordon H, Frooman B, Lavie P (1982) Shift in cognitive asymmetries between wakings from REM and NREM sleep. Neuropsychologica 20:99–103

Gordon C, Lavie P (1986) The role of the sympathetic nervous system in the regulation of ultradian rhythms in urine excretions. Physiol Behav 38:307–313

Gorton B (1957) The physiology of hypnosis. I. J Soc Psychosom Dent 4(3):86–103

Gorton B (1958) The physiology of hypnosis: vasomotor activity in hypnosis. J Am Soc Psychosom Dent 5:20–28

Green R, Green M (1987) Relaxation increases salivary immunoglobulin A. Psychol Rep 61:623–629

Grevert P, Goldstein A (1985) Placebo analgesia, naloxone, and the role of endogenous opioids. In: White L, Tursky B, Schwartz G (eds) Placebo: theory, research, and mechanisms. Guilford Press, New York, pp 332–350

Hall H (1982–1983) Hypnosis and the immune system: a review with implications for cancer and the psychology of healing. Am J Clin Hypn 25:92–103.

Hartmann E (1968) The 90-minute sleep–dream cycle. Arch Gen Psychiatry 18:280–286

Hasegawa M, Kem E (1978) Variations in nasal resistance in man: a rhinomanometric study of the nasal cycle in 50 human subjects. Rhinology 16:19–29

Hayes D, Cobb L (1979) Ultradian biorhythms in social interaction. In: Siegman A, Feldstein F (eds) On time and speech. Lawrence Erlbaum Associates, Hillsdale, NJ, pp 57–70

Heetderks D (1927) Observations on the reaction of normal nasal mucosa membrane. Am J Med Sci 174:231–244

Hilgard E (1965) Hypnotic susceptibility. Harcourt, Brace & World, New York

Hilgard E (1982) Hypnotic susceptibility and implications for measurement. Int J Clin Exp Hypn 30:394–403

Hilgard E, Hilgard J (1983) Hypnosis in the relief of pain. William Kaufmann, Los Altos, CA

Hobson J (1988) The dreaming brain. Basic Books, New York

Holley D, Winget C, DeRoshia C (1981) Effects of circadian rhythm phase alteration on psysiological and psychological variables: implications to pilot performance. NASA Tech Mem 81277

Horne J, Whitehead M (1976) Ultradian and other rhythms in human respiration rate. Experientia 32:1165–1167

Hunsaker W, Reiser B, Wolynetz M (1977) Vaginal temperature rhythms in sheep. Int J Chronobiol 4:151–162

Iranmanesh A, Lizarradle G, Johnson M, Veldhuis J (1989) Circadian ultradian, and episodic release of β-endorphin in men, and its temporal coupling with cortisol. J Clin Endocrinol Metab 68:1019–1025

Jewett J, Kronauer R, Czeisler A (1991) Light-induced suppression of endogenous circadian amplitude in humans. Nature 350:59–62

Jouvet M (1962) Recherches sur les structures nerveuses et les mechanismes responsables des différentes phases du sommeil physiologique. Arch Ital Biol 100:125–206

Jouvet M (1973) Telencephalic and rhonbencephalic sleep in the cat. In: Webb W (ed) Sleep: an active process. Scott Foresman & Co., Glenview, IL, pp 12–32

Kandel E (1989) Genes, nerve cells, and the remembrance of things past. J Neuropsychiatry 1:103–125

Kandel E, Schwartz G (1985) Principles of neural science, 2nd edn. Elsevier, New York

Katz R (1980) The temporal structure of motivation. III. Identification and ecological significance of ultradian rhythms of intracranial reinforcement. Behav Neur Biol 30:148–159

Kayser R (1895) Die exacte Messung der Luftdurchgangigkeit der Nasa. Arch Laryngol Rhinol 3:101–120

Keuning J (1968) On the nasal cycle. Int Rhinol 6:99–136

Kiecolt-Glaser J, Glaser R (1986) Psychological influences on immunity. Psychosomatics 27:621–624

Kirschbaum C, Hellhammer D (1989) Salivary cortisol in psychobiological research: an overview. Biol Pharmacopsychol 22:150–169

Klein R, Armitage R (1979) Rhythms in human performance: $\frac{1}{2}$ hour oscillations in cognitive style. Science 204:1326–1328

Klein R, Pilon D, Prosser S, Shannahoff-Khalsa D (1986) Nasal airflow asymmetries and human performance. Biol Psychol 23:127–137

Kleitman N (1963) Sleep and wakefulness as alternating phases in the cycle of existence. University of Chicago Press, Chicago, IL

Kleitman N (1969) Basic rest–activity cycle in relation to sleep and wakefulness. In: Kales A (ed) Sleep: physiology and pathology. Lippincott, Philadelphia, PA, pp 33–38

Kleitman N (1970) Implications of the rest–activity cycle: implications for organizing activity. In: Hartmann E (ed) Sleep and dreaming. Little, Brown, Boston, pp 13–14

Kleitman N (1982) Basic rest–activity cycle – 22 years later. Sleep 5:311–315

Klevecz R, Braly P (1987) Circadian and ultradian rhythms of proliferation in human ovarian cancer. Chronobiol Int 4:513–523

Knobil E, Hotchkiss J (1985) The circhoral gonadotropin releasing hormone (GnRH) pulse generator of the hypothalamus and its physiological significance. In: Schulz H, Lavie P (eds) Ultradian rhythms in physiology and behavior. Springer, Berlin Heidelberg New York, pp 32–40

Kosunen K, Kuoppasalmi K, N'averi H, Rehunen S, N'arv'anen S, Adlercreutz H (1977) Plasma renin activity, angiotensin II, and aldosterone during the hypnotic suggestion of running. Scand J Clin Lab Invest 37:99–103

Kripke D (1972) An ultradian biological rhythm associated with perceptual deprivation and REM sleep. Psychosom Med 34:221–234

Kripke D (1984) Critical interval hypothesis for depression. Chronobiol Int 1:73–81

Kripke D, Sonnenschein D (1978) A biologic rhythm in waking fantasy. In: Pope K, Stringer J (eds) The stream of consciousness. Plenum Press, New York, pp 321–332

Krynicki V (1975) Time trends and periodic cycles in REM sleep eye movements. Electroencephalogr Clin Neurophysiol 39:507–513

Kupfer D, Monk T, Barchas J (1988) Biological rhythms and mental disorders. Guilford Press, New York

Lavie P (1976) Ultradian rhythms in the perception of two apparent motions. Chronobiologia 3:214–218

Lavie P (1977) Nonstationarity in human perceptual ultradian rhythms. Chronobiologia, 4:38–48

Lavie P (1979) Ultradian rhythms in alertness – a pupillometric study. Biol Psychol 9:49–62

Lavie P, Kripke D (1977) Ultradian rhythms in urine flow in waking humans. Nature 269:142–144

Lavie P, Kripke D (1981) Ultradian circa $\frac{1}{2}$ hour rhythms: a multioscillatory system. Life Sci 29:2445–2450

Levin B, Goldstein A, Natelson B (1978) Ultradian rhythm of plasma noradrenaline in rhesus monkeys. Nature 272:164–166

Levitt E, den Breeijen A, Persky H (1960) The induction of clinical anxiety by means of a standardized hypnotic technique. Am J Clin Hypn 2:206–214

Lewis B, Kripke D, Bowden D (1977) Ultradian rhythms in hand–mouth behavior of the rhesus monkey. Physiol Behav 18:283–286

Lippincott B (1990) Testing predictions of the ultradian theory of therapeutic hypnosis. Paper presented at the 32nd Annual Scientific Meeting of The American Society of Clinical Hypnosis, 24–28 March 1990, Orlando, FL

Lippincott B (1991) Owls and larks in hypnosis: an experimental validation of the ultradian theory of hypnotic susceptibility. Paper presented at the 33rd Annual Scientific Meeting of The American Society of Clinical Hypnosis, 14–18 April 1991, St Louis, MI

Lippincott B (1992) Owls and larks in hypnosis: individual differences in hypnotic susceptibility relating to biological rhythms. Am J Clin Hypn 34:185–192

Lissoni P, Marelli O, Mauri R et al. (1986) Ultradian chronomodulation by melatonin of a placebo effect upon human killer cell activity. Chronobiologia 13:339–343

Lloyd D, Edwards SW (1984) Epigenetic oscillators during the cell cycles of lower eucaryotes are coupled to a clock: life's slow dance to the music of time. In: Edmunds L (ed) Cell cycle clocks. Marcel Dekker, New York, pp 27–46

Lloyd D, Edwards SW (1987) Temperature-compensated ultradian rhythms in lower eukaryotes: timers for cell cycles and circadian events? In: Pauly J, Scheving L (eds) Advances in chronobiology, pt A. Alan R. Liss, New York, pp 131–151

Lovett J (1976) Two biological rhythms of perception distinguishing between intact and relatively damaged brain function in man. Int J Chronobiol 4:39–49

Lovett J (1980) Sinus tachycardia and abnormal cardiac rate variation in schizophrenia. Neuropsychobiology 6:305–312

Lovett J, Podnieks I (1975) Comparison between some biological clocks regulating sensory and psychomotor aspects of perception in man. Neuropsychobiology 1:261–266

Lovett J, Payne W, Podnieks I (1978) An ultradian rhythm of reaction time measurements in man. Neuropsychobiology 4:93–98

Luce G (1970) Biological rhythms in psychiatry and medicine. US Dept. of Health, Education and Welfare, Washington, DC

Lydic R (1987) State-dependent aspects of regulatory physiology. FASEB J 1:6–15

Mason A, Black S (1958) Allergic skin responses abolished under treatment of asthma and hayfever by hypnosis. Lancet 1:877–880

Migaly P (1987) Integrated approach of hypnotic theories and hallucinatory states. Paper presented at Fourth European Congress of Hypnosis in Psychotherapy and Psychosomatic Medicine, 11–17 July, Oxford

Millard W, Reppert S, Sagar S, Martin J (1981) Light–dark entrainment of the growth hormone ultradian rhythm in the rat is mediated by the arcuate nucleus. Endocrinology 108:2394–2396

Moore-Ede M, Czeisler C, Richardson G (1983) Circadian timekeeping in health and disease. Part I: Basic properties of circadian pacemakers. N Engl J Med 309:469–476

Murray A, Kirschner M (1989) Cyclin synthesis drives the early embryonic cell cycle. Nature 339:275–280

Murray A, Kirschner M (1991) What controls the cell cycle? Sci Am 266:56–63

Murray A, Solomon M, Kirschner M (1989) The role of cyclin synthesis and degradation in the control of maturation promoting factor activity. Nature 339:280–286

Naish P (ed) (1986) What is hypnosis? Open University Press, Milton Keynes, Philadelphia

Naitoh P (1982) Chronobiological approach for optimizing human performance. In: Brown F, Graeber R (eds) Rhythmic aspects of behavior. Lawrence Erlbaum Associates, Hillsdale, NJ, pp 41–103

Olness K, Conroy M (1985) A pilot study of voluntary control of transcutaneous PO by children: a brief communication. Int J Clin Exp Hypn 33(15):1–5

Olness K, Wain H, Ng L (1980) Pilot study of blood endorphin levels in children using self-hypnosis to control pain. Dev Behav Pediatr 1:187–188

Olness K, Culbert T, Uden D (1989) Self-regulation of salivary immunoglubulin A by children. Pediatrics 83:66–71

Orne M (1972) On the simulating subject as a quasi-control group in hypnosis research: what, why, and how. In: Fromm E, Shor R (eds) Hypnosis: research developments and perspectives. Aldine, Chicago, IL, pp 399–443

Orne M, Evans F (1966) Inadvertent termination of hypnosis with hypnotized and simulating subjects. Int J Clin Exp Hypn 14:61–78

Orr W, Hoffman H, Hegge F (1974) Ultradian rhythms in extended performance. Aerosp Med 45:995–1000

Osowiec D (1991) Ultradian rhythms in self-actualization, anxiety, and stress-related somatic symptoms. PhD dissertation, California Institute of Integral Studies

Perry C, Laurence J, D'eon J, Tallant B (1988) Hypnotic age regression techniques in the elicitation of memories: applied uses and abuses. In: Pettinati H (ed) Hypnosis and memory. Guilford Press, New York, pp 128–154

Pert C, Ruff M, Weber R, Herkenham M (1985) Neuropeptides and their receptors: a psychosomatic network. J Immunol 135:820s–826s

Pert C, Ruff M, Spencer D, Rossi E (1989) Self-reflective molecular psychology. Psychol Perspect 20:213–221

Pettinati H (1988) Hypnosis and memory. Guilford Press, New York

Poirel C (1982) Circadian rhythms in behavior and experimental psychopathology. In: Brown F, Graeber R (eds) Rhythmic aspects of behavior. Lawrence Erlbaum Associates, Hillsdale, NJ, pp 363–398

Putnam F (1985) Dissociation as a response to extreme trauma. In: Kluft R (ed) Childhood antecedents of multiple personality. American Psychiatric Press, Washington, DC

Rama S, Ballentine R, Ajaya S (1976) Yoga and psychotherapy: the evolution of consciousness. Himalayan International Institute of Yoga Science and Philosophy, Honesdale, PA

Rapp P (1987) Why are so many biological systems periodic? Prog Neurobiol 29:261–273

Rasmussen D (1986) Physiological interactions of the basic rest–activity cycle of the brain: pulsatile luteinizing hormone secretion as a model. Psychoneuroendocrinology 2:389–405

Rasmussen D, Malven P (1981) Relationship between rhythmic motor activity and plasma luteinizing hormone in ovariectomized sheep. Neuroendocrinology 32:364–369

Reid A, Curtsinger G (1968) Physiological changes associated with hypnosis: the affect of hypnosis on temperature. Int J Clin Exp Hypn 16:111–119

Reinberg A, Smolensky M (1983) Biological rhythms and medicine. Springer, Berlin Heidelberg New York

Ritterman M (1983) Using hypnosis in family therapy. Jossey-Bass, San Francisco, CA

Romano S, Gizdulich P (1980) Suggestion of ultradian rhythm in peripheral blood flow. Chronobiology 7:259–261

Rosbash M, Hall J (1989) The molecular biology of circadian rhythms. Neuron 3:387–397

Rose K (1988) The body in time. Wiley & Sons, New York

Rossi E (ed) (1980) The collected papers of Milton H. Erickson on hypnosis: vol I, The nature of hypnosis and suggestion; vol II, Hypnotic alteration of sensory, perceptual, and psychophysical processes; vol III, Hypnotic investigation of psychodynamic processes; vol IV, Innovative hypnotherapy. Irvington, New York

Rossi E (1982) Hypnosis and ultradian cycles: a new state(s) theory of hypnosis? Am J Clin Hypn 25:21–32

Rossi E (1986a) The psychobiology of mind–body healing. Norton, New York (German edn: Die Psychobioloqie der Seele–Körper-Heilung. 1991, Synthesis-Verlag, Essen)

Rossi E (1986b) Altered states of consciousness in everyday life: the ultradian rhythms. In: Wolman B, Ullman M (eds) Handbook of altered states of consciousness. Van Nostrand, New York, pp 97–132

Rossi E (1986c) Hypnosis and ultradian rhythms. In: Zilbergeld B, Edelstien M, Araoz D (eds) Hypnosis: questions and answers. W. W. Norton, New York, pp 17–21

Rossi E (1986d) The indirect trance assessment scale (ITAS): a preliminary outline and learning tool. In: Yapko M (ed) Hypnotic and strategic interventions: principles and practice. Irvington, New York, pp 1–29

Rossi E (1987) From mind to molecule: a state-dependent memory, learning, and behavior theory of mind–body healing. Advances 4:46–60

Rossi E (1989) Mind–body healing, not suggestion, is the essence of hypnosis. Am J Clin Hypn 32:14–15

Rossi E (1990a) A clinical–experimental assessment of the ultradian theory of therapeutic suggestion. Paper presented at the 32nd Annual Scientific Meeting and Workshops on Clinical Hypnosis, 24–28 March 1990, Orlando, FL

Rossi E (1990b) The new yoga of the west: natural rhythms of mind–body healing. Psychol Perspect 22:146–161

Rossi E (1990c) The eternal quest: hidden rhythms of stress and healing in everyday life. Psychol Perspect 22:6–23

Rossi E (1990d) Mind–molecular communication: can we really talk to our genes? Hypnos 17:3–14

Rossi E (1990e) From mind to molecule: more than a metaphor. In: Zeig J, Gilligan S (eds) Brief therapy: myths, methods and metaphors. Brunner/Mazel, New York, pp 445–472

Rossi E (1991a) The wave nature of consciousness. Psychol Perspect 24:1–10

Rossi E (1991b) The twenty minute break: the ultradian healing response. Jeremy Tarcher, Los Angeles

Rossi E (1992a) Periodicity in self-hypnosis and the ultradian healing response: a pilot study. Hypnos 19:4–13

Rossi E (1992b) A clinical–experimental exploration of Erickson's naturalistic approach: ultradian time and trance phenomena. Hypnos (in press)

Rossi E, Cheek D (1988) Mind–body therapy: ideodynamic healing in hypnosis. W. W. Norton, New York

Rossi E, Nimmons D (1991) The 20-minute break: using the new science of ultradian rhythms. Tarcher, Los Angeles, CA

Rossi E, Ryan M (eds) (1986) Mind–body communication in hypnosis, vol. 3, The seminars, workshops, and lectures of Milton H. Erickson. Irvington, New York

Rossi E, Smith M (1990) The eternal quest: hidden rhythms of mind–body healing in everyday life. Psychol Perspect 22:6–23

Roth J, Le Roith D, Collier E et al. (1985) Evolutionary origins of neuropeptides, hormones, and receptors: possible applications to immunology. J Immunol 135:816s–819s

Sabourin M (1982) Hypnosis and brain function: EEG correlates of state–trait differences. Res Commun Psychol Psychiatr Behav 7:149–168

Sachar E, Fishman J, Mason J (1965) Influence of the hypnotic trance on plasma 17-hydroxycorticosteroid concentration. Psychosomat Med 27:330–341

Sanders S (1991) Self-hypnosis and ultradian states: are they related? Paper presented at the 33rd Annual Scientific Meeting of The American Society of Clinical Hypnosis, 14–18 April, St Louis, MO

Sarbin T (1976) Hypnosis as role enactment: the model of Theodore R. Sarbin. In: Sheehan P, Perry C (eds) Methodologies of hypnosis. Lawrence Erlbaum Associates, Hillsdale, NJ, pp 123–152

Sarbin T, Slagle R (1972) Hypnosis and psychophysiological outcomes. In: Fromm E, Shor R (eds) Hypnosis: research developments and perspectives. Aldine, Chicago, IL, pp 185–214

Schmitt F (1984) Molecular regulators of brain function: a new view. Neuroscience 13:991–1001

Schmitt F (1986) Chemical information processing in the brain: prospect from retrospect. In: Iversen L, Goodman E (eds) Fast and slow signalling in the nervous system. Oxford University Press, New York, pp 239–243

Schulz H, Lavie P (1985) Ultradian rhythms in physiology and behavior. Springer, Berlin Heidelberg New York

Shaffer J, Schmidt C, Zlotowitz H et al. (1978) Biorhythms and highway crashes: are they related? Arch Gen Psychiatry 35:41–46

Sheehan P, McConkey K (1982) Hypnosis and experience: the exploration of phenomena and process. Erlbaum, New York

Sheehan P, Perry C (1976) Methodologies of hypnosis. Lawence Erlbaum Associates, Hillsdale, NJ

Shiotsuka R, Jovonovich J, Jovonovich J (1974) In vitro data on drug sensitivity: circadian and ultradian corticosterone rhythms in adrenal organ cultures. In: Schoff J et al. (eds) Chronobiological aspects of endocrinology. Schattauer, Stuttgart, pp 225–267

Simon C, Brandenberger G, Follenius M (1987) Ultradian oscillations of plasma glucose, insulin, and c-peptide in man during continuous enteral nutrition. J Clin Endocrinol Metab 64:669–674

Simons R, Ervin F, Prince R (1988) The psychobiology of trance. Transcult Psychiatr Res Rev 25:249–284

Smolensky M, Reinberg A (eds) (1977) Chronobiology in allergy and immunology. C. C. Thomas, Springfield, IL

Sommer C (1990) The ultradian rhythm and the common everyday trance. Paper presented at the 32nd Annual Scientific Meeting of The American Society of Clinical Hypnosis, 24–28 March, Orlando, FL

Soubeyrand L (1964) Action des medicaments vasomoteurs sur le cycle nasal et la fonction ciliare. Rev Laryngol Oto-Rhonol 85:43–113

Spanos N (1986) Hypnotic behavior: a social-psychology interpretation of amnesia, analgesia, and "trance logic". Behav Brain Sci 9:449–503

Spanos N, Cobb P, Gorassini D (1985) Failing to resist hypnotic test suggestions: a strategy for self-presenting as deeply hypnotized. Psychiatry 48:282–292

Spiegel D, Detrick D, Frischholz E (1982) Hypnotizability and psychopathology. Am J Psychiatr 139:431–437

Spiegel D, Hunt T, Dondershine H (1988) Dissociation and hypnotizability in posttraumatic stress disorder. Am J Psychiatry 145:301–305

Spiegel D, Bloom J, Kraemer H, Gottheil E (1989) Effect of psychosocial treatment on survival of patients with metastatic breast cancer. Lancet October 14:888–891

Spiegel H, Spiegel D (1978) Trance and treatment: clinical uses of hypnosis. Basic Books, New York

Steiner R, Peterson A, Yu J, Conner H, Gilbert M, Penning B, Brenner W (1980) Ultradian luteinizing hormone and testosterone rhythms in the adult male monkey, *Macaca fasicularis*. Endocrinology 107:1489–1493

Stephenson J (1978) Reversal of hypnosis-induced analgesia by naloxone. Lancet 2:991–992

Stoksted P (1953) Rhinometric measurements for determination of the nasal cycle. Acta Otolaryngol 109 [Suppl]:159–175

Stroebel C (1969) Biologic rhythm correlates of disturbed behavior and Rhesus monkey. In: Rohles F (ed) Circadian rhythms in non-human primates. Karger, New York, pp 19–105

Tart C (1983) States of consciousness. Psychological Processes, El Cerrito, CA

Tepas D (1982) Work/sleep time schedules and performance. In: Webb W (ed) Biological rhythms, sleep, and performance. John Wiley & Sons, New York, pp 175–204

Timney B, Barber T (1969) Hypnotic induction and oral temperature. Int J Clin Exp Hypn 17:121–132

Tinterow M (1970) Foundations of hypnosis. C. C. Thomas, Springfield, IL

Todorov I (1990) How cells maintain stability. Sci Am 263:66–75

Ullner R (1974) On the development of ultradian rhythms: the rapid eye movement activity in premature children. In: Scheving LE, Halberg F, Pauly JE (eds) Chronobiology. Igaku Shoin Ltd, Tokyo, pp 478–481

Van Cauter E, Desir D, Decoster C, Fery F, Balasse E (1989) Nocturnal decrease in glucose tolerance during constant glucose infusion. J Clin Endocr Metab 69:604–611

Veldhuis J, Johnson M (1988) Operating characteristics of the hypothalamo-pituitary–gonadal axis in men: circadian, ultradian, and pulsatile release of prolactin and its temporal coupling with luteinizing hormone. J Clin Endocrinol Metab 67:116–123

Veldhuis J, King J, Urban R et al. (1987) Operating characteristics of the male hypothalamo-pituitary–gonadal axis: pulsatile release of testosterone and follicle-stimulating hormone and their temporal coupling with luteinizing hormone. J Clin Endocrinol Metab 65:929–941

Wagstaff G (1986) Hypnosis as compliance and belief: a socio-cognitive view. In: Naish P (ed) What is hypnosis? Current theories and research. Open University Press, Milton Keynes Philadelphia, pp 57–84

Wehr T (1982) Circadian rhythm disturbances in depression and mania. In: Brown F, Graeber R (eds) Rhythmic aspects of behavior. Lawrence Erlbaum Associates, Hillsdale, NJ, pp 46–74

Wehr T, Goodwin F (1981) Biological rhythms and psychiatry. Am Handbook Psychiatry 7:46–74

Wehr T, Goodwin F (eds) (1983) Rhythms in psychiatry. Boxwood Press, Pacific Grove, CA

Weitzenhoffer A (1971) Ocular changes associated with passive hypnotic behavior. Am J Clin Hypn 14:102–121

Werntz D (1981) Cerebral hemispheric activity and autonomic nervous function. Doctoral thesis, University of California, San Diego

Werntz D, Bickford R, Bloom F, Shannahoff-Khalsa D (1982a) Alternating cerebral hemispheric activity and lateralization of autonomic nervous function. Human Neurobiol 2:225–229

Werntz D, Bickford R, Bloom F, Shannahoff-Khalsa D (1982b) Selective hemispheric stimulation by unilateral forced nostril breathing. Human Neurobiol 6:165–171

Wester W, Smith A (eds) (1984) Clinical hypnosis. J. B. Lippincott, Philadelphia, PA

Wever R (1989) Light effects on human circadian rhythms: a review of recent Andechs experiments. J Biol Rhythms 4:161–185

Wickramasekera I (1987) Risk factors leading to chronic stress-related symptoms. Advances 4:9–35

Williams R, Kraus L, Inbar M, Dubey D, Yunis E, Halberg F (1981) Circadian bioperiodicity of natural killer cell activity in human blood (individually assessed). In: Walker C, Winget C, Soliman K (eds) Chronopharmacology and Chronotherapeutics, pp 269–273

Wilson S, Barber T (1978) The creative imagination scale as a measure of hypnotic responsiveness: applications to experimental and clinical hypnosis. Am J Clin Hypn 20:235–249

Winfree A (1980) The geometry of biological time. Springer, Berlin Heidelberg New York

Winfree A (1987) The timing of biological clocks. Scientific American Library, New York

Wolcott J, McMeekin R, Burgin R et al. (1977) Correlation of occurrences of aircraft accidents with biorhythmic criticality and cycle phase in U.S. Air Force, U.S. Army, and civil aviation pilots. Aviat Space Environ Med 48:976–983

Young M (1988) The metronomic society: natural rhythms and human timetables. Harvard University Press, Cambridge, MA

Zimbardo P, Maslach C, Marshall G (1972) Hypnosis and the psychology of cognitive and behavioral control. In: Fromm E, Shor R (eds) (1972) Hypnosis: research developments and perspectives. Aldine, Chicago, IL, pp 539–571

Epilogue: The Unification Hypothesis of Chronobiology – Psychobiology from Molecule to Mind

D. Lloyd and E. L. Rossi

It is the theory which decides what we can observe.

Einstein in Werner Heisenberg, *Physics and beyond*

Anyone whose disposition leads him to attach more weight to unexplained difficulties than the explanation of a certain number of facts will certainly reject my theory.

Darwin, *Origin of species*

While many seeds of the unification hypothesis of chronobiology have been germinating quietly and independently for more than a generation, it is only with the assembled research of this volume that its outlines become clearly evident.

At the molecular–genetic–cellular level intracellular coordination and time-based synchronization of processes has been perfected during the evolutionary development of extant organisms: many clock-associated mechanisms have ancient origins and are highly conserved. As well as fast intramolecular motions and conformational changes within macromolecules, the reversible binding of small molecules at specific recognition sites on DNA, proteins and membranes are all processes required for molecular transformation and change. All involve cycles; all are potentially oscillatory. Electrons, ions, small molecules, macro-molecules, membranes, organelles, cells, organisms and populations are ascending levels in a hierarchy. As we ascend, reaction time constants become larger; we go from solid state processes, to hydrophobic then aqueous phase reactions, from catalytic turnover, to biosynthetic and degradative turnover, to cell division. ("What are called structures are slow processes of long duration; functions are quick processes of short duration" (von Bertalanffy 1960, p. 134)). Periodic behaviour abounds on every time scale, so that the whole temporal hierarchy shows self-similarities over 22 decades of time as measured in seconds.

At the neuroendocrinal and developmental levels a network of interlaced temporal controls reiterate and subsume those at the molecular level, but now intermolecular communication involves transport of informational substances

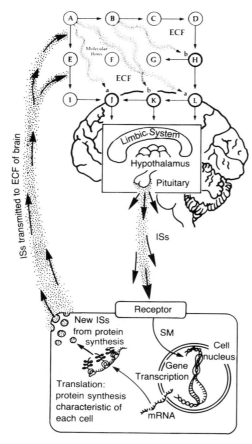

I THE MIND–BRAIN CONNECTION

1. Neural networks of the brain's cortical-limbic systems encode state-dependent memory, learning and behaviour of "mind" (words, images, sensations, perceptions, etc.) with the help of cybernetic information substances from cells of the body.

II THE BRAIN–BODY CONNECTION

2. Neuroendocrinal information transduction in the limbic-hypothalamic-pituitary system of the brain. The information in neural networks of the brain is transduced into molecular (hormonal) information substances of the body.

3. Information substances (ISs) travel to cells of body with appropriate receptors.

III THE CELL–GENE CONNECTION

4. Cellular receptors binding ISs.

5. Intracellular secondary messengers (SM) lead to activation of "housekeeping" genes.

6. Transcription of genetic information into mRNA.

7. Translation: protein synthesis characteristic of each cell.

8. New ISs from proteins flow to brain to cybernetically encode state-dependent aspects of mind and behaviour.

Fig. 19.1. The unification hypothesis of chronobiology: three levels of the mind–gene–molecular cybernetic network of information transduction hypothesized to be operative in the psychobiological processes of chronobiology. On all levels, ultradian rhythms up to 90–120 min (e.g. basic rest–activity cycle) integrate and modulate psychobiological processes such as memory, learning, stress, and human performance variables. ECF, extracellular fluid. (Reproduced, with permission, from Rossi and Cheek 1988.)

between the otherwise disparate cells, tissues and organs of the metazoan assemblage. Integrated rhythmic function constitutes the emergent observable output. These principles, obscured by an excessively zealous adherence to the homeostatic postulates of Claude Bernard for more than a century, have suddenly (in less than a decade) dawned upon a new generation of endocrinologists to transform our understanding of basic physiology and will soon revolutionize clinical pharmacology and medicine.

For the first time psychobiological processes of information transduction and communication that encompass behavioural and psychosocial levels become both clearly delineated and firmly based within the molecular, genetic and cellular dynamics of the whole organism. Integrated function at the highest level (i.e. self-conscious thought and creativity) shows the most complex time dependencies, as underlying is a concatenation of nested time domains. Human health, performance, stress and illness show dependence on intricate time structures; both life styles and therapies need to take heed of these new chronobiological insights. Fig. 19.1 shows the pathways of information flow involved in this new paradigm of non-steady state (oscillatory) molecule–mind communication. Time has for too long been the neglected parameter of biology and psychology. Our current dynamic perspectives replace the static snapshots of the past to propose a scientifically based holistic synthesis of coordinated temporal behaviour from amoeba to human.

References

Rossi E, Cheek D (1988) Mind–body therapy: ideodynamic healing in hypnosis. W.W. Norton, New York
von Bertalanffy L (1960) Problems of life. Harper & Rew, New York

Index